ADAPTIVE STRATEGIES FOR SMALL-HANDED PIANISTS

ADAPTIVE STRATEGIES FOR SMALL-HANDED PIANISTS

Lora Deahl
and
Brenda Wristen

OXFORD
UNIVERSITY PRESS

OXFORD
UNIVERSITY PRESS

Oxford University Press is a department of the University of Oxford. It furthers the University's objective of excellence in research, scholarship, and education by publishing worldwide. Oxford is a registered trade mark of Oxford University Press in the UK and certain other countries.

Published in the United States of America by Oxford University Press
198 Madison Avenue, New York, NY 10016, United States of America.

Library of Congress Cataloging-in-Publication Data
Names: Deahl, Lora. | Wristen, Brenda G.
Title: Adaptive strategies for small-handed pianists / Lora Deahl
and Brenda Wristen.
Description: New York, NY : Oxford University Press, [2017] |
Includes bibliographical references and index.
Identifiers: LCCN 2017014830 | ISBN 9780190616847 (hardcover : alk. paper) |
ISBN 9780190616854 (pbk. : alk. paper)
Subjects: LCSH: Piano—Performance—Physiological aspects. | Hand.
Classification: LCC MT221 .D43 2017 | DDC 786.2/19365—dc23
LC record available at https://lccn.loc.gov/2017014830

Contents

Preface xi

Acknowledgments xv

Key to Abbreviations and Graphic Symbols xix

1 Understanding Small-Handedness 1

Defining the Small Hand 2

Demographics of Small-Handedness 5

The Evolution of the Modern Piano and Its Relationship to Small-Handedness 8
Historical Background 8
Ergonomically Scaled (Reduced-Size) Piano Keyboards 12

Pedagogical Misconceptions about Small Hands 14
Social Bias 14
Attempts to Stretch the Small Hand 15

Small-Handedness and Injury 18

The Role of the Teacher 20

Conclusion 21

Reflection Points 22

2 Fundamental Principles of Movement and the Playing Apparatus 23

Principles of Work Efficiency 23

Principles of Movement 28
Leverage 28
Newton's Laws and Piano Technique 31
Muscular Contraction and Release 32
Additional Characteristics of Movement Economy 35
Planes of Motion 35
Neutral Position at the Piano 37

The Parts of the Playing Apparatus 37
Torso 37
Shoulder and Upper Arm 40
Forearm 42
Wrist 43
Hand and Fingers 47

Conclusion 52

Reflection Points 52

3 Technique and the Small-Handed Pianist **54**

Maladaptive Tendencies 55
Fatigue 55
Power 56
Reach 57

Lexicon of Fundamental Facilitative Techniques for the Small-Handed Pianist 58
Avoiding Fatigue 59
Timing Muscular Releases 59
Using Levers Appropriately 59
Cultivating Finger-to-Elbow Alignment 60
Combining Basic Movements into Larger Gestures 61
Reducing the Amplitude of Movements and Gestures 67
Producing Power 67
Maximizing Key Speed 68
Transmitting Force through Skeletal Structures 69
Using Arm Impulses 69
Addressing Large Reaches and Distances 71
Using the Arm to Navigate Distances 71
Using Passive Spread 72
Using Vertical Spanning 75

Boundaries 77

Conclusion 80

Reflection Points 82

4 Redistribution **83**

Uncrossing Parts 84

Eliminating Stretches in Chords and Arpeggios 86

Facilitating Leaps or Hand Shifts 95

Increasing Accuracy, Power, and Control 98

Maintaining More Neutral Hand and Wrist Positions 100

Facilitating Trills and Tremolos 101

Maintaining Legato and Line 102

Projecting Harmony 104

Conclusion 104

Application Problems 105

5 Refingering **106**

Fingering and Hand Size 108

Encouraging Compact Hand Positions 110
- Chords 111
- Arpeggios, Broken Chords, and Melodic Intervals 112
- Consecutive Double Notes and Octaves 117
- Readjusting the Hand to a More Compact Shape 120
- Playing Seconds with the Thumb Alone 120
- Using Non-Consecutive Fingerings 121
- Playing Trills 123

Repeating the Same Finger to Recover Space in the Hand 124

Optimizing Sound Production and Voicing 128
- Repeated Use of the Fifth Finger and Thumb to Project Melodies 128
- Aligning the Hand and Forearm in a Skeletal Position 130

Moving Longer Fingers over Shorter Fingers 131

Substituting the Thumb for Inner Fingers 133

Using Positional Fingerings to Achieve Maximum Speed 134

Substituting One Finger for Another on the Same Key 138

Retaking Released Notes Silently 138

Conclusion 139

Application Problems 140

6 Maximizing Reach and Power 142

Covering Large Distances 143
  Applications of Either Oval or Fan-Shaped Gestures 144
  Applications of Oval Gestures 146
  Applications of Fan-Shaped Gestures 147

Facilitating Large Reaches 154
  Applications of Passive Spread 154
  Applications of Vertical Spanning 155
  Applications of Passive Spread in Conjunction with Vertical Spanning 158

Dividing or Rolling Chords 160
  Consecutive and Non-Consecutive Divisions or Rolls 161
  Simultaneous Divisions or Rolls 166

Releasing Notes 167
  Releasing Notes in Chords and Multi-Layered Textures 168
  Releasing Notes in Melodic Intervals, Arpeggios, and Broken Chords 174
  Releasing Notes in Broken Octaves 177

Using Rotation for Power 178

Taking Advantage of Skeletal Support 180

Playing Staccato Chords and Octaves 181
  Repeated Chords and Octaves 181
  Traveling Staccato Chords and Octaves 184
  Legato Octaves 185

Grouping Notes into Larger Gestures 187

Omitting and Revising Notes 189

Conclusion 201

Application Problems 201

7 Musical Solutions for Technical Problems 202

Legato 203
  Legato in Widespread Cantabile Melodies and Arpeggiated Accompaniments 203
  Legato in Octave and Double-Note Passages 206

Fortissimo Playing 208

Pedaling 211
Damper Pedal 211
Sostenuto Pedal 215

Voicing 219

Rhythmic Inflection, Tempo Modification, and Rubato 220

Gestural Shaping 226

Conclusion 230

Reflection Points 230

8 Integration and Exploration **231**

Diagnosis and Integration 232

Exploration 237

Closing Thoughts 245

Glossary 247

Bibliography 251

List of Activities 255

List of Examples 257

List of Figures 265

List of Tables 267

Index 269

Preface

Touch is the language of sensation, the membrane through which feeling, meaning, and emotion seamlessly pass.[1] Touch is what draws us to the piano and what we strive to master through years of dedicated study. For pianists, it is the portal to music itself. Through touch, a pianist tests new sounds and means of expression. In the words of Ravel, it is how "he finds his most beautiful rhymes . . . where he measures . . . differing degrees of seductiveness or astringency, where he develops his alchemy . . . in which is distilled his most intoxicating perfumes."[2] Liszt, the performer par excellence, elevated virtuosity of touch to an even higher calling, proclaiming it as "no act of empty receptivity" but rather as the means by which musical ideas are made manifest and transferred "out of the fringe of incorporeal abstraction into the tangible, visible world."[3]

Sadly, many small-handed pianists literally and figuratively lose touch with the piano when they advance beyond the first years of study and begin to encounter passages that strain or exceed the capacities of their hands. Hours are spent in fruitless pursuit of unreachable goals. Injuries may result, and motivation declines. At the heart of this book lie memories of our own epic battles at the keyboard. One author sustained a devastating musculoskeletal injury after spending hours stretching her hands to reach large chords and drilling fingerings that never seemed to work. The other, assigned by her teacher to learn the first pages of all the Chopin etudes, was dismayed to find that she could not

---

[1] Adam Gopnik, "Feel Me: What the New Science of Touch Says about Ourselves," *New Yorker* (May 16, 2016): 56-67.

[2] Paul Roberts, *Reflections: The Piano Music of Maurice Ravel* (Milwaukee, WI: Amadeus Press, 2012), 91.

[3] Franz Liszt, *Gesammelte Schriften* IV: 194, as quoted in Victoria A. von Arx, *Piano Lessons with Claudio Arrau* (London: Oxford University Press, 2014), 29.

play the notes of the first measure of the very first etude. Perceptions about our physical inadequacies rapidly escalated into personal indictments—perhaps we were just not good enough. Paradoxically, understanding and acknowledging the inherent difficulties of small-handedness freed us to hear our musical voices more clearly and empowered us to find effective ways to meet these challenges. We have seen our experiences replicated in the lives of our students, especially those with hands even smaller than our own. Most are unaware of the fundamental impact of hand size on performance. When struggles persist over time, many embrace the punishing notion that they should work harder, without stopping to consider whether there might be different paths to the same goals.

Hands come in many different sizes and shapes, and some pianistic tasks are simply easier or more difficult for some than for others. Small-handed pianists are not the only ones experiencing challenges at the piano. On the other side of the coin, we have noticed that players with larger hands and fingers often experience difficulty playing in the spaces between the black keys or executing rapid scales, trills, and ornaments. Therefore, it would be a mistake to categorically attribute shortfalls in execution to a lack of talent or effort. It is equally possible that pianists are struggling because they simply do not have the appropriate technical tools for the task, tools that work for their own individual hands.

This book is unique in bringing together information from ergonomics, physics, anatomy, medicine, music pedagogy, and our years of professional experience in the field to focus on the subject of small-handedness. The first chapter presents an overview of small-handedness from historical, anatomical, and pedagogical perspectives. The second establishes a basic understanding of work efficiency and the human anatomy, moves on to general observations about piano playing and the constraints of physics, and explains the principles of healthy movement. The third is a focused analysis of technique as it relates to small-handedness. Chapters 2 and 3 are not intended as comprehensive explorations of piano technique. Rather, they provide the theoretical foundation for the detailed discussions of facilitative approaches contained in the remaining chapters.

Since individuals are unique in physical size, dimension, and the way their brains perceive, process, and execute information, the strategies we have proposed may not work equally well for all. Some methods may initially strike the reader as unorthodox or counterintuitive. However, our hope is that this book will enable or restore higher levels of music-making by helping small-handed pianists to establish healthy physical and mental boundaries, by illuminating many possible alternative pathways to musical and technical goals, and by directing them away from errant choices or dead ends. Furthermore, although finding physical efficiency is a fundamental goal, we seek to avoid facilitative strategies that ignore or contravene musical values.

The illustrations provided fall into three categories: Figures, Activities, and Examples. Figures are line drawings, graphs, or pictures provided to deepen understanding of textual descriptions. Activities are intended as elemental experiences that direct the reader's attention to kinesthetic sensations of movements described in the text. They are not technical exercises to be practiced repetitively to the point of mental and/or physical fatigue. In fact, readers are strongly cautioned against engaging in mindless practice that is divorced from musical content. Finally, the examples represent concrete applications within the piano literature, demonstrating how adaptive strategies can be implemented in a way that maintains musical integrity. Since the examples are necessarily brief, readers are encouraged to consult reputable editions to gain perspective of the broader musical context.

Many of the thoughts expressed here are not new and reflect the accumulated wisdom of scholars, colleagues, teachers, and students who have shared our musical journey. We value perspectives gained from the close study of varied approaches to technique written by those who have come before us. In particular, we often refer to Otto Ortmann's seminal 1929 treatise for two primary reasons.[4] Ortmann was the first to examine the work of piano playing from an objective viewpoint grounded in an understanding of physics and anatomy; his study may thus be regarded as the foundation for all subsequent biomechanical approaches to piano technique. Ortmann was also the first to specifically acknowledge the inherent challenges of small-handedness, although his treatment of the subject was limited. We have used Ortmann's findings, however, only on a selective basis to verify particular strategies or concepts. Our book should therefore not be read as a wholesale endorsement of any particular pedagogical approach or system of piano playing but rather as an investigation of the unique challenges faced by a particular group of pianists. Since all writers about piano playing describe the same tasks, there will always be points of similarity that are freely discoverable and shared by all. Furthermore, while we have tried to make this book accessible to any pianist who wishes to learn about small-handedness, we acknowledge that words can only go so far. Writing about movement is notoriously difficult, and we acknowledge that the expression of our ideas may be incompletely communicated to the reader. It is our hope that adherents of many different approaches will find points of agreement and application within these pages and that the strategies proposed here will inspire ingenuity in thinking about the many ways pianists can use their own unique hands at the keyboard.

[4] Otto Ortmann, *The Physiological Mechanics of Piano Technique* (Toronto: E. P. Dutton, 1929; reprint, 1962).

Acknowledgments

This book, which was decades in the making, would not have been possible without the extraordinary support of many people. Lora Deahl and Brenda Wristen would jointly like to thank the following:

E. L. Lancaster, editor-in-chief at Alfred Music Company, for his candid feedback and support at the outset of this project.

Our editor at Oxford University Press Suzanne Ryan, editorial assistants Andrew Maillet and Victoria Kouznetsov, Damian Penfold, and indexer Rachel Lyon, and the entire production staff for their unwavering enthusiasm for this project and their patient guidance in working through the myriad details of publication. We are grateful for the help of Samara Stob, marketing manager for music, and many others at Oxford whom we do not know by name who assisted with the production of this book.

Vanessa Cornett, director of keyboard studies and associate professor of piano and piano pedagogy at the University of St. Thomas, and Catherine Kautsky, professor of music and chair of the keyboard department at Lawrence University. Their willingness to interrupt the progress of their own manuscripts to offer invaluable written feedback on ours meant more than we can express, and their comments strongly influenced the final shape of this book.

Justin Houser, for expert preparation of the musical examples in the text and for multiple revisions of the same; Matt Bailey for graphic design and illustration; Justin Mohling, for photographic services; Ella Durham for photo editing; Greg Bashford, biomedical engineer and associate professor, Department of Biological Systems Engineering at the University of Nebraska-Lincoln, and Thad Buster, chief research analyst in the Movement and Neurosciences Center, Institute for Rehabilitation Science and Engineering at

Madonna Rehabilitation Hospital in Lincoln, Nebraska for surface electromyography figures; and Erica Deahl for graphic design consultation.

Numerous piano teachers who have attended our lectures and/or read our articles over the years and who have provided stimulating discussion that furthered our understanding of this topic, in particular, members of the Music Teachers National Association, the Texas Music Teachers Association, Plano Music Teachers Association, and the Nebraska Music Teachers Association. We would like to specifically acknowledge Louis Nagel, Martha Thomas, Timothy Woolsey, Catherine Kautsky, and Vanessa Cornett in this regard.

Other researchers and teachers who have shed light on the special needs of small-handed pianists, including Patricia Eger, Christopher Donnison, David Steinbuhler, Rhonda and Robin Boyle, Carol Leone, Erin Booker, and Eri Yoshimura.

Our students, especially those with hands even smaller than our own, whose struggles and victories helped shape our thinking about adaptive piano technique.

Lora Deahl would additionally like to thank the following:

My collaborator, Brenda Wristen, whose brilliant dissertation on the biomechanics of piano technique marked the beginning of this epic journey and whose wiser judgment prevailed on countless occasions when it needed to do so.

Texas Tech University for a 2012 Creative Arts, Humanities, and Social Sciences Grant and 2016 Internal Competitive Funding Opportunity Scholarship Catalyst Program Grant awarded in support of this research.

William Ballenger, former director of the Texas Tech University School of Music, and the entire faculty and staff past and present, whose boundless creativity, intellectual generosity, and devotion to the music profession are a constant source of inspiration. I am especially grateful for the support and encouragement of members of the piano division, including Carla Cash, William Westney, and Jane Ann Wilson.

Beloved teachers who profoundly changed the way I thought about music and the arts, especially David Renner, Douglass Green, Hanns-Berthold Dietz, Elliott Antokoletz, Josef Gingold, Inda Howland, Ellen Johnson, and Melvin Gallagher.

Catherine Kautsky, Timothy Woolsey, Amy and David Troyansky, Martha Thomas, Lea Schmidt Rogers, Sho-mei Pelletier and Dwight Shambley, Betty and Harry Mallard, Sujatri Reisinger, Rodrigo Davies, Toni and Thomas McGovern, Ellen Ching and Michael Belles, and Steven Ching and Lisa Robson—comrades in music and in life.

My husband, Robert Deahl, who did not survive to see the publication of this book but whose unwavering support and love are woven into every page, and our beloved daughter, Erica, the light of our lives.

And finally, the composers who make all our efforts worth the while.

Brenda Wristen would additionally like to thank the following:

Above all, Lora Deahl, collaborator and cherished friend, whose keen musical insight, diligence, and unwavering devotion to her friends, her students, and her profession continue to be an inspiration to me.

The Hixson-Lied Foundation, the Hixson-Lied College of Fine and Performing Arts, and the Glenn Korff School of Music at the University of Nebraska-Lincoln for publication support grant funding.

Mark Clinton, treasured colleague and friend, whose constant encouragement helped me see this book through to its fruition.

The innumerable teachers and professors in various fields who inspired and shaped me over the years. There are far too many to mention each by name, but I am especially grateful to Lora Deahl, Jane Ann Wilson, Ruth J. Holmes, and Lavoy Leach.

John Richmond, director emeritus of the Glenn Korff School of Music at the University of Nebraska-Lincoln, who enthusiastically supported this project from its inception, and my colleagues, who gave freely of their expertise and advice.

Fellow teachers and scholars in the larger piano and piano pedagogy communities whose creativity and energy constantly engage and challenge me. For specific advice regarding this book, I am thankful to Sylvia Coats, Chris Fisher, and Vanessa Cornett. The list of those who have provided professional mentorship and support over the years is a long one, but Lora Deahl, Joan Reist, and Gail Berenson merit special mention.

My students, who daily inspire me to be the best musician and teacher I can be.

Finally, all of my family and friends who continue to remind me that there is life outside academia.

Key to Abbreviations and Graphic Symbols

Term	Abbreviation	Symbol
Down/Up or Drop/Lift		
Fan shape		
Left hand	LH	
Oval		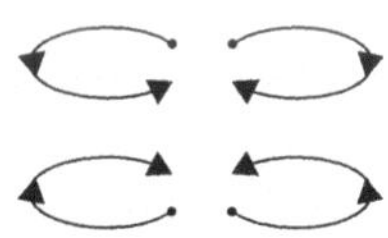
Overarc	OA	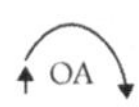
Passive spread	PS	
Redistribution		
Release, caesura, breath		,
Right hand	RH	

Term	**Abbreviation**	**Symbol**
Rotation		
Rotation with preparatory stroke		
Underarc	UA	
Vertical spanning	VS	

ADAPTIVE STRATEGIES FOR SMALL-HANDED PIANISTS

ONE

Understanding Small-Handedness

SMALL-HANDEDNESS WAS NOT A SERIOUS CONCERN FOR PIANISTS WHO performed on early fortepianos during the 1700s. Unlike the imposing black machines of today, Cristofori's instruments were as light and easy to play as Italian harpsichords. Lawrence Libin notes that

> Keys of instruments made before 1800 tend to be slightly narrower than modern piano keys, and hand stretches beyond an octave were rare; for the most part then, hand size was not a limiting technical factor, as it was to become in some Romantic piano music. The relatively light "touch" of 18th-century keyboard instruments, excepting some large organs, fostered a relaxed, fleet, well-articulated technique that depended less on weight and strength than does modern piano technique. Indeed, proper performance practice—for example, correct fingering for crisp ornaments and audible inner voices—was premised on a delicate, responsive touch; a sluggish "action" . . . would have been incompatible with textural clarity.[1]

Not until the nineteenth century did pianos become ubiquitous in middle-class European households. For many children of this era, music-making in the home centered around the piano, and learning to play it was an integral part of growing up. Large numbers of small-handed players now entered into the musical worlds of Beethoven, Chopin, Schumann, Liszt, and Brahms, where massive octaves and chords replaced delicate scales and ornaments. The plight of small-handedness was born.

[1] Lawrence Libin, "The Instruments," in *Eighteenth-Century Keyboard Music*, ed. Robert Marshall (New York: Schirmer, 1994), 8.

From that time to the present, small-handedness has been overlooked, discounted, or stigmatized as a disability. Pianists are often reluctant to admit to themselves or others that they have small hands, fearing that their work ethic, talent, or expectations for success will be questioned. As rigid boundaries circumscribing participation in music begin to crumble, however, it is possible to take a more clear-minded look at small-handedness. Musicians are more willing to concede that pure virtuosity is not the only mark of a fine pianist and that musicians of all descriptions—including amateurs—have something to offer. They have seen that even famous concert artists can suffer from piano-related injuries and that difficulties connected with playing the instrument do not necessarily reflect on one's musical talent. In studies dating from 1984 and 1988, Christoph Wagner documented that people with many different hand sizes and shapes can play the piano well and observed that no single hand type is a predictor of pianistic ability.[2]

Small-handed players are often unaware that many of their struggles arise directly from the size of the piano keys, wrongly attributing their difficulties to imagined personal deficits. In fact, psychological barriers can be even more detrimental than the actual physical limitations of small-handedness. Because hand size cannot be self-determined and because there is no proven way to increase it, ignoring the strengths and limitations of one's hands only stands in the way of technical and musical progress. Realistic assessment and creative problem-solving can open new paths to success.

DEFINING THE SMALL HAND

In a seminal study of piano technique published in 1929, Otto Ortmann documented many different hand sizes and dimensions among his students, noting that hand size and shape clearly influenced pianistic use.[3] Because children have small hands, Ortmann pointed out the necessity of adapting fingerings and means of tonal production to fit their needs.[4] He opined that

> to attempt to force all hands into one standard position—the normal arched position—a position, by the way, that is not even a physiological norm . . . is seriously to restrict pianistic freedom and, in my estimation, is unwise pedagogy. It is as if we obliged each person to walk with a step of standard length, regardless of the length of the leg.[5]

---

[2] Christoph Wagner, "Success and Failure in Musical Performance," in *The Biology of Music Making: Proceedings of the 1984 Denver Conference*, ed. Franz Roehmann and Frank Wilson (St. Louis: MMB, 1988), 154–179; Christoph Wagner, "The Pianist's Hand: Anthropometry and Biomechanics," *Ergonomics* 31, no. 1 (1988): 97–131.

[3] Otto Ortmann, *The Physiological Mechanics of Piano Technique* (1929; reprint New York: E. P. Dutton, 1962), 318–319.

[4] *Ibid.*, 300.

[5] *Ibid.*, 311.

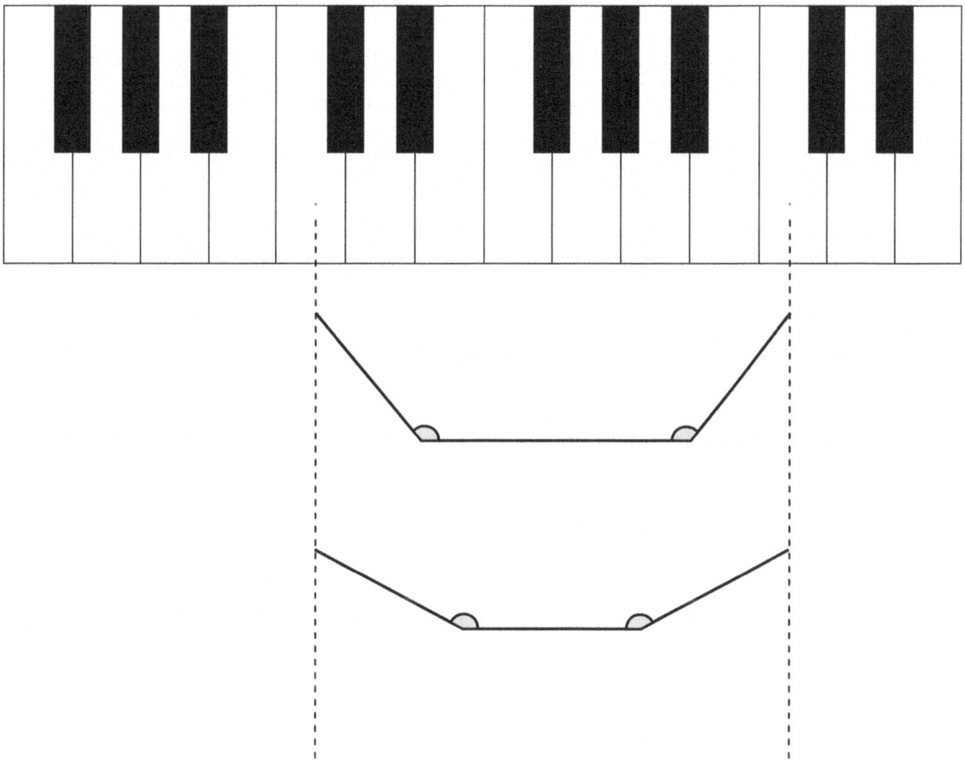

FIGURE 1.1. *Impact of palm size on angles of finger abduction. Adapted from Otto Ortmann,* The Physiological Mechanics of Piano Technique *(1929; reprint New York: E. P. Dutton, 1962), 312. Credit: Matthew Bailey, graphic artist.*

Ortmann went on to say that three main physiological parameters determine the ability of the hand to cover distance at the keyboard: the length of the fingers, the width of the palm, and the maximum spread (angles of abduction) between all fingers. He observed that large size in one or two of these dimensions could offset small size in another. For example, in hands with the same finger length, having a larger palm would result in smaller joint angles between the fingers to cover the same keyspan (Figure 1.1). Conversely, having longer fingers would compensate for a smaller palm.[6]

Ortmann furthermore attempted to classify the adult hand as being subnormal, normal, or supernormal in size based on measurements he took of pianists' finger lengths, hand widths, and angles between adjacent fingers when fully spread apart. These data cannot serve as norms for modern-day pianists, however, because Ortmann did not identify the precise anatomical landmarks he used as points of measurement.[7] More

---

[6] *Ibid.*, 311–312.

[7] For example, it is unclear whether Ortmann's measurements were taken from the outer edges or the midpoints of the joints. This distinction can easily change the determination of whether a particular hand is classified as subnormal, normal, or supernormal.

important, Ortmann provided almost no information about the participants in his study. Arnold Schultz, author of the introduction to the 1962 reissue of Ortmann's book, stated that Ortmann's subjects were expert pianists, but it is unclear whether these subjects were randomly sampled from a representative population.[8] Even if they were, selection bias was likely at play because women and other small-handed players were rarely numbered among "expert" performers in Ortmann's day. Applicability of his measurements aside, Ortmann clearly recognized that small-handedness was defined by a combination of different dimensions, not by a single measurement, and he understood that having a small hand imposed distinct challenges. Unfortunately, he also concluded that small hand size would limit professional success.

More recently, the inventor of the reduced-size piano keyboard, David Steinbuhler, defined a small hand as having a thumb-to-fifth-finger span of 8 inches or less. This measurement was used to define small-handedness for research purposes in studies by Wristen and Hallbeck in 2009 and Boyle in 2012.[9] In contrast, Kalmosiri defined small hands as those unable to comfortably play an octave and/or reach a ninth, and Farias as those unable to reach a tenth.[10] Boyle addressed the span between the second and fifth fingers as an important determinant of small-handedness, observing that if the pianist cannot easily reach the interval of a sixth with these fingers, he or she will be seriously hampered in playing large chord shapes.[11] Having a small span between the third and fourth fingers of the right hand has been implicated as another marker, since this condition has been associated with higher rates of reported pain arising from piano playing.[12]

The spread between the thumb and fifth finger is therefore not the only criterion for small-handedness. Other dimensions factor into how the hand fits the keyboard, even for hands that do not appear to be small at first glance. Characteristics such as range of motion at all the joints, webbing between the fingers, amount of cushioning on the fingertip, location and shape of the fingernails, length of the forearm and upper arm, and

---

[8] Arnold Schultz, Introduction to *The Physiological Mechanics of Piano Technique*, by Ortmann, xvii.

[9] Brenda Wristen and M. Susan Hallbeck, "The 7/8 Piano Keyboard: An Attractive Alternative for Small-Handed Players," *Update: Applications of Research in Music Education* 28 (2009): 9–16; Rhonda Boyle, "The Experience of Playing Reduced-Size Piano Keyboards: A Survey of Pianists," *Music Teachers National Association e-Journal* 3, no. 4 (April 2012): 3.

[10] Tretip Kamolsiri, "Even a Star Shines in the Darkness: Overcoming Challenges Faced by Pianists with Small Hands" (DMA dissertation, West Virginia University, 2002); J. Farias et al., "Anthropometrical Analysis of the Hand as a Repetitive Strain Injury (RSI) Predictive Method among Pianists," *Italian Journal of Anatomy and Embryology* 107, no. 4 (2002): 225–231.

[11] Boyle, "The Experience of Playing Reduced-Size Piano Keyboards," 3.

[12] Eri Yoshimura, Pamela Mia Paul, Cyriel Aerts, and Kris Chesky, "Risk Factors for Playing-Related Pain among College Students," *Medical Problems of Performing Artists* (September 2006): 123.

width of the shoulders are also important determinants. Additionally, one cannot make assumptions about hand size based on overall size or body type. Many pianists who are very short have small hands, but this trait might be shared by individuals of average or even tall stature.

For the purposes of this book, the authors consider small-handedness to be self-defined. If pianists notice that the size of their hands restricts their performances, they are effectually small-handed. Even someone with an average or large hand span may be constrained by certain dimensions, such as a short reach between individual fingers. However, many people who identify as being small-handed can reach no more than a ninth. They may struggle to play the thick chords, extended arpeggios, octaves, and tenths that are staples of Romantic and modern piano works and the broken octaves and contrapuntal textures of Classical and Baroque music. Furthermore, while it is true that many hallmark difficulties turn up in the advanced piano literature, challenges also surface in pedagogical repertoire, but on a more limited basis.

It bears mentioning, however, that there are some tasks to which a small hand is ideally suited. In tight scalar or chromatic passages, it is agile and compact enough for the fingers to work easily in between the keys; large-handed pianists may report feeling "cramped" in similar circumstances. Having short fingers may enable players to work in the black-key area without hitting the fallboard. Finally, the small hand may find it easier to produce the small movements needed for quick trills and ornaments.

DEMOGRAPHICS OF SMALL-HANDEDNESS

In 2009, Boyle and Boyle measured thumb-to-fifth-finger spans and documented the largest white-key interval 25 adult pianists could play with the thumb and fifth finger on a conventionally sized keyboard.[13] They further noted whether these pianists retained the ability to comfortably slide the thumb and fifth finger in toward the black keys or whether players were forced to stay at the very edges of the white keys. Boyle and Boyle compared their results to two previous studies documenting hand size—Wagner's 1988 study and David Steinbuhler's 2004 unpublished data.[14] Although neither Wagner's nor Steinbuhler's data were observed from scientifically based random samples, the results of all three showed that the average hand span (means and medians) of adult males was approximately one

[13] Rhonda Boyle and Robin Boyle, "Hand Size and the Piano Keyboard: Literature Review and a Survey of the Technical and Musical Benefits for Pianists using Reduced-Size Keyboards in North America," *Proceedings of the Australasian Piano Pedagogy Conference Association*, North Parramatta, Sydney, July 13–17, 2009, www.appca.com.au (accessed August 5, 2015): 5.

[14] Wagner, "The Pianist's Hand"; David Steinbuhler, "Measuring Hand Span," *Steinbuhler & Company*, accessed July 2015, www.steinbuhler.com.

inch greater than that of adult females.[15] This measurement is slightly greater than the width of one standard piano key. Boyle and Boyle concluded that while most females can reach an octave, a significant majority cannot play a ninth or tenth comfortably, even when playing "on the edge" of the keys. On the other hand, a significant majority of males can reach a ninth comfortably, and most can also play a tenth "on the edge" of the keys.

In 2015, Boyle, Boyle, and Booker measured differences in hand sizes among a larger sample of pianists and again compared their findings to those of Wagner and Steinbuhler. These three studies used slightly different measurement protocols: Boyle, Boyle, and Booker and Steinbuhler measured the thumb-to-fifth-finger hand span by measuring from the outside edges of the thumb and fifth finger when fully extended while Wagner measured from the midpoint of the fingertips.[16] Even so, all three studies again confirmed a marked difference in hand size between males and females (Table 1.1).

TABLE 1.1 Comparison of pianist's thumb to fifth finger span data in inches (centimeters)

	Males			Females		
	Steinbuhler (Mix of R and L hands)	Wagner (RH)	Boyle et al. (RH)	Steinbuhler (Mix of R and L hands)	Wagner (RH)	Boyle et al. (RH)
Sample size	66	111	159	90	105	314
Minimum	7.7 (19.6)	7.91 (20.1)	7.8 (19.7)	7.0 (17.8)	7.01 (17.8)	6.4 (16.3)
Maximum	10.2 (25.9)	10.24 (26.0)	10.8 (27.4)	8.9 (22.6)	9.33 (23.7)	9.5 (24.1)
Arithmetic Mean	8.9 (22.6)	8.93 (22.7)	8.9 (22.6)	7.9 (20.1)	8.17 (20.7)	7.9 (20.1)
Median	8.9 (22.6)	8.86 (22.5)	8.9 (22.6)	7.9 (20.1)	8.11 (20.6)	7.9 (20.1)
First Quartile	8.5 (21.6)	8.62 (21.9)	8.5 (21.6)	7.5 (19.1)	7.83 (19.9)	7.6 (19.2)
Third Quartile	9.3 (23.6)	9.21 (23.4)	9.2 (23.4)	8.2 (20.8)	8.50 (21.6)	8.3 (21.0)
Range	2.5 (6.4)	2.32 (5.9)	3.0 (7.7)	1.9 (4.8)	2.32 (5.9)	3.1 (7.9)
Inter-Quartile Range	0.8 (2.0)	0.59 (1.5)	0.7 (1.8)	0.7 (1.8)	0.67 (1.7)	0.7 (1.8)
Standard Deviation	0.56 (1.42)	0.45 (1.14)	0.56 (1.43)	0.47 (1.19)	0.45 (1.14)	0.53 (1.35)

Source: Boyle, Boyle, and Booker, "Pianist Hand Spans," 37. Reprinted with permission from Rhonda Boyle.

[15] Boyle and Boyle, "Hand Size and the Piano Keyboard," 4.

[16] Rhonda Boyle, Robin Boyle, and Erica Booker, "Pianist Hand Spans: Gender and Ethnic Differences and Implications for Piano Playing," presented to the Australasian Piano Pedagogy Conference Association, Melbourne, July 2015, www.appca.com.au (accessed August 5, 2015): 39.

Boyle, Boyle, and Booker observed that a startling 87.1% of women in their study had small hands, defined by having a thumb-to-fifth-finger span of less than 8.5 inches.[17] There was a noticeable gender gap with regard to hand size, but small-handedness was not limited to the female gender. As shown in Table 1.2, a significant number of men (23.8%) in Boyle, Boyle, and Booker's study also had small hands.

In addition to hand-size dissimilarities between genders, differing hand sizes have also been noted between various ethnic groups. Nag, Nag, and Desai measured the hands of Indian women, finding them to be significantly smaller than those of British, American, and West Indian women.[18] Sakai noted that Japanese women were likely at a hand-size disadvantage in comparison with Europeans and Americans.[19] In Boyle, Boyle, and Booker's study, a greater proportion of both Asian males and females had smaller hands than Caucasians (Table 1.2). Caucasian male pianists in their study had an average (arithmetic mean) hand span of 9.0 inches, while Asian males averaged a hand span of 8.7 inches; Caucasian females had an average hand span of 8.0 inches, compared to 7.8 inches for Asian females.[20] Boyle, Boyle, and Booker went on to say that these proportions may well underestimate the global magnitude of small-handedness among pianists.[21]

TABLE 1.2 Proportions of adult pianists with "small" or "large" hands reported by Boyle, Boyle, and Booker

	"Small" hands: 1–5 spans less than 8.5 in.	"Large" hands: 1–5 spans 8.5 in. or more
All males*	23.8%	76.2%
All females*	87.1%	12.9%
Caucasian males	20.2%	79.8%
Caucasian females	82.3%	17.7%
Asian males	29.9%	70.1%
Asian females	94.0%	6.0%

**Based on the ethnic mix in this particular sample of adult pianists*

Source: Boyle, Boyle, and Booker, "Pianist Hand Spans," 59. Reprinted with permission from Rhonda Boyle.

[17] *Ibid.*, 59.

[18] Anjali Nag, P. K. Nag, and Hina Desai, "Hand Anthropometry of Indian Women," *Indian Journal of Medical Research 117* (June 2003): 264; Boyle and Boyle, "Hand Size and the Piano Keyboard," 4.

[19] Naotaka Sakai, "Hand Pain Related to Keyboard Techniques in Pianists," *Medical Problems of Performing Artists* 7 (1992): 65.

[20] Boyle, Boyle, and Booker, "Pianist Hand Spans: Gender and Ethnic Differences," 56.

[21] *Ibid.*, 60.

The small-handed population also includes children and aging adults. Most children have small hands relative to the conventionally sized piano keyboard. Their challenges may persist for only a short time until their hands grow, or for many years, inhibiting their musical and technical development. At the other end of the life spectrum, some older adults whose hands once fit comfortably on the keyboard may find their hand span and flexibility diminishing over time due to arthritis or other impacts of aging. It is apparent that small hand size is not limited to one gender, ethnic group, or age.

THE EVOLUTION OF THE MODERN PIANO AND ITS RELATIONSHIP TO SMALL-HANDEDNESS

Before discussing particular challenges posed by small-handedness, some underlying assumptions should be examined and dispelled. One supposition is that the size of the keyboard has been fixed through time. On the contrary, the piano has continued to evolve since its invention by Bartolomeo Cristofori at the beginning of the eighteenth century. The width and length of the keys, height of the black keys over the white keys, and depth of key travel have all generally increased over time, along with the heaviness of the action.[22] Even now, there are subtle variations in these dimensions from one manufacturer to the next.

Historical Background

In 2008, Naotaka Sakai investigated differences in octave spans from C4 to C5 among 120 historical and modern instruments, including harpsichords, clavichords, spinets, virginals, pianofortes, and square pianos (Table 1.3).[23] The oldest harpsichords and the earliest Cristofori piano that he measured had average octave spans equivalent to that of the modern piano.[24] However, between 1784 and 1850, the average octave spans of both the harpsichord and grand piano actually decreased. This means that many of the

[22] Hunter J. H. Fry, "Overuse Syndrome in Musicians—100 Years Ago: An Historical Review," *Medical Journal of Australia* 145 (December 1986): 623.

[23] Naotaka Sakai, "Keyboard Span in Old Musical Instruments: Concerning Hand Span and Overuse Problems in Pianists," *Medical Problems of Performing Artists* 23 (2008): 169–171.

[24] The Society for Piano Keyboard Standards suggests measuring from the left side of the piano's first C key to the left side of the seventh C key and dividing the result by seven as a very accurate means for obtaining the measurement of the octave. This yields a number representing the distance from the middle of one C key to the middle of the next C key. Octaves on modern conventionally sized keyboards typically measure 6.50 ± 0.04 inches using this method. Sakai measured octaves differently but was nonetheless consistent in measuring from one instrument to the next, so his observations about trends in key sizes over time may be regarded as valid.

masterworks in the piano literature—including works by Mozart, Beethoven, Schubert, Chopin, Schumann, and Liszt—may have been composed on smaller keyboards. Sakai noted that the mean span of the octave increased significantly after 1850, and it has remained essentially unchanged ever since (Table 1.3).

In addition to a dramatic increase in the size of the octave during the second half of the nineteenth century, the touch of the piano evolved markedly in response to the public demand for instruments of greater brilliance and power. The introduction of cast iron frames allowed makers to build pianos that could withstand exponentially increased string tension so as to project greater volumes of sound into larger performance halls. The larger octaves, thicker strings, bigger hammers, and heavier and deeper actions (Table 1.4) exacerbated problems for small-handed players.

TABLE 1.3 Summary of Sakai's measurements of octave spans on historical instruments

Year of manufacture	Mean distance of octave span (C4 to C5)
1720 (Cristofori)	7.40 in. (188 mm.)
1784–1800	7.16 ± 0.14 in. (181.9 ± 3.6 mm.)
1801–1825	7.16 ± 0.07 in. (182 ± 1.8 mm.)
1826–1850	7.24 ± 0.12 in. (183.8 ± 3.0 mm.)
1851–1875	7.37 ± 0.04 in. (187.3 ± 1.0 mm.)*
After 1876	7.39 ± 0.01 in. (187.8 ± 0.3 mm.)

**Shows statistically significant increase*

Source: Based on information contained in Sakai, "Keyboard Span in Old Musical Instruments," 169–170.

TABLE 1.4 A comparison of nineteenth-century grand pianos with modern Steinway grand pianos

Piano	Travel depth	Weight for full key depression (rounded to nearest hundredth)	String tension in middle register (rounded to nearest hundredth)
Early nineteenth-century grand	0.24 in. (6 mm.)	0.81 oz. (23 g.)	26.46–33.07 lb. (12–15 kg.)
Late nineteenth-century grand	0.35 in. (9 mm.)	1.59 oz. (45 g.)	176.34 lb. (80 kg.)
Modern Steinway grand	0.41 in. (10.5 mm.)	1.76–2.12 oz. (50–60 g.)	198.42 lb. (90 kg.)

Source: Based on information contained in Fry, "Overuse Syndrome in Musicians," 623.

Lora Deahl compared the horizontal and vertical dimensions of a historical range of original and reconstructed keyboards from the collection of the Schubert Club in St. Paul, Minnesota, playing a test chord from Schumann's *Träumerei* on each (Example 1.1). The instruments examined are summarized in Table 1.5. In addition to octave span, Deahl measured the length of the black keys as well as the length of the white keys in front of the black keys (Figure 1.2). Like Sakai, she found that the size of the octave in the oldest pianos equaled that of modern twentieth-century instruments, but that this measurement decreased during a period dating from the last quarter of the eighteenth century

EXAMPLE 1.1. *Schumann test chord, "Träumerei," from* Kinderszenen, *op. 15, m. 22*

TABLE 1.5 Historical/reconstructed instruments measured by Deahl

Data point	Brand/reconstruction	Type	Year
1	Annibale dei Rossi	Spinet	1542
2	Cristofori/David Sutherland	Grand	1726
3	Silbermann/David Sutherland	Grand	1749
4	Zumpe	Square Grand	1768
5	Stein/Richard Sorenson	Grand	1784
6	Johann Xaver Christoph	Grand	1790
7	Broadwood	Grand	1795
8	Schantz/T. and B. Wolf	Grand	1800
9	Broadwood	Square Grand	1807
10	Streicher née Stein/T. and B. Wolf	Grand	1815
11	Anonymous	Lyre	1825
12	H. Kisting	Grand	1830
13	Pleyel	Grand	1839
14	Erard	Upright	1844
15	Steinway	Upright	1866
16	J. B. Streicher	Grand	1869
17	Bechstein	Grand	1878
18	Erard	Grand	1899
19	Virgil	Silent Keyboard	1909
20	Wurlitzer	Grand	1935

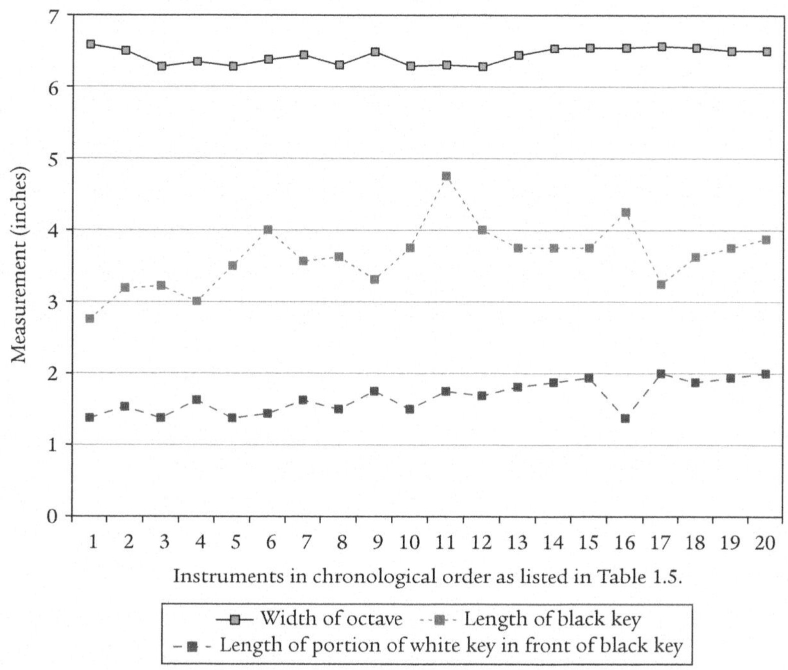

FIGURE 1.2. *Deahl's comparison of changes in horizontal dimensions*

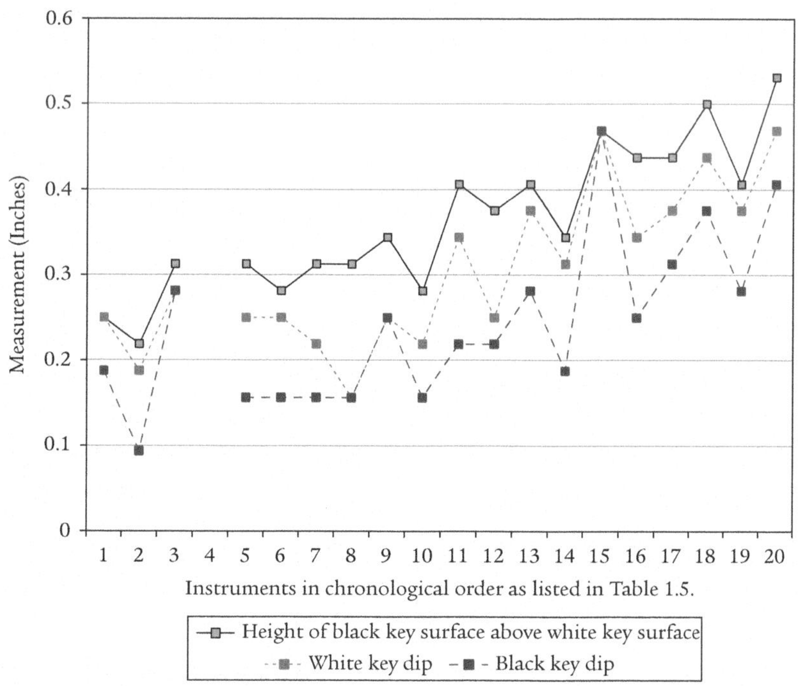

FIGURE 1.3. *Deahl's comparison of changes in vertical dimensions*

through the first half of the nineteenth century. Her data also document a gradual trend toward greater key length. In addition, she collected data concerning vertical dimensions (Figure 1.3), including key depression depth of both black and white keys and the height of the black keys over the white keys. These data also reflect an upward trend, generally increasing over time. The compound result of these changes in horizontal and vertical dimensions was that Deahl could easily play the Schumann test chord on an 1830 Kisting grand that had been played by Robert and Clara Schumann, Mendelssohn, and Brahms, but could not reach the same chord on an 1878 Bechstein that had been played by Liszt, Brahms, and Mahler.

Ergonomically Scaled (Reduced-Size) Piano Keyboards

Increased awareness of ergonomic principles has motivated employers to adapt work environments to the dictates of tasks and the design of the human body. The wide availability of redesigned tools such as desk chairs and computer keyboards in the marketplace testifies to a growing concern about occupational safety. With respect to the modern piano, however, a "one-size-fits-all" mentality has remained pervasive. This is not the case with other instruments, particularly stringed instruments, which are available in many sizes to suit the player. Other instrumentalists also have access to numerous customizations. Violinists and violists often use freely adjustable chin and shoulder rests. Woodwind and brass instruments are frequently modified through customizing grips, mouthpieces, or key extensions.

Recently, there has been a growing interest in sizing piano keyboards to fit small-handed players. These keyboards, currently referred to as Ergonomically Scaled Piano Keyboards (ESPKs), are manufactured in several sizes, all with octaves smaller than the conventional keyboard. It is important to realize, however, that while the octaves are narrower on ESPKs, other dimensions that may impact small-handed players, such as the length of the white keys in front of the black keys or the vertical key dip, remain unchanged. As measured by David Steinbuhler, the leading manufacturer of ESPKs, the conventional piano has an octave measuring 6.5 inches. By contrast, the more common ESPKs have octaves measuring 6.0 inches (15/16 size) and 5.5 inches (7/8 size).[25] The narrower keys of ESPKs require less hand extension to play large chords and octaves. A study by Wristen and others measured the movements of the thumb and fifth finger in two pianists as they performed the first minute and a half of the Tchaikovsky first piano concerto. Each pianist performed on both the conventional and 7/8-size keyboard. On average, playing the 7/8 keyboard required five degrees less

[25] David Steinbuhler, "Our Research," *Steinbuhler & Company*, accessed July 2015, www.steinbuhler.com.

thumb abduction (movement away from the midline of the hand) and 10 degrees less fifth-finger abduction.[26]

A number of studies have examined muscular activity and perceptions among small-handed pianists regarding comfort on the 7/8 piano as compared with conventionally sized piano keyboards. A study by Wristen and Hallbeck found that the majority of participants expressed a clear preference for the 7/8 keyboard.[27] On a similar note, Yoshimura reported that small-handed piano majors at the University of North Texas anecdotally observed that playing on ESPKs reduced pain and discomfort and allowed them to perform repertoire that would otherwise be problematic.[28] Carol Leone noted that the ability to produce beautiful legato and powerful sound were primary benefits of the 7/8 keyboard.[29] In a study conducted by Boyle, 100% of the small-handed participants cited the ability to sustain notes for full value and an increased feeling of power as advantages when playing on reduced-size keyboards. Other benefits included greater ease in executing fast passages of octaves, large chords, broken octaves, and arpeggios; less time required to master technically difficult passages or learn new repertoire; an increased overall feeling of security; fewer awkward or non-ideal fingerings, hand shifts, or leaps; increased accuracy; and better control of legato and melodic projection. Many participants also commented that they could play repertoire that was previously beyond their grasp.[30] Boyle suggested that, based on hand sizes, approximately 50% of adult women "will find advantages in reduced-size keyboards including creating technical ease, comfort and security, relief from pain and injury, reduced learning times, expanded repertoire, musical improvements and most importantly, greater enjoyment." Boyle contended that 70% to 80% of adult women and 20% of males would prefer a reduced-size keyboard if their aim was to play diverse repertoire, pointing out that a pianist with an 8.5-inch hand span can barely accommodate a tenth and that in performing much Romantic and twentieth-century repertoire, "there are significant advantages in being above this threshold." She further maintained that since there is approximately a one-inch difference in the average spans of adult males and females, a 7/8 keyboard essentially places males and females on a level playing field.[31]

---

[26] B. G. Wristen, M-C. Jung, A. K. G. Wismer, and M. S. Hallbeck, "Assessment of Muscle Activity and Joint Angles in Small-Handed Pianists: A Pilot Study on the 7/8 Keyboard versus the Full-Sized Keyboard," *Medical Problems of Performing Artists* 21 (2006): 3–9.

[27] Wristen and Hallbeck, "The 7/8 Piano Keyboard," 9–16.

[28] Eri Yoshimura, Anncristine Fjellman-Wiklund, Pamela Mia Paul, Cyriel Aerts, and Kris Chesky, "Risk Factors for Playing-Related Pain among Piano Teachers," *Medical Problems of Performing Artists* 23, no. 3 (September 2008): 112.

[29] Carol Leone, "Goldilocks Had a Choice," *American Music Teacher* 52, no. 6 (June/July 2003): 26–29.

[30] Boyle, "The Experience of Playing Reduced-Size Piano Keyboards," 2–20.

[31] *Ibid.*, 8.

Increasing public acceptance of ergonomically modified keyboards represents an exciting development in addressing small-handedness. Advocates are working toward greater availability of ESPKs not only in universities and private settings but also in concert halls and competitions. The use of smaller keyboards is not without precedent: the famous pianist Josef Hofmann regularly toured with a Steinway custom built to his smaller hands. The size of the octave on Hofmann's piano purportedly spanned a distance closer to a seventh on the standard keyboard of the day, something Hofmann claimed was the norm for Russian and Polish instruments.[32] In March 2014, the first piano competition offering competitors a choice between piano keyboard sizes was held in Dallas, Texas. Until ergonomic keyboards become universally available, however, there is a need for heightened awareness about the specialized technical problems imposed on small-handed players by the size of the conventional keyboard.

PEDAGOGICAL MISCONCEPTIONS ABOUT SMALL HANDS

The "one-size-fits-all" mentality that has prevailed since the piano evolved to larger dimensions was likely influenced by the innovations of mass production. The ergonomic perspective that the work interface should be adapted to fit the worker is a relatively recent development, and the history of piano pedagogy is rife with assumptions that small-handed players are simply inadequate. Thus, in addition to coping with the demands of the instrument itself, small-handed players have been subject to many pedagogical misconceptions.

Social Bias

The piano rose to prominence in late eighteenth-century musical life precisely because playing it became a fashionable and expected skill among women. In fact, all upper-class women were expected to be able to play solos or accompany themselves as they sang in domestic settings.[33] However, women did not dare to play the piano in public; such behavior was well outside acceptable gender roles of the time. They were nevertheless expected to aspire to the same level of virtuosity exhibited by men.[34]

[32] Patricia Eger, "A Study of the Problems Encountered by the Pianist with Small Hands and a Compendium of Practical Solutions" (DA dissertation, Ball State University, 1982), 37.

[33] James Parakalis, "Austen and the Domestic Life of the Piano," in *Piano Roles: Three Hundred Years of Life with the Piano* (New Haven: Yale University Press, 1999), 96.

[34] *Ibid.*, 97 and 99.

Parakalis points out, however, that technological developments in piano construction affected males, females, and particularly children differently, given the marked disparity in their hand sizes. "[W]ith pedals appearing at floor level, the keyboard expanding in both directions, and passages in octaves appearing regularly in the newest repertory, the piano became, for children who learned it, a daunting daily reminder of how far they had to go to measure up in the adult world. . . .. [M]any more girls than boys have felt that they would never have the instrument—or the part of the repertory created by big-handed men—within their reach."[35]

Even though many more girls studied the piano than boys, nineteenth-century male virtuosos with large hands strongly influenced the trajectory and expectation for what the hand should be able to reach and how big the keyboard should be. In keeping with the fascination for standardization during the Machine Age, attitudes about how the piano should be played, including norms for hand and wrist position, fingerings, and posture, were codified in writings and etudes by pedagogues and composers such as Cramer, Clementi, Chopin, and Czerny.

In the twentieth century, having a small hand was still typically viewed as an exclusionary barrier to serious pianism. Many influential piano teachers including Ernö Dohnányi, Alfred Cortot, Isidor Philipp, Heinrich Neuhaus, Otto Ortmann, Louis Kentner, Adele Marcus, Ruth Slenczynska, and Josef Hofmann commented or wrote about the "problem" of small hands. For example, Ortmann observed that he knew of "numerous instances where diminutive hands definitely terminated musical careers."[36] Even though Ortmann devoted a chapter in his groundbreaking book to discussing individual differences of the hand, he had very few suggestions for adapting technique for a small hand. He felt that the coping strategies he observed among small-handed pianists were inadequate solutions to untenable problems. The few recommendations he made were intended as stop-gap measures for children whose hands were still growing. Among these recommendations were refingering, redistribution, sliding with one finger from a black to a white key, division of chord tones between hands, and using the thumb for two adjacent white or black keys.[37]

Attempts to Stretch the Small Hand

Until quite recently, the prevailing interest of pedagogues was to try to remodel the small hand to fit the piano rather than to consider how the pianistic task or the piano itself might be adapted to fit the hand. Some pedagogues prescribed relatively benign treatments such as kneading and massaging the joints and fingers or soaking the hands in hot

---

35 *Ibid.*, 151.

36 Ortmann, *Physiological Mechanics of Piano Technique*, 313.

37 *Ibid.*, 300.

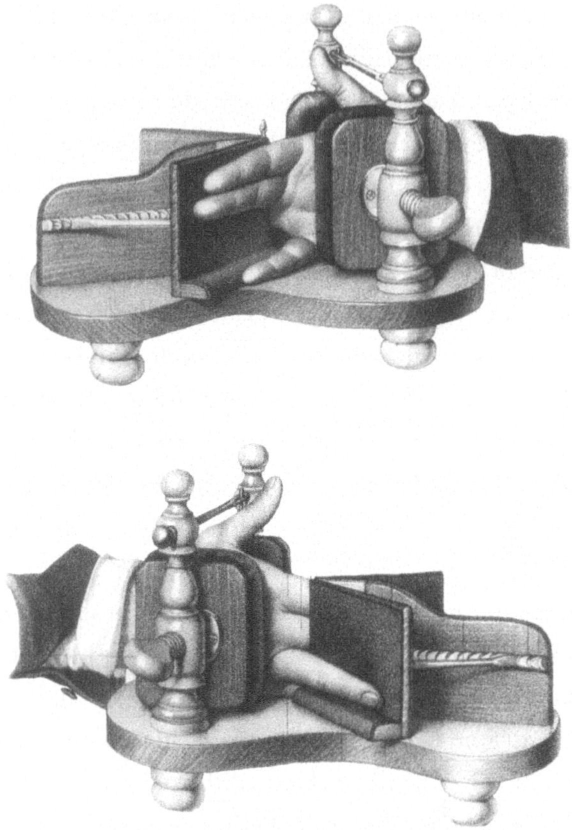

FIGURE 1.4. *Stretching device recommended by Sigismond Thalberg. Credit: Lithograph after a drawing by Emile Beau from* L'anatomie de la main, *Paris 1846, as reproduced in Ernst Burger,* Robert Schumann: A Chronicle of his Life in Pictures and Documents *from* Robert Schumann: New Edition of the Complete Works, *Series VIII: Supplements, Vol. 1 (Main: Schott, 1998), 105.*

water to loosen joints.[38] However, others recommended regimens that were potentially dangerous—finger gymnastics for strength conditioning, lifting weights with the fingers, or using devices to exercise or stretch the hand. Small-handed pianists often resorted to using mechanical devices, usually requiring the participant to force fingers over sized wedges. For example, the "Hand Extender" designed by Frederick Crane of Massachusetts in 1899 featured a threaded rod located between two finger slings which could force apart any pair of fingers (not necessarily adjacent) for extended periods of time.[39] Figure 1.4 shows an apparatus that was highly recommended by Sigismond Thalberg, the famous virtuoso who was a contemporary of Liszt. Ernst Burger reports that Friedrich Wieck,

---

[38] Recommended by Thomas Fielden and Hofmann as discussed in Eger, "A Study of the Problems Encountered by the Pianist with Small Hands," 52–53.

[39] Robert E. Sheldon, "Keyboard Practice and Exercise Aids," in *Encyclopedia of Keyboard Instruments*, Vol. 1: *The Piano*, ed. Robert and Margaret Palmieri (New York: Garland, 1994), 197.

the teacher and father-in-law of Robert Schumann, sold similar devices, many of which Wieck constructed together with his son Alwin. All of these devices were ineffective and had the potential for doing serious damage.

The twentieth century also witnessed a proliferation of exercises that purportedly stretched the fingers apart.[40] Typical exercises prescribed by Dohnányi, Pischna, Marcus, and Philipp—among many others—advocated holding down the notes of a widely spaced chord, lifting only one or two fingers high above the keys, and then sounding those fingers as loudly as possible while keeping the elbow close to the body and the wrist low (Example 1.2).

EXAMPLE 1.2. *Excerpt of exercise from Isidore Philipp's* School of Technic for the Piano *(1908)*

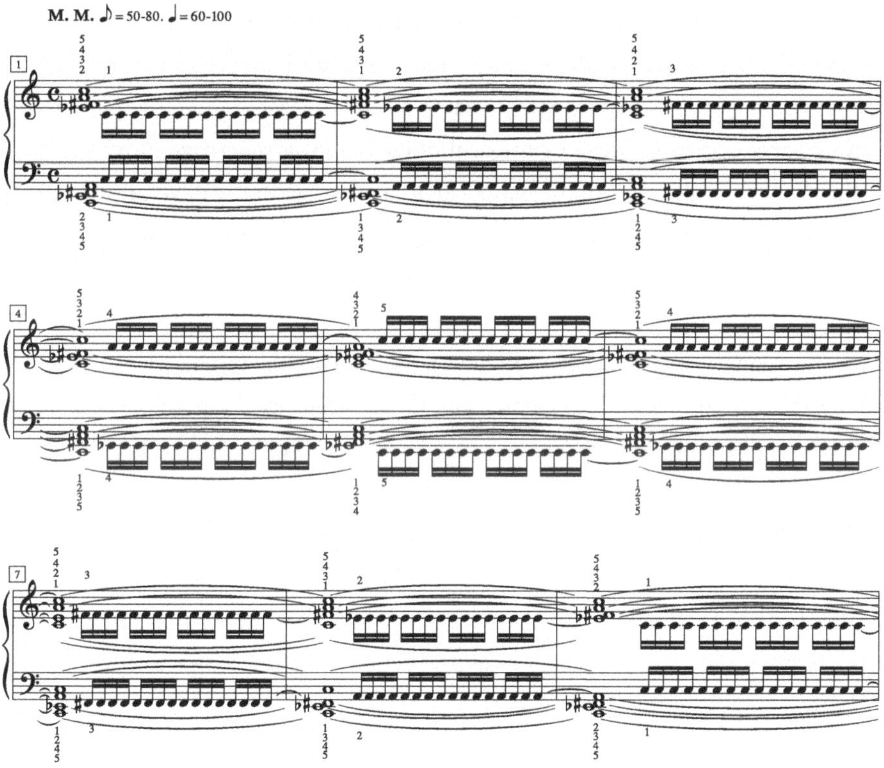

The reason that stretching the hand by means of devices and exercises of this specific type is especially problematic for small-handed pianists is due to the anatomical design of the hand. The tendon that lifts the fingers off the key surface, the extensor digitorum, is divided into strands that attach to the fingers.[41] However, the extensor attachments to the

[40] De Larrocha and Marcus recommended stretching exercises at the piano; Ortmann, Gát, and others recommended gymnastic exercises for strengthening away from the piano, as discussed by Eger, "A Study of the Problems Encountered by the Pianist with Small Hands," 53–54 and 67–70.

[41] See the section on the anatomy of the hand in Chapter 2 for a fuller explanation.

third, fourth, and fifth fingers are interconnected so that these fingers are limited by design in their ability to lift independently. This limitation is compounded when the hand is opened to its limit. Furthermore, there is some evidence that playing with the hand fully stretched may be a risk factor contributing to injury, especially among small-handed pianists.[42]

While the body can change to accommodate tasks performed over time, the extent of physical remodeling is not only impossible to qualify or quantify but varies widely according to individual physical limits. Despite anecdotal accounts of individuals who express confidence in the value of stretching exercises or devices, there is no empirical evidence demonstrating that these exercises or devices can physically extend the reach of the hand. In fact, routines or mechanical appliances of this kind have been implicated in a number of devastating injuries, most famously the case of Robert Schumann. They ought to be viewed with extreme caution and are perhaps best avoided entirely.

Discretion is probably warranted any time the small-handed pianist is engaged in playing repetitive patterns or exercises, such as those exercises where the objective is to inculcate a particular movement or musical objective. Any repetitive movements, even those that lie comfortably within the hand, can increase risk for the development of repetitive strain injuries, as will be discussed in Chapter 2.[43] Alice Brandfonbrener, one of the founders of performing arts medicine in the United States, noted:

> If the musician, trying to master a difficult passage, plays the same series of notes in an unvarying rhythm without interruption, those tissues used repetitiously without adequate opportunity for rest and recovery become subject to injury. While repetition is essential to most musical learning, without modifications it can be hazardous.[44]

The danger may be even higher if the pianist's attention wanders due to the unchanging nature of a drill or exercise. If pianists choose to practice exercises or repetitive patterns of any kind, they should carefully weigh potential benefits against risks and always be attentive to both mental and muscular fatigue.

SMALL-HANDEDNESS AND INJURY

Several research studies have implicated hand size as a distinct factor in the development of physical pain and injury among pianists. In a study of 66 pianists, De Smet, Ghyselen,

[42] Sakai, "Hand Pain Related to Keyboard Techniques in Pianists," 63–65.

[43] Susan J. Hall, *Basic Biomechanics*, 3rd ed. (Boston: WCB McGraw-Hill, 1999), 78–79.

[44] Alice G. Brandfonbrener, "The Etiologies of Medical Problems of Performing Artists," in *Performing Arts Medicine*, 2nd ed., eds. Robert Thayer Sataloff, Alice G. Brandfonbrener, and Richard J. Lederman (San Diego: Singular, 1998), 28.

and Lysens found that females with an injury had a shorter hand length (as measured from the crease of the wrist to the tip of the third finger) than females without injury.[45] Farias and others studied 341 professional and student pianists. Of the 222 players who reported experiencing a repetitive strain injury, 60% had hands that were unable to reach a tenth.[46] A study by Yoshimura and others of 35 university piano majors in 2006 identified small-handedness, particularly having a small span between the third and fourth fingers, as a distinct risk factor for the development of performance injuries.[47] In 2008, Yoshimura studied 47 piano teachers and again found that having a smaller hand size and less hand strength were strongly correlated with reports of pain.[48]

Sakai studied the relationship of injury to specific hand motions on the piano. In one study, he found that 30 of 40 pianists reported that they had been practicing a particular skill at the onset of their injuries.[49] In a follow-up study, he found that 70 of 200 pianists attributed their injury to a specific technique.[50] In both studies, octave and chord playing accounted for approximately two-thirds of these attributable injuries, with *fortissimo* playing, scales, arpeggios, and wide extended passages accounting for the rest.[51] Sakai thus concluded that playing with full extension of the thumb and fifth finger could lead to pain and injury, especially when done repetitively. This observation is particularly relevant to pianists with small hands, who generally must extend their fingers more than a pianist of average hand size.[52] In a separate study, Sakai and others observed that small-handed pianists played with their thumbs outstretched for longer periods of time than pianists with larger hands and also used more lateral deviations at the wrist. Because these wrist positions increase the stress on the interior compartments in the wrist, he postulated that using them frequently could contribute to development of an injury.[53]

Pianists with small hands may also be subject to higher joint and tendon forces. Hillberry pointed out that joint and tendon stresses can be determined by dividing the forces by the cross-sectional area receiving the force. The force necessary to depress a

---

[45] L. De Smet, H. Ghyselen, and R. Lysens, "Incidence of Overuse Syndromes of the Upper Limb in Young Pianists and Its Correlation with Hand Size, Hypermobility, and Playing Habits," *Annals of Hand and Upper Limb Surgery* 17 (1998): 312.

[46] Farias et al., "Anthropometrical Analysis," 229.

[47] Yoshimura et al., "Risk Factors for Playing-Related Pain among College Students," 123.

[48] Yoshimura et al., "Risk Factors for Playing-Related Pain among Piano Teachers," 107–113.

[49] Sakai, "Hand Pain Related to Keyboard Techniques in Pianists," 63–65.

[50] N. Sakai, "Hand Pain Attributed to Overuse among Professional Pianists," *Medical Problems of Performing Artists* 17 (2002): 178–180.

[51] Octave and chord playing accounted for 77% of all cases of injury in the 1992 study and 74% in the 2002 study.

[52] Sakai, "Hand Pain Related to Keyboard Techniques in Pianists," 65.

[53] N. Sakai, M. C. Liu, F. C. Su, A. T. Bishop, and K. N. An, "Hand Span and Digital Motion on the Keyboard: Concerns of Overuse Syndrome in Musicians," *Journal of Hand Surgery* 31A (2006): 831–835.

piano key is the same for a small-handed pianist as it is for a pianist with larger hands; however, a small-handed pianist will have smaller joint areas and thus higher joint stresses.[54]

In the world of music, the injured often deny their pain because they fear repercussions from a culture that tolerates no weakness.[55] Many small-handed players suffer in silence, overtaxing themselves to succeed on a keyboard that does not fit their hands. While additional research is still needed, findings from studies to date do point to small-handedness as a discrete factor for developing playing-related pain, fatigue, and injury.

THE ROLE OF THE TEACHER

Many pianists struggle needlessly to master skills more suited to larger hands. As primary purveyors of information about piano technique, teachers perform a critical role in helping young students learn the basic principles of healthy movement from their first lessons. Early training is important because kinesthetic sensations of movement develop from the onset of piano study, and maladaptive habits tend to be ingrained over time. It can be difficult to diagnose the reasons for physical discomfort in children because incoordination can stem from multiple sources such as small-handedness, a lack of fine motor control, or other developmental issues. Understanding the principles of physics and body movement discussed in Chapter 2 will improve a teacher's ability to pinpoint problems and to prioritize and sequence technical development. Fortunately, most elementary and early-intermediate pedagogical literature is written with a small hand in mind. However, the development of a physically coordinated technique cannot wait until more difficult repertoire is encountered.

Having laid the groundwork for efficient movement, the teacher should help students learn to recognize and solve problems. It is vital to remember that students are often completely unaware that their difficulties relate to their hand size. Since a student's sensory expectations regarding how it should feel to play the piano are still being formed, he or she may not experience inefficient movements or postures as awkward or fatiguing. It is up to the teacher to intervene and establish limits to the physical and mental struggle that is an intrinsic and necessary part of learning. Students who are taught about the inherent challenges of small-handedness and guided in the exploration of potential solutions can then be empowered to seek answers of their own.

Developing problem-solving skills goes hand in hand with teaching students healthy practice behaviors. Unfortunately, numerous studies indicate that most students do

---

[54] Benjamin M. Hillberry, "Dynamic Effects of Work on Musculoskeletal Loading," in *Repetitive Motion Disorders of the Upper Extremity* (Rosemont, IL: Academy of Orthopaedic Surgeons, 1995), 106.

[55] Parakalis, "A History of Lessons and Practicing," in *Piano Roles*, 152.

not receive explicit practice instructions and therefore develop an overreliance on repetition.[56] Since mindless repetition is ineffective and corrosive to motivation, teachers need to provide students with concrete practice strategies and establish reasonable guidelines on how much to practice each day, particularly in a single sitting. Teachers should also help students to monitor fatigue, perhaps by asking students to assign a scaled number to the fatigue levels felt in specific parts of the playing apparatus. Above all, students should be taught to identify pain with the extreme end of the fatigue continuum and to stop before that point is reached. Because no formula fits all pianists equally well and because students typically lack the background to assess these challenges, the teacher's guidance is critical.

Choice of repertoire represents another major concern. Repertoire that benefits the small-handed pianist in a physical sense and allows him or her to play the music with integrity should be selected. Pieces filled with extended *fortissimo* passages, large chords, and/or stretches that are just barely within reach of the hand should be modified or perhaps avoided entirely. However, the teacher should guard against assuming that a student's inability to master a passage is due to a lack of talent or insufficient practice. Problems that persist in spite of the small-handed student's best effort likely call for a different approach.

As pianists progress to playing intermediate and advanced literature, problems related to small-handedness increase in both number and complexity. Enthusiastic learners might be tempted to try to expand the reach of the hand by using mechanical devices or stretching exercises. However, the history of the piano is replete with accounts where this approach yielded disastrous results. Choosing appropriate repertoire and implementing adaptive strategies such as those recommended in this book are safer alternatives.

CONCLUSION

The authors have known many gifted performers who faced great difficulties playing standard-sized keyboards. Their special needs have been frequently ignored, misunderstood, or underestimated by those who have never had to grapple with issues of hand size. In the past, there has been little conversation among pianists about how to accommodate small hands at the instrument. However, as musicians increasingly accept that physical injury can arise in even the most talented of players, attention has turned to defining the healthful boundaries of piano playing. There is more interest in exploring and addressing

[56] Gary E. McPherson and James M. Renwick, "A Longitudinal Study of Self-Regulation in Children's Musical Practice," *Music Education Research* 3, no. 2 (2001): 169–186; Sarah J. Bartolome, "Naturally Emerging Self-Regulated Practice Behaviors among Highly Successful Beginning Recorder Students," *Research Studies in Music Education* 31, no. 1 (2009): 37–51.

the challenges of small-handedness. Even though these hurdles cannot be solved in the space of a few paragraphs or even this entire book, they are not insurmountable. Finding workable physical and musical solutions will require ingenuity and continuing diligence on the part of all.

REFLECTION POINTS

1. Considering the various criteria discussed in this chapter, do you think you have small hands?
2. What criteria can you use to determine whether a particular piece fits your hand?
3. In terms of the repertoire you prefer to play, what pianistic tasks or techniques come naturally to you? What tasks or techniques are the most challenging?
4. Have you experienced fatigue or strain while practicing or performing? What piece were you playing, or what technique were you using at the time? Have you played on instruments with keyboards that were either smaller or larger than the conventional keyboard? How did it affect your power and accuracy?

TWO

Fundamental Principles of Movement and the Playing Apparatus

TECHNIQUE IS THE PHYSICAL MEANS BY WHICH MUSICIANS TRANSLATE notated or imagined music into sound. There are, however, as many ways of playing the piano as there are pianists. This chapter is not intended as an exhaustive treatise on technique or an endorsement of a particular school of playing. Rather, it aims to establish a basis for understanding the general physical and anatomical principles that govern movement at the piano so that the challenges of small-handedness can be clearly analyzed.

PRINCIPLES OF WORK EFFICIENCY

For the first 200 years of the piano's existence, technical training was informed by the experiential knowledge of well-known pianists and handed down from teacher to student through subsequent generations. Beginning in the nineteenth century, many famous virtuosos, including Clementi, Hummel, Czerny, and Kalkbrenner, began to publish pedagogical methods that were widely disseminated throughout the Western world. As the twentieth century dawned, pianists increasingly turned to science, hoping to discover truths that would be universally generalizable. Otto Ortmann, a pioneer in this regard, sought to examine piano technique from an empirical rather than a subjective perspective, bringing to bear principles from anatomy and physics as well as the nascent science of biomechanics. Ortmann's 1929 book, *The Physiological Mechanics of Piano Technique*, was sharply criticized because it challenged long-standing pedagogical practices.[1] Many

[1] Otto Ortmann, *The Physiological Mechanics of Piano Technique* (1929; reprint Toronto: E. P. Dutton, 1962). The reprint was made at the request of Egon Petri and contained a foreword by Arnold Schultz,

pianists felt that the objectification of technique was antithetical to making good music. Since Ortmann's time, however, pianists have become more accepting of the idea that science can provide valuable observations about the limitations and possibilities inherent in physical bodies. As the music world awakened to the reality of high rates of occupational injury, pianists began to examine technique from a biomechanical standpoint to discover more healthful and efficient ways of playing.

With this goal in mind, several scientific disciplines are germane to a study of piano technique relative to the human body. Ergonomics is the study of interactions between humans, the objects they use, and the environments in which they function.[2] Most of the ergonomic problems that arise from playing the piano are cognitive/psychomotor, musculoskeletal, or anthropometric in nature. Cognitive/psychomotor problems pertain to how the brain takes in and processes information and how this impacts execution; an example of a psychomotor problem is maintaining adequate attention during piano practice or performance. Musculoskeletal problems can arise if biological limits are exceeded; solving pianistic musculoskeletal problems involves redesigning the task to fit the player's capabilities. Anthropometry concerns itself with how well the human body fits with the work interface.[3] Anthropometry is typically expressed in terms of linear measurements of body dimensions, including length, weight, and volume. Some anthropometric considerations for pianists include distance between the bench and keyboard, height of the music rack relative to gaze, and, of course, the size of the keyboard relative to the size of the hand.

Like ergonomics, the science of biomechanics considers the functional design and limits of the human body. Biomechanics uses traditional engineering sciences, including applied mechanics, continuum mechanics, mechanism analysis, and structural analysis, to analyze biological systems. Additionally, both kinesiology and kinematics play prominent roles in biomechanics. Kinesiology is the study of human movement, and kinematics is a branch of classical mechanics that describes the motion of points, objects, and groups of objects.[4] The basic principles of Newtonian physics underlying all of these disciplines are treated later in this chapter.

Each discipline has contributed to the growing body of knowledge about factors that can facilitate or interfere with the body's ability to do work and offers useful lenses through which to evaluate piano technique. Table 2.1 summarizes these factors, several of which are explored in greater detail below.

---

who published a book in 1969 called *The Riddle of the Pianist's Finger* that was highly indebted to Ortmann's study.

[2] B. Mustafa Pulat, *Fundamentals of Industrial Ergonomics* (Englewood Cliffs: Prentice Hall, 1992), 3.

[3] *Ibid.*, 9.

[4] Kinesiology, the study of movement, is not to be confused with Applied Kinesiology, which is an alternative health-care method of diagnosis and treatment based on the belief that weakness in a particular muscle signals a problem with an associated internal organ, gland, blood supply, nerve, or chemical imbalance.

TABLE 2.1 General work factors that impact piano playing

Category	Examples and Descriptions

Physics	• Force loading
	  • Amount of force applied
	  • Direction/vector of force (linear or rotational)
	  • Manner of application:
	    • Static (gripping, holding) versus dynamic (force that results in visible movement) loading
	    • Smooth application versus "jerk" (sudden linear changes of direction), impact
	• Inertia and momentum: Friction, resistance, and inertia must be overcome by momentum for movement to take place.

Biomechanics/ Anatomy	• Work-rest cycles: Frequency, repetition, and duration of rest periods between work cycles
	• Posture
	  • Positional postures maintained without movement can be more tiring than dynamic or flexible ones.
	  • Bending and other awkward or asymmetrical postures are more stressful than neutral postures.
	• Individual variations, such as endurance or strength

Cognitive Factors	• Motivation and attention • Complexity of task: Difficult/complex tasks require more attention and are more taxing than easier or more familiar tasks. • Focus and concentration: Uninterrupted work for long periods of time without a break tends to corrode attention and focus as do attempts to do more than one task at a time. • Critical thinking and problem solving: Ability to break complex problems into smaller parts and to choose appropriate strategies to target problems • Self-regulation and persistence: Ability to plan and carry out work, even in the face of challenge or frustration
Psychosocial Factors	• Perception of reasonableness of expectations • Concerns about evaluation of work • Control over timing and speed of work • Perception of being overwhelmed by work • Having tight or frequently changing deadlines • Having proper training to complete a task
Environmental Factors	• Light • Noise • Temperature (Colder environments tax the body more than warm ones.) • Distance of music from eyes and angle of viewing
Task-Specific Factors	• Parameters or requirements of a specific task at the piano (for example, opening the hand to play a large chord)

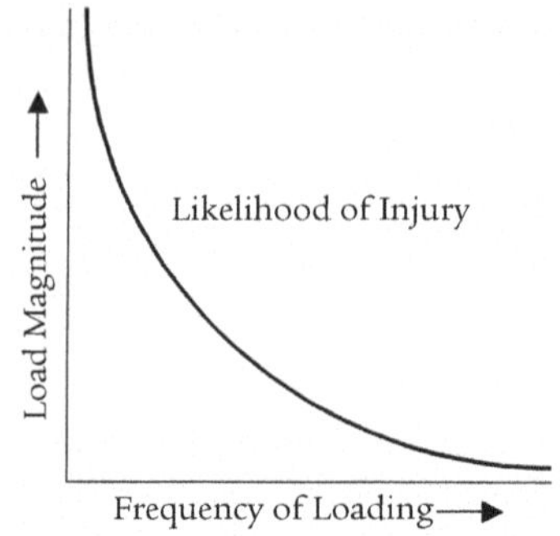

FIGURE 2.1. *Relationship between load magnitude, repetition, and likelihood of injury. Credit: Susan J. Hall,* Basic Biomechanics, *3rd ed. (1999), 79, © McGraw-Hill Education. Reproduced by permission.*

Interactions among force load, frequency, and fatigue are a major factor in how work impacts the body. Too much of a sudden single force—a fall down the stairs, a football tackle, or an automobile accident—can inflict acute traumatic damage to tissues. However, injury can also result from cumulative wear by lower forces that are applied repeatedly.[5] Injuries associated with piano playing tend to develop over time. The forces generated by piano playing are not sufficient to cause acute tissue failure in a single application; however, with enough repetition, even small force loads can cause pain or injury, as illustrated by Figure 2.1.

Repetitive Strain Injuries (RSIs) occur at a high rate in occupations requiring long hours of repetitive work—typing on computer keyboards, packing or unpacking boxes in warehouses, operating machines on assembly lines, or indeed, playing the piano. RSIs typically involve the soft tissues of the body. They most commonly affect the tendons and nerves of the shoulder, upper arm, forearm, wrist, and fingers but may also impact joints or muscles in these areas. Though noted as early as 1717 by Ramazzini, RSIs began receiving much more attention in the 1980s. Ergonomic experts identify the following as primary risk factors:

- Unnatural joint posture, including deviations at the wrist. Risk of injury increases notably when extreme joint positions are used.
- Application of force through hinge joints, such as the wrist. Pivot joints such as the elbow are also at risk, but to a lesser extent.
- Repetition of highly repetitive tasks of short duration. Over time, risk of injury increases exponentially.
- Individual factors such neuropathy, arthritis, circulatory issues, reduced estrogen levels, and small hand or wrist size.

---

[5] Susan J. Hall, *Basic Biomechanics,*3rd ed. (Boston: WCG McGraw-Hill, 1999), 78–79.

When these factors are combined, the risk of injury rises dramatically.[6]

Ergonomics has also shed light on how musculoskeletal fatigue relates to injury. Muscle fatigue may be simply defined as a decline in the ability to maintain a desired intensity or power.[7] It has been identified as a risk factor for injury, though its precise role is not completely understood. For one thing, fatigue can be very difficult to quantify and describe. Additionally, as previously mentioned, injuries can develop over time, even when the forces involved are small. Researchers believe that fatigue contributes to a process of progressive damage when healthy tissue in the body is subjected to higher levels of force or more repetitions than can be successfully repaired through normal recovery processes. Once this threshold is exceeded, microscopic tears develop. If the damaged tissue then continues to be exposed to forces or repetitions, the tears will expand, worsening the injury.[8]

The issue of how much force a body can bear is also difficult to quantify. How much force constitutes "high" force? In ergonomic studies, force is often contextualized as a percentage of effort relative to the failure limit of tissue. However, failure limits cannot be defined with a hard-and-fast numerical value and vary somewhat from one individual to the next. According to fatigue failure theory, the maximum stress a body tissue can withstand is the amount of force that results in complete tearing or rupture with just one loading cycle. If the stress load is lowered, the material should be able to withstand more cycles before failing. Carrying this reasoning further, there is a limit at which repeated loading will not lead to failure; this safe zone is generally accepted to be around 30 percent of ultimate stress.[9] There is thus an ideal range in which repetitive loading will not result in fatigue of the material.

However, it is unclear where this range lies with regard to piano playing. First, living tissue can repair itself, but not at a steady rate. Rather, healing is influenced by a number of factors and is highly individual. Second, it would be unethical to induce complete tissue failure in living human beings, and thus it is impossible to determine ultimate limits. Finally, since in reality the body functions as a system of parts, it would be very difficult to quantify force in a single part or to determine what percentage of the ultimate failure limit is attributable to a particular motion. For these reasons, the ultimate fatigue limits of the various segments of the playing apparatus cannot be precisely determined, nor

---

[6] Pulat, *Fundamentals of Industrial Ergonomics*, 49.

[7] Sheila A. Dugan and Walter R. Frontera, "Muscle Fatigue and Muscle Injury," *Physical Medicine and Rehabilitation Clinics of North America Journal* 11, no. 2 (May 2000): 391.

[8] Sean Gallagher and John R. Heberger, "Examining the Interaction of Force and Repetition on Musculoskeletal Disorder Risk: A Systematic Literature Review," *Human Factors: The Journal of the Human Factors and Ergonomics Society* 55, no. 1 (February 2013): 108–124.

[9] *Ibid.*, 117.

is it possible to quantify how much force application or repetition lies within the ideal safe range. Clearly the forces used to play the piano do not exceed ultimate stress limits, or no pianist would be able to do so successfully more than once. It is apparent, however, that fatigue influences work efficiency and can lead to the development of playing-related injuries. Furthermore, having a small hand may predispose a pianist to fatigue more quickly, especially when playing large shapes on the piano. Thus, it is important to consider how much muscular effort is truly necessary in a functional piano technique and how this effort can be minimized and sustained over time.

PRINCIPLES OF MOVEMENT

To build an understanding of piano technique, small-handed pianists should develop a familiarity with the factors that govern work efficiency as well as a basic biomechanical understanding of how the body accomplishes work.

Leverage

The skeleton of the body is a system of levers moving around joints. Joints serve as a fulcrum for motion, or they can be stabilized or fixated by muscles and ligaments in order to bridge smaller levers together to create longer levers. Joint fixation as it is used when playing the piano does not entail a prolonged state of joint rigidity or stiffness; rather, joint fixation should be released as soon as it is practical to do so.

Pianists commonly speak of the wrist or elbow leading a motion, but this is a misnomer because both are joints. Joints are acted upon during movement; they do not initiate movement. While the wrist and elbow trace distinct paths through space and are thus useful landmarks for observing motion, producing and controlling movement is the job of muscles working at the command of the nervous system.

Opposing muscle-tendon groups located on either side of each lever cross over and under the joints and cooperate to move the levers of the skeleton much like ropes around a pulley. Generally, the muscles primarily responsible for movement of a particular lever lie on either side of the next lever closer to the center of the body. For example, the muscles controlling hand movements that are visible at the wrist lie in the forearm, while the muscles that lift and lower the forearm at the elbow lie in the upper arm.

Muscles act on the levers of the body through the development of tension. In the music world, "tension" often carries a negative connotation, but in reality, muscle tension is necessary to create movement; playing the piano is impossible without muscular tension. To produce a loud chord for instance, muscles must contract to produce the energy for tone production. Then other muscles must contract to stabilize the joints of the fingers, hand, and wrist so that they will not collapse from the impact against the keys. When pianists refer to playing with tension in the negative sense, they are really commenting on the accumulation of muscular tension over time in a way that is not conducive to movement. The timing of contraction and release is off, resulting in visible rigidity. For instance, if both the biceps on the front of the upper arm and the triceps on the back of the upper arm are activated to the same degree and in direct opposition, the upper arm cannot move through space. The goal of efficient piano playing is not to avoid tension per se but rather to activate the right muscles at the right time only up to the necessary level of intensity so that muscular effort is maximized and fatigue is avoided.

Depending on the movement being produced, most of the voluntary muscles in the body are capable of acting as agonists, antagonists, neutralizers, and stabilizers. Different forms of muscular tension are used in these varied capacities.[10] The muscles that most directly cause movement are functioning as agonists. At the same time, antagonists, generally the opposing muscles on the other side of the lever, either relax to allow the agonists to work unopposed or contract to slow, control, or stop the movement. Muscles may also work as neutralizers, preventing undesirable movements that may occur when agonists contract. For example, when lifting the arm to prepare for a chord attack, the biceps muscle of the upper arm acts as an agonist, pulling the forearm toward the upper arm while the triceps muscles on the opposite side of the upper arm function as an antagonist, relaxing to allow the biceps to lift the forearm unopposed. However, flexing the biceps also rotates the forearm slightly toward the fifth finger, so in order to keep the thumb and fingers properly positioned, additional muscles must be contracted to neutralize forearm rotation.[11] Muscles can also contract to stabilize other muscles, bones, or joints. If opposing muscles generate the same amount of force at the same time, muscle length does not change, so no movement occurs. This is known as isometric contraction. For example, when playing hand staccatos at the piano, muscles on both sides of the knuckles and finger joints isometrically contract to stabilize these joints so that force can be efficiently transmitted to the key. Healthy use of isometric contraction is a matter of degree and timing rather than avoidance; it is only undesirable when applied to a part of the anatomy that does not need to be stabilized, when it reaches an excessive degree, or when the pianist fails to release it as soon as it is no longer needed.

When considering the various motions of piano technique, it is useful to know which muscles are serving as the primary actors and which muscles are playing supporting roles.

---

[10] Concentric contraction occurs when a muscle shortens as it develops tension. Eccentric contraction occurs when a muscle lengthens as it is tensed. The third type of muscular tension, isometric contraction, is explained later in this paragraph.

[11] Hall, *Basic Biomechanics*, 165.

Sometimes different muscles share the same function. Furthermore, even when there is no visible movement, muscles may still be contracting. Because piano playing is so complex, completely describing the activity and role of every muscle would be not only impossible but impractical. Instead, this book focuses on the most active agents that initiate various movements.

Muscles direct and control the levers, which can be used singly or combined with other levers. Finger segments represent the smallest levers used to play the piano. Smaller levers can be combined with their nearest neighbors to form longer levers. For example, all three bones of a particular finger can be used together with their adjacent hand bone to form one lever with the fulcrum at the wrist. Similarly, all the bones of the finger, hand, and forearm can function together as a single lever with the fulcrum at the elbow. This is a simplification, however, because levers are capable of working against a fixed base of movement as in the example above, or against a moving base. For instance, the finger can work with the hand as a base, but with the forearm moving in a different direction.

In any case, not all levers and muscles are equally suited for a particular task. Smaller, lighter levers like the fingers work well in fast-moving passagework, but they are not able to produce powerful chords without assistance from the larger levers. Conversely, the longer levers are useful for producing power, but they cannot move through space as quickly. The same can be observed regarding the muscles of the body. Smaller muscles, such as the intrinsic muscles of the hand, are suitable for light, quick motions that are limited in range of movement and power, but they quickly tire when exposed to higher forces. Larger muscles, such as those found in the upper back, can generate power and resist fatigue at slow tempos but cannot be activated in quick succession. Thus, piano technique necessitates a constant negotiation of activity between smaller and longer levers and muscles.

As the force required for a task increases, muscular activity spreads up the arm to successively larger muscle groups. Ortmann observed that additional joints and levers are brought into play and that this transfer of activity must be implemented gradually in order to produce smooth, unbroken, multi-dimensional movements.[12] However, spread of activity can work against the pianist if he or she uses a lever that is longer than necessary, especially when rapid changes of direction or lightness of touch are needed.[13] For instance, contracting the muscles of the upper arm would work against a pianist's ability to play individual keystrokes in fast scalar passages. Thus, the ability to achieve the right balance between small and large playing units is critical.

---

[12] Ortmann, *Physiological Mechanics of Piano Technique*, 34 and 114.

[13] *Ibid.*, 289.

Newton's Laws and Piano Technique

As mentioned above, the science of biomechanics relies heavily upon the laws of Newtonian mechanics in order to analyze and describe movements made by the human body. Piano playing is governed by these principles, and understanding them is beneficial in cultivating effective and efficient piano technique. Newton's three laws, which form the basis for classical mechanics, are

- Law of inertia: An object in motion remains in motion unless external force acts upon it.
- Law of acceleration: Force is proportional to the acceleration of mass.
- Law of reaction: For every action there is an equal and opposite reaction.

Newton's first law, the law of inertia, states that an object in motion tends to remain in motion in the same direction of travel, and that an object at rest tends to remain at rest unless acted on by an external force. Inertia is the property by which objects continue in their existing state. The greater the mass of an object, the more inertia must be overcome to move or stop it. In a leverage system, longer levers produce more power, but they must overcome greater inertia. In the context of piano playing, the forearm and whole arm are typically used as power levers. However, they have more mass and respond more slowly. On the other hand, fingers have less mass and move more quickly but are limited in power. Therefore, larger tone-producing units are desirable for slow, sustained tone production, but the fingers are more suitable as primary levers where speed is needed.

The law of inertia also has implications for the vector of motions used to play the piano. Vector is defined as a quantity having direction as well as magnitude in determining the position of one point in space to another. Once a lever is set in motion, it continues along the same path until either the keybed or another muscular contraction redirects or stops it. Because the levers of the body move around joints, it is impossible to make a smooth, steady movement in more than one plane without using a curve.[14] Thus, a lever moving through space purely by its inertia will describe a curve rather than a straight line. Angular gestures that stop and start are normally less coordinated than curves because momentum must first be actively interrupted and then additional muscular contraction must take place to change direction.[15] In common parlance, smooth curvilinear movements are more economical in piano playing because less work is required to redirect motion than to regenerate it. For example, when playing a smooth ascending and

[14] *Ibid.*, 29–30.

[15] Physics describes this phenomenon as "jerk" (also known as "lurch" or "jolt").

descending four-octave arpeggio, the forearm will describe a curvilinear shape, especially when the arpeggio changes direction.

Newton's second law, the law of acceleration, also has profound implications for pianists, small-handed pianists in particular. This fundamental law of physics describes interactions among force, mass, and acceleration (or momentum) and is typically expressed as $f=ma$. Force and acceleration are directly related, as are force and mass. More power at the piano can be produced by increasing acceleration and/or adding mass by using longer levers, a choice that will be influenced by the musical context.[16]

Newton's third law states that for every action there is an equal and opposite reaction. This means that movement can only result from unbalanced forces, where force applied in one direction is greater than its resistance from the opposite direction. If the force applied to the piano key is sufficient to overcome the resistance of the key, then the key is depressed. The amount of force and the speed at which it is applied affects how far and fast the key is depressed, which in turn produces louder or softer sounds. The direction in which force is applied also matters. Greatest power is achieved when force is applied at a 90-degree angle to the lever. Less power is produced as the angle decreases until it reaches zero at zero-degree force application angle. The speed effect is the exact reverse: speed is least when force acts at right angles to the lever arm and increases as the angle of force application decreases.[17] The angle at which a muscle acts, therefore, helps determine the work it can accomplish. As the resistance to be overcome increases, parts of the body are repositioned so that force can be applied more advantageously.[18] For example, fingers deliver more force when they are curved (working at an angle approaching 90 degrees in relation to the key) than when they are flat. The basic precepts of Newtonian physics underscore the importance of choosing the right combination of levers, applying them in the correct manner, and favoring curvilinear over angular gestures.

Muscular Contraction and Release

Coordinate piano technique depends on the proper timing of contractions and releases. As discussed in the section on leverage, muscles arranged on either side of the bones produce movement; the brain sends signals to the muscles via the nerves, and the muscles respond by contracting. Electromyography (EMG), a tool used in scientific studies and biofeedback, measures muscle activity in the body and produces a tracing useful for visualizing this activity. The surface EMG tracings shown in Figures 2.2a and 2.2b

[16] The relative merits of using acceleration and/or mass to produce power are discussed in Chapter 3.

[17] Ortmann, *Physiological Mechanics of Piano Technique*, 6–7.

[18] *Ibid.*, 113.

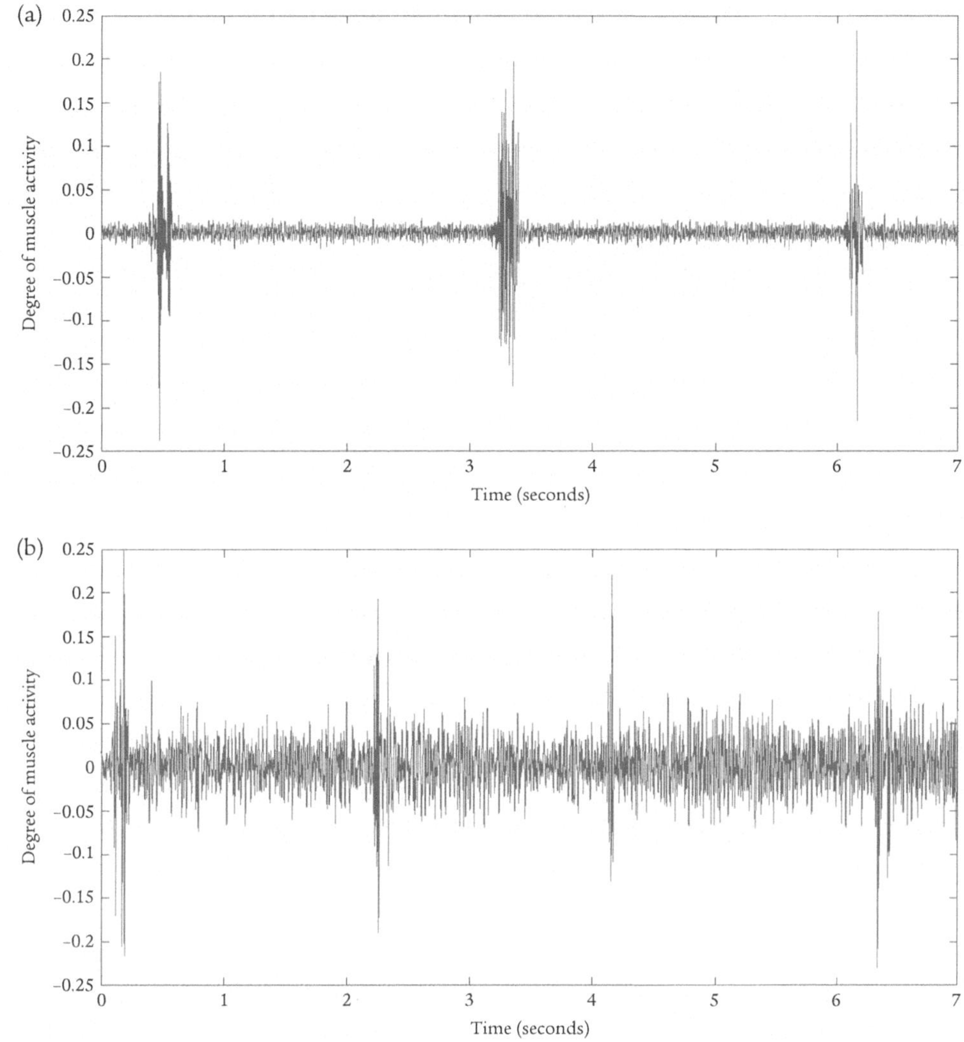

FIGURE 2.2. *Surface EMG tracing of muscle activity. Credit: Greg Bashford. Used by permission.*
a. Efficient dynamic muscle contraction
b. Inefficient dynamic muscle contraction

contrast two types of muscular contraction when the pianist is playing *fortissimo* chords. Figure 2.2a shows efficient dynamic contraction—the muscle contracts vigorously to play the chord and then quickly releases to its initial state of rest, referred to as the baseline, prior to the next chord attack. Returning to the baseline provides the maximum potential for the next muscle contraction. In contrast, Figure 2.2b illustrates inefficient contraction. Discrete muscular contractions are still visible coinciding with each chord attack; however, the muscle does not release completely to its resting baseline between contractions but instead continues to exert isometric tension. This extraneous muscular tension does not contribute to, and in fact interferes with, producing the movements necessary for the task.

Some isometric loading of the muscles of the neck, shoulder, and upper back is necessary to maintain a seated position at the piano with the arms in playing position. However, static contraction can be counterproductive where body movement is necessary. Pianists may subconsciously and unnecessarily statically contract larger muscles to compensate for fatigue in other body parts or in reaction to the stresses of performance. These excessive or prolonged states of contraction restrict blood flow, and without a steady or adequate supply of oxygen, muscle tissues deplete their internal reserves of fuel and become tired. Waste products created by the muscles build up, leading to soreness and stiffness. On the other hand, efficient dynamic muscular effort that alternates between contraction and relaxation actually allows more blood to flow than if the muscle had remained at rest. Static effort without sufficient rest tends to result in more "wear and tear" than dynamic loading.[19] Thus, a primary goal of coordinate piano technique is to optimize the timing and release of muscle contractions.

Counterintuitively, it is the rate of muscular release rather than the rate of contraction that is implicated in the fatigue process. As a muscle tires, it is no longer able to release accumulated tension back down to baseline before the next stimulus is received and thus becomes progressively more exhausted. Additional time is required for the overworked muscle to release accumulated tension. At this point, the muscle has less potential for contraction because it is not starting from baseline. To avoid this vicious cycle, muscle contraction should take place a mere moment before, or just as, the resistance is met and should cease once the resistance is removed. Thus, after playing a *fortissimo* chord, the muscles of the arm should release as soon as the point of escapement is reached. Ortmann felt so strongly on this point that he added, "I am inclined to believe that the readiness with which relaxation sets in between movements, be they movements of fingers, hand, or arm, is a fair index of kinesthetic talent as applied to the piano."[20] While the ability to release muscular contraction immediately after tone production is a feature of coordinate movement, incoordinate technique is characterized by muscular contraction that is sustained far past the point of tone production, resulting in static loading.[21]

There may not be enough time to release muscular contractions after each and every key attack, particularly in loud or rapid passages. In this case, movements should be combined into larger physical gestures, and players should plan release points to correspond with structural or phrase groupings in the music. Pianists should cultivate an acute

---

[19] Pulat, *Fundamentals of Industrial Ergonomics*, 31–32; Mark S. Sanders and Ernest J. McCormick, *Human Factors in Engineering and Design*, 7th ed. (New York: McGraw-Hill, 1993), 297.

[20] Ortmann, *Physiological Mechanics of Piano Technique*, 120.

[21] *Ibid.*, 285.

awareness of muscular work and release relative to baseline so that the work of playing can be managed within comfortable limits. Developing the ability to regulate contraction and release is a priority for all pianists, but it is crucial for small-handed pianists.

Additional Characteristics of Movement Economy

In addition to managing muscular contraction and release, there are other principles of movement economy that apply to piano playing. One is cultivating curved rather than angular movements as previously discussed. Another is keeping the amplitude of these curved movements as small as possible, staying close to the keys. There are, of course, many occasions when pianists may make large dramatic gestures if there is sufficient time. However, keeping the hand close to the piano keys is advisable when fast tempos or quick changes of direction are called for.

Another important consideration is that equal and opposite motion through space does not entail equal and opposite muscular effort. Moving in toward the center of the body tends to require less effort than moving outward. Closing the hand is easier than opening it, and moving the arm in toward the torso is easier than the opposite. For instance, when playing contrary motion scales starting in front of the body, moving away from the body's center as the right hand goes up and left hand goes down takes more work than the reverse. The same asymmetry is true of rotational movements of the forearm toward or away from the body. This is discussed in more detail in the section dedicated to the movements of the forearm later in this chapter.

Finally, the position of the hands at the keyboard affects physiological comfort. Playing with the hands directly in front of the body or at the extreme outer registers is more awkward than playing with the hands neutrally situated on either side of the torso.

Planes of Motion

The human body moves in three dimensions. The field of biomechanics typically describes movements as occurring on and across three cardinal planes of motion: the frontal plane (movements observed while standing directly in front of or behind the player), the sagittal plane (movements observed from the side of the player), and the transverse plane (movements observed from any other vantage point, especially from above) (Figure 2.3). The frontal plane splits the body vertically into front and back while the sagittal plane divides the body vertically into right and left. The transverse plane separates the body horizontally into top and bottom. When observing or discussing piano technique, these perspectives are valuable to keep in mind as some movements may be visible in one plane, but not in another.

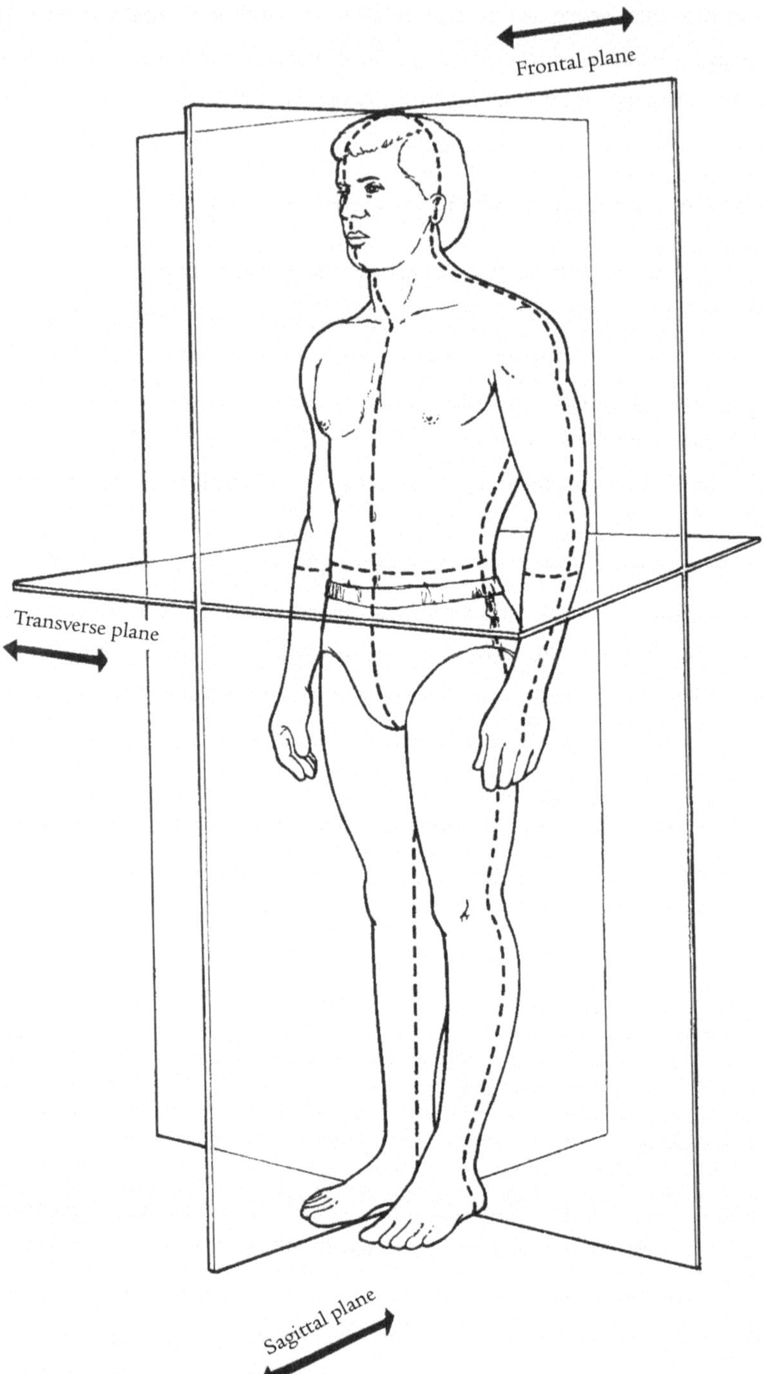

FIGURE 2.3. *The cardinal planes of motion. Credit: Susan J. Hall,* Basic Biomechanics, *3rd ed. (1999), 31, © McGraw-Hill Education. Reproduced by permission.*

Neutral Position at the Piano

In biomechanics and ergonomics, the standard neutral position used as a reference point for describing motion is one in which the body is standing face forward, arms down by the side with the palms facing forward, and thumbs pointing away from the body. Of course, this position is not really neutral in terms of muscular effort as the upper arms are rotated significantly outward and the muscles of the legs and core are contracted to support the body in standing position. Additionally, this biomechanical reference position is not particularly relevant to playing the piano. Thus, in this book, a task-specific neutral position that requires the least muscular effort to maintain has been adopted: the pianist is seated at the piano with the fingertips resting on the keys, ready for movement, but not yet moving. Pianists should seek to return or get as close as possible to this neutral position as soon and as often as they can.

THE PARTS OF THE PLAYING APPARATUS

Many books about piano technique focus exclusively on the arm, hand, and fingers. While these parts are important, the entire body is involved in piano playing, from the feet through the core to the fingertips. In addition to understanding the structure and function of the anatomy, all musicians need to cultivate an awareness of how to move their bodies efficiently and effectively.[22]

Torso

The spine is the central support column for the weight of the body, like a steel girder holding up a building. It also contains the nerves that transmit signals for movement from the brain to the rest of the body. Thus, spinal alignment is important for optimal function. In a neutral spinal alignment, all the natural curves—cervical, thoracic, and lumbar—are maintained (Figure 2.4). Exaggerated postures—slumping over or sitting excessively straight—are not neutral postures. Twisting the spine, leaning to either side, or leaning

[22] A list of some somatic training resources currently available, though by no means exhaustive, includes the Alexander Technique; the Taubman Approach; Barbara Conable and Benjamin Conable, *What Every Musician Needs to Know about the Body*; Thomas Mark, *What Every Pianist Needs to Know about the Body;* Julie Lyonn Liebermann, *You Are Your Instrument*; Barbara Lister-Sink, *Freeing the Caged Bird* (DVD); Paul Wirth, *The Gravi-DVD* (DVD); Seymour Fink, *Mastering Piano Technique: A Guide for Students, Teachers, and Performers*; Alan Fraser, *The Craft of Piano Playing: A New Approach to Piano Technique* (book and DVD); and Fred Karpoff, *3-D Piano Method*. Biofeedback, a technique that uses electromyography in real time to teach subjects to monitor the overall level of muscle tension, has also been adapted as a pedagogical tool for use by pianists, notably by Kathleen Riley.

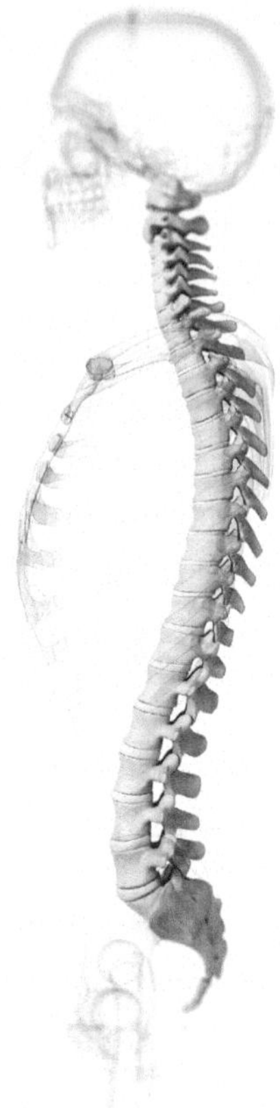

FIGURE 2.4. *Neutral curves of the spine. Credit: Sebastian Kaulitzki/Shutterstock.com.*

forward or backward also takes the spine out of neutral alignment. In the course of playing the piano, the spine may need to move out of its neutral position, but the goal is to return to neutral as often and as soon as possible.

Keeping the spine in its neutral position while seated requires balancing the body's weight evenly on the knobby bones called ischial tuberosities at the base of the buttocks. Yoga teachers sometimes refer to these as the sitting bones. Contrary to popular belief, the tailbone, or coccyx, does not come in contact with the seat in a balanced spinal alignment (Figure 2.5)

With the neck in neutral position, the head is balanced on top of the spinal column. When pianists hunch over with the chin jutting forward, a tendency often observed when playing *pianissimo*, static loading in the trapezius muscle that crosses the neck, shoulders, and upper back increases markedly, subjecting that muscle to fatigue. Tipping the head backward from its balance point also imposes extra work on the trapezius. This calls for somatic awareness as well as attention to ergonomic concerns, such as the placement of the piano bench and the music rack. Pianists usually center themselves in front of the keyboard for maximum access. However, on occasion, they may wish to sit closer to the top or bottom

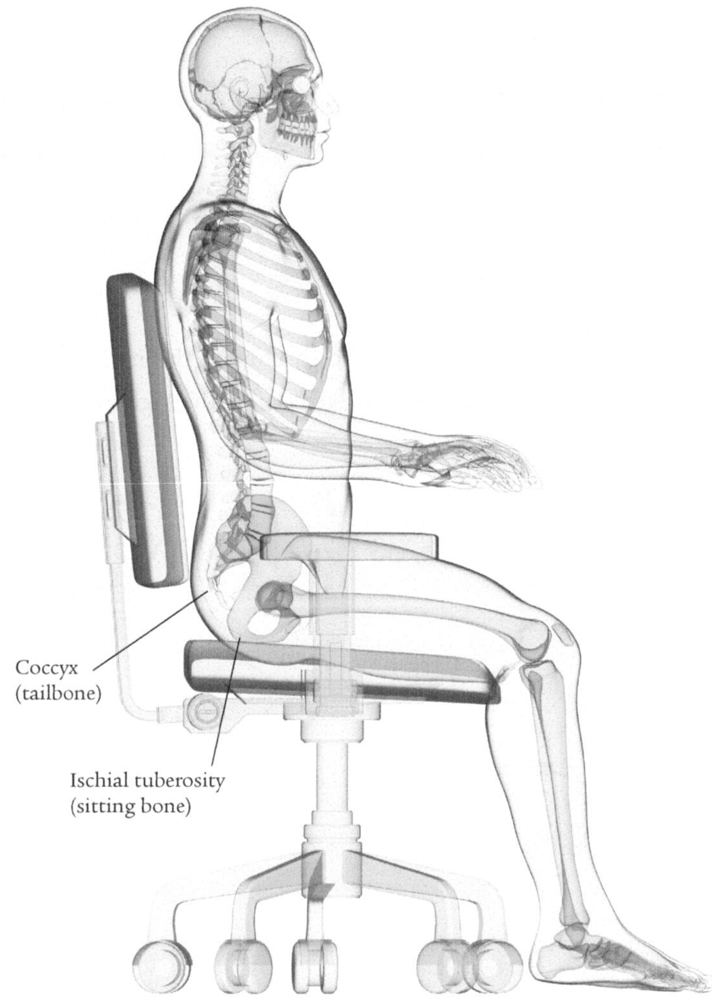

FIGURE 2.5. *Neutral seated position. Credit: Sebastian Kaulitzki/Shutterstock.com.*

end of the piano. For example, in pedagogical works where both right and left hands are notated in the treble clef, it makes more sense to sit near the treble side. Similarly, if rehearsing the secondo part of a duet, players should place themselves closer to the bass registers.

Pianists tend to overlook the importance of their feet in maintaining a balanced seating position. If the feet dangle, they can tug on the lower back, pulling the lumbar spine out of alignment. If the feet maintain contact with the floor, however, they provide stability for the lower body, which in turn grounds torso movement. This stability is especially important when the pianist needs to play at the extremes of the keyboard. Furthermore, the feet are essential for pedaling. Providing raised foot supports may be necessary for children and adults of short stature.

The torso provides a stable base of support for motion, and it may also be used—very selectively—as a lever itself. When combined with the entire arm, the torso can provide power in the form of increased mass. An example of this would be playing a single, powerful chord at the end of a section or a piece. Here, the torso and arms are thrown forward from the waist, and muscular effort is released immediately afterward. Of course, using the torso this way is best confined to instances when the pianist needs a single burst of power in relative isolation.

Shoulder and Upper Arm

The shoulder is the most complex joint in the body, allowing the arm to move in multiple planes (Figure 2.6). The arm can be flexed or extended in the sagittal plane, or abducted

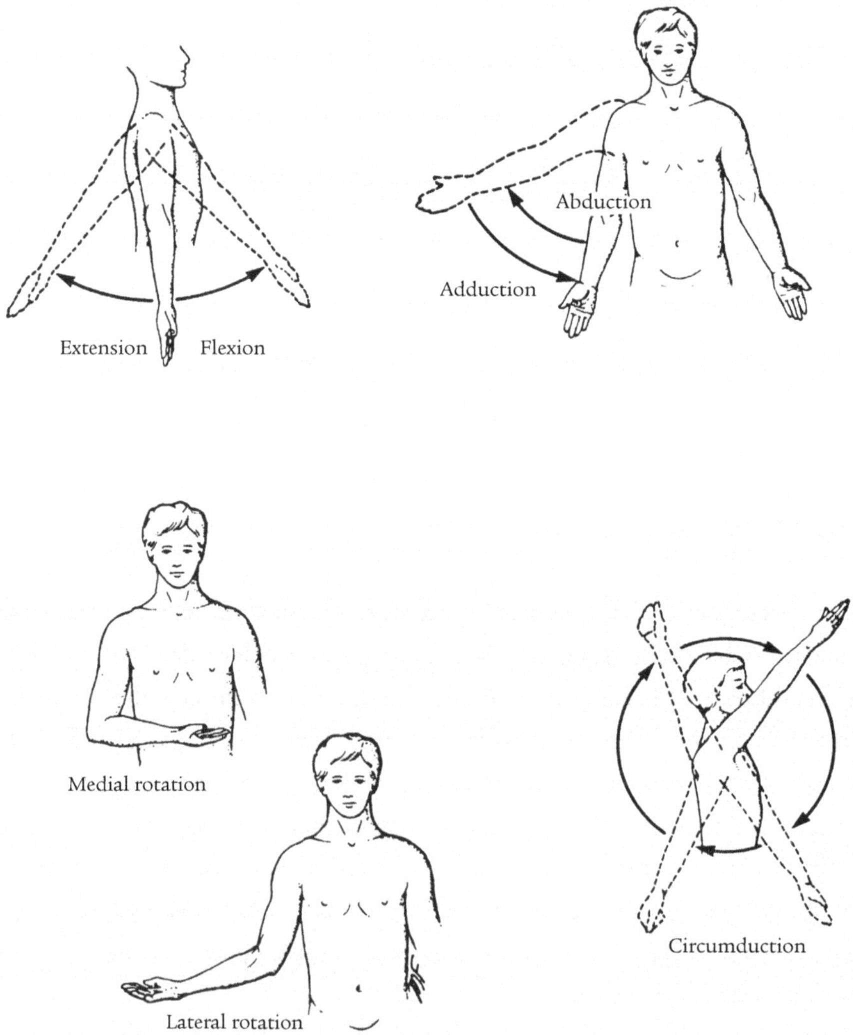

FIGURE 2.6. *Movements at the shoulder. Credit: Richard S. Snell,* Clinical Anatomy for Medical Students, *5th ed. (1995), 410, © Little, Brown and Company. Reproduced by permission.*

(moved away from the center of the body) or adducted (moved back toward the body's center) in the frontal plane. The bone of the upper arm—the humerus—can also be rotated in the shoulder joint across multiple planes of motion.

The muscles responsible for moving the upper arm lie between the base of the skull and the shoulder joint. The trapezius, a large, kite-shaped muscle running along the back of the body between the neck and arm, is one of the most active. It connects to the base of the skull, extends down across both shoulders, and finally attaches to the vertebrae behind the chest cavity. When shrugging the shoulders, it can be felt with the hand between the neck and shoulder joint. Raising the shoulders even slightly while playing can lead to excess static loading of the trapezius. It may also contract involuntarily in response to psychological stress; this tension is then added to workloads that are already present.[23]

The trapezius helps support posture. It suspends the shoulder girdle from the skull and spine and helps the neck support the weight of the head, especially when the head is tipped forward and is thus not balanced on top of the spine. The trapezius is also responsible for moving, rotating, and stabilizing the shoulder blade.[24] It can also activate the whole arm as a single lever to produce powerful large chords, although a price must be paid in terms of the increased mass that must be moved and the extra time needed to reset between consecutive uses. Due to its role in activating the arm and its tendency to develop static loading, the trapezius muscle is a frequent site of discomfort among pianists. It is thus important to constantly monitor the level of muscle tension in the trapezius and to release tension back down to muscular baseline at frequent intervals.

In neutral playing position, the arms should hang naturally from the shoulders with the elbows falling slightly in front of the rib cage, not held away from the torso. Rotating the upper arm forward in the shoulder girdle or lifting it away from the torso should be avoided unless needed for a particular purpose, such as reaching for the outer registers of the piano keyboard. Pianists should return to neutral position in between movements to avoid causing a high degree of static loading in the muscles of the shoulder girdle. The entire upper body, including the collarbones and shoulder blades, should be kept flexible so that the pianist can move the upper arm in or out, produce a good tone when playing in front of the body, or rotate from the shoulder joint to produce loud *fortissimo* chords.

---

[23] Ulf Lundberg, Roland Kadefors, Bo Melin, Gunnar Palmerud, Pater Hassmen, Margareta Engstrom, and Ingela Elfsberg Dohns, "Psychophysiological Stress and EMG Activity of the Trapezius Muscle," *International Journal of Behavioral Medicine* 1, no. 4 (1994): 354–370.

[24] Richard S. Snell, *Clinical Anatomy for Medical Students*, 5th ed. (New York: Little, Brown, 1995), 400–401.

Forearm

The forearm plays a very active role in piano technique. The forearm moves freely up and down using the elbow as a hinge joint, but it is limited in its side-to-side movement without the cooperation of the upper arm. The forearm generates or transmits force necessary for key depression and also plays a crucial role in navigating distances.

The muscles with primary control for the movements of the forearm are located in the upper arm. They generate power for keystrokes, help lift the weight of the forearm out of the keys, and serve as stabilizers for forearm motion. These muscles can be used to generate impulses that travel down the arm through a flexible wrist to produce bouncing motions of the hand. They are also involved in positioning the arm in space when navigating wide spans across the keyboard. When their capacity is exhausted, the trapezius muscle must then be recruited to help, which can result in exhaustion and rigidity if not monitored and controlled.

The forearm can join with the upper arm, hand, and fingers to form the powerful whole arm lever. It can also link up just with the hand and fingers to produce forearm staccato, a strong touch whose slightly percussive effect is caused by the impact of the fingers on the keys. The wrist joint is stabilized during the preparatory phase of movement and at the moment of key impact to allow the mass of the arm to be transmitted directly to the key. Playing with the whole arm or using forearm staccato involves longer levers and thus requires time.

If desired, the forearm can simply transmit impulses from the upper arm to the fingers. In this use, varying degrees of the forearm's mass are suspended by the muscles of the upper arm and shoulder, and impulses from the upper arm are sent through a pliable wrist so that the hand bounces in response yet the fingers remain on or close to the key surfaces. Impulses can be timed selectively—one per note or chord, or if the tempo is too fast for that, one per phrase or grouping. The force and repetition speed of these impulses can be tailored along a continuum to suit the musical context.

Forearm rotation is one of the most valuable resources in the pianist's technical toolkit, useful both for covering distances and for generating powerful sound. Rotation is the motion of the forearm used in turning water faucets, door knobs, or screw drivers. It is most visible at the wrist, but contrary to popular belief, it is not produced by the wrist. Rather, rotation results from crossing and uncrossing the long forearm bones of the radius attached on the thumb side of the wrist and the ulna attached on the fifth-finger side (Figure 2.7). Rotating toward the thumb (radial) side of the forearm is called pronation, and rotation toward the fifth finger (ulnar) side of the forearm is called supination.

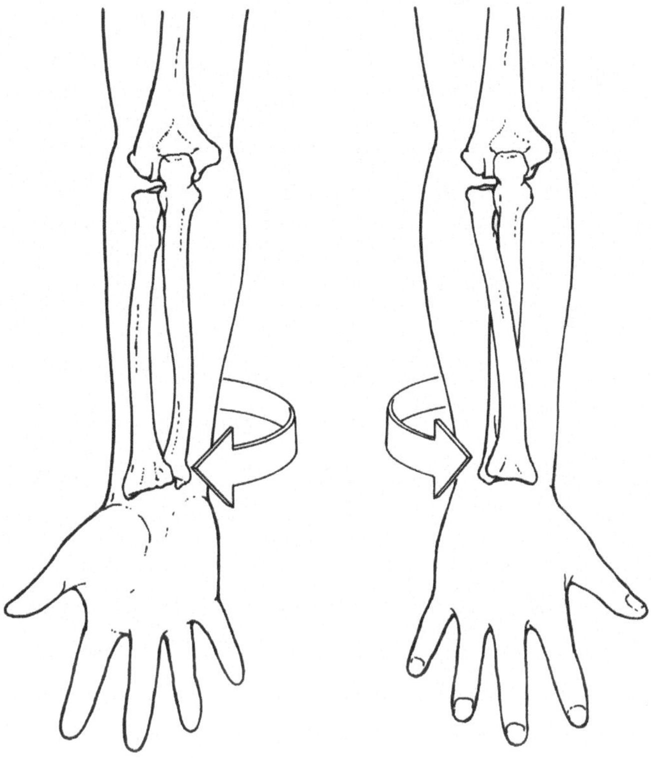

FIGURE 2.7. *Forearm rotation. Credit: Susan J. Hall,* Basic Biomechanics, *1st ed. (1991), 160, © McGraw-Hill Education. Reproduced by permission.*

Although supination and pronation are opposite motions, they are not, strictly speaking, equal motions. As Ortmann observed, rotating the forearm while using the fifth finger as a pivot to play the thumb is easier than rotating the forearm using the thumb as a pivot to play the fifth finger.[25] This in effect makes the thumb the most powerful digit at the keyboard. Movements produced by the forearm may often be visualized by tracking the movement of the wrist through space.

Wrist

To speak of moving the wrist—or any other joint—can cause confusion. The wrist does not act directly but rather is acted upon by muscles originating in the forearm and ending in the hand or fingers. Even though it does not generate motion, however, the wrist visibly describes motion through space as muscles of the forearm are activated. Thus, it can be a useful point of reference in making observations about piano technique.

[25] Ortmann, *Physiological Mechanics of Piano Technique,* 36.

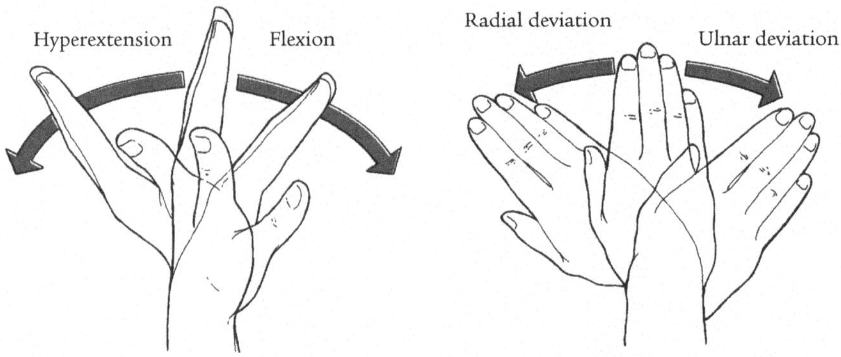

FIGURE 2.8. *Movements at the wrist. Credit: Susan J. Hall,* Basic Biomechanics, *3rd ed. (1999), 218, © McGraw-Hill Education. Reproduced by permission.*

The wrist allows the hand to move in multiple planes, as shown in Figure 2.8. Flexion and extension occur in the sagittal (vertical, observable from the side) plane of motion, while ulnar and radial deviation occur in the transverse (horizontal, observable from the top) plane of motion. Wrist flexion occurs when the hand is bent downward toward the underside of the forearm; extension occurs when the hand is bent back to neutral from a starting position of flexion. Hyperextension occurs if the hand bends beyond a neutral position toward the back of the forearm. Radial deviation is a position in which the hand is turned laterally so that the fingers point toward the thumb side of the wrist, while ulnar deviation is a position in which the hand is turned so that the fingers angle toward the fifth-finger side of the wrist. Vertical movements at the wrist (flexion and extension) are associated with key depression and release, having to do with force moving in and out of the keyboard. Lateral motions at the wrist (radial and ulnar deviations) are necessary to position the hand over chords or groups of notes. Lateral and vertical movements at the wrist can also be combined to allow for movement across multiple planes.

The vertical movements of flexion/extension have a greater possible range of motion than lateral movements (radial/ulnar deviation). Like forearm rotation, opposite lateral motions of the wrist (radial/ulnar deviations) are not equal. When resting the hand and forearm flat on a surface and moving the hand toward the thumb and then the fifth finger, the wrist has a greater range of lateral motion toward the fifth finger than toward the thumb because the forearm bone on the fifth finger side is shorter. True lateral motion at the wrist is very limited if the wrist is also flexed or extended. In this case, any lateral movement usually involves rotation.

Excessive wrist motion has been linked to occupational injury. Epidemiological, biomechanical, and ergonomic studies have determined that degree, repetitiveness, and forcefulness of wrist motions are factors in the development of tendinitis, carpal tunnel syndrome, and other nerve entrapments. For example, a common motion that can injure the tendons and/or their protective sheaths is the forceful rotational movement used to wring wet clothes, drive screws, and loop wire with pliers.[26] Repeated or forceful flexion/extension is also suspect. When the wrist is flexed, the tendons bunch up in the carpal tunnel, putting pressure on the median nerve.[27] Wrist pressure in the average person increases by a factor of two to ten with passive flexion or extension.[28] Moreover, Rempel observed that the relationship of carpal tunnel pressure to wrist angle is parabolic, increasing sharply with greater degrees of wrist hyperextension or flexion.[29] The greater the degree of active wrist motion, the higher the carpal tunnel pressure. While the wrist is designed to flex and extend, there are limits to its tolerance. Pressure on the median nerve is normally relieved with a return to neutral position, but with excess or repetitive use, the tendons can become so inflamed that the median nerve becomes compressed by either the tendons themselves or by excess pressure from fluid in the wrist, resulting in carpal tunnel syndrome (CTS).[30] Since increased carpal tunnel pressure has been theoretically linked to the development of CTS, pianists should carefully consider the degree, forcefulness, and repetitiveness of their wrist use. For example, using forceful consecutive vertical strokes of the forearm is one way to produce a powerful sound, but it might overstress tendons and muscles.

Lateral motion at the wrist can also cause problems. As with flexion and extension, Rempel observed that carpal tunnel pressure increases dramatically with greater degrees of lateral wrist deviation.[31] One key rule in designing hand tools is to avoid ulnar deviation of the wrist.[32] This has led to redesigns featuring bent handles in brooms, hammers, knives, axes, hoes, shears, snow shovels, and even sports equipment. A softball bat with a bent handle is now legal for the game.[33] Radial deviation is similarly problematic and has been linked to the development of "tennis elbow."[34] In-Seol Chung documented a much greater range of lateral wrist motion in piano playing than that which falls within normal functional range in the performance of both technical exercises and repertoire.[35] While all pianists should reflect on the degree, force, and repetitiveness of their wrist movements, this issue is of special concern for small-handed pianists because the smaller the

---

[26] Sanders and McCormick, *Human Factors in Engineering and Design*, 387.

[27] *Ibid.*

[28] R. H. Gelberman et al., "The Carpal Tunnel Syndrome: A Study of Carpal Canal Pressures," *Journal of Bone and Joint Surgery* 63A (1981): 380–383.

[29] David Rempel, "Musculoskeletal Loading and Carpal Tunnel Pressure," in *Repetitive Motion Disorders of the Upper Extremity* (Rosemont: American Academy of Orthopedic Surgeons, 1995), 125.

[30] Robert M. Szabo and Michael Madison, "Carpal Tunnel Syndrome as a Work-Related Disorder," in *Repetitive Motion Disorders of the Upper Extremity* (Rosemont: American Academy of Orthopedic Surgeons, 1995), 426–427.

[31] Rempel, "Musculoskeletal Loading and Carpal Tunnel Pressure," 125.

[32] Sanders and McCormick, *Human Factors in Engineering and Design*, 387.

[33] *Ibid.*, 388–389.

[34] *Ibid.*, 390.

[35] In-Seol Chung et al., "Wrist Motion Analysis in Pianists," *Medical Problems of Performing Artists* 7 (March 1992): 5.

hand, the greater the necessity for using larger degrees of both ulnar/ radial deviation and flexion/extension at the wrist.

The easiest way to experience neutral wrist position is to allow the heavy arm to hang loosely from the shoulder while standing so that the hand is in a straight line with the wrist and forearm in all planes (Figure 2.9). This position can then be transferred to the keyboard. When playing the piano, it is important to return the wrist to neutral position as soon and as frequently as possible (Figure 2.10).

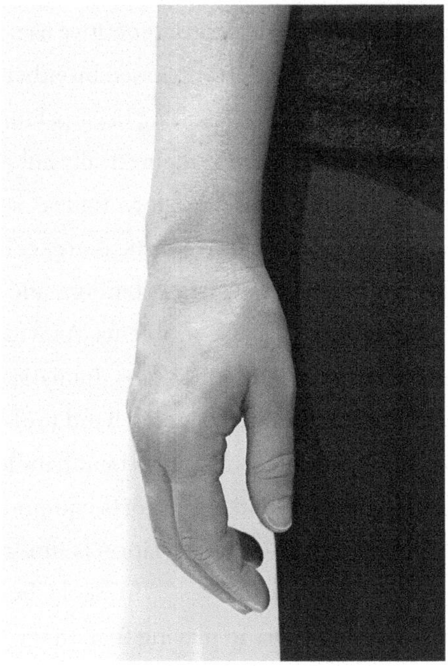

FIGURE 2.9. *Neutral wrist and hand position while standing. Credit: Justin Mohling, photographer, and Ella Durham, photo editor.*

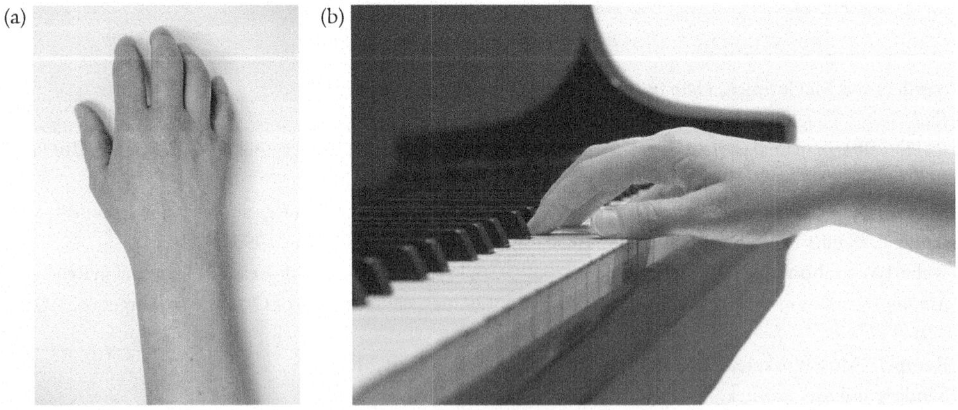

FIGURE 2.10. *Neutral wrist and hand position at the piano. Credit: Justin Mohling, photographer and Ella Durham, photo editor.*
a. Viewed from above (transverse plane)
b. Viewed from the side (sagittal plane)

Hand and Fingers

While each individual hand looks slightly different in this regard, there are noticeable arches observable from the wrist down to the tip of each finger when the hand is allowed to hang naturally at the side of the torso (Figure 2.9). Another arch forms across the knuckles of the hand at what is often called the bridge. At the keyboard, the fingertips should rest lightly on the keys with the hand maintained in all its naturally supported arches (Figure 2.10b). However, many motions are necessary to actually play the keys, and the fingers and hand must frequently move out of neutral position. The goal, as with other parts of the playing apparatus, is to return to neutral as soon and as often as possible.

The fingers move in multiple planes. During piano playing, flexion and extension can be seen in the sagittal (side, vertical) plane, and spreading the fingers away from each other and then moving them back together (abduction and adduction) are visible in the transverse (horizontal) plane (Figure 2.11). The thumb functions unlike the other digits. Not only is it obviously positioned farther back on the hand but it also has additional sets of muscles that allow it to move in opposition to the fifth finger (Figure 2.12).

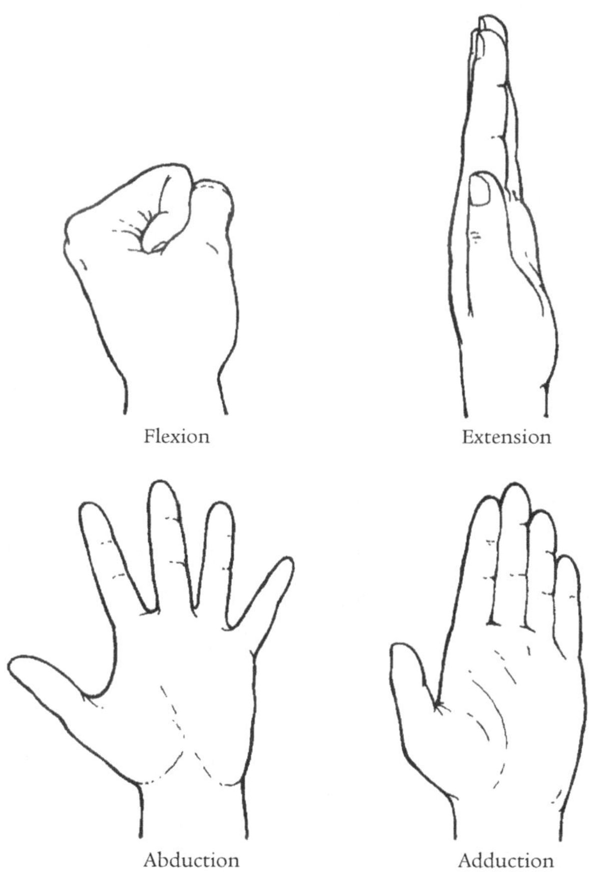

FIGURE 2.11. *Finger movements. Credit: Susan J. Hall,* Basic Biomechanics, *3rd ed. (1999), 223, © McGraw-Hill Education. Reproduced by permission.*

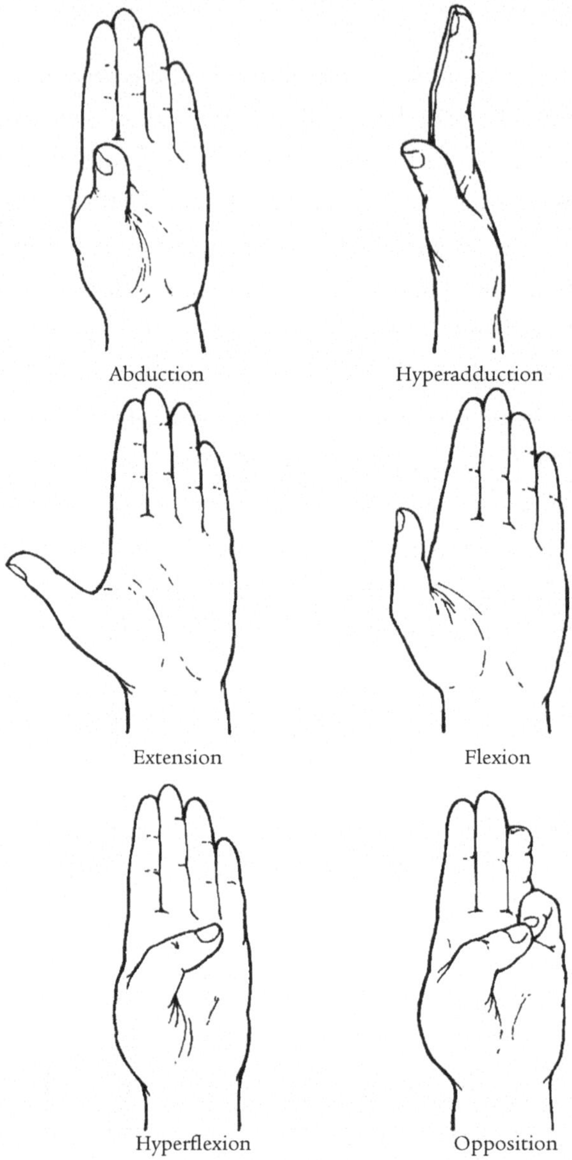

FIGURE 2.12. *Thumb movements. Credit: Susan J. Hall,* Basic Biomechanics, *3rd ed. (1999), 222, © McGraw-Hill Education. Reproduced by permission.*

The fingers and hand are made up of a collection of smaller bones: the phalanges of the fingers and the metacarpals of the hand (Figure 2.13). The phalanges are joined by two interphalangeal joints—the proximal interphalangeal (PIP) joint and distal interphalangeal (DIP) joint—and the fingers are connected to the metacarpals of the hand via the metacarpophalangeal (MCP) joints, often called the hand knuckles. The metacarpal bones can serve as a base against which fingers move.

The fingers and hand can be used together as one lever joined by stabilized MCP and interphalangeal joints. Muscles of the forearm flex and extend the hand lever. When

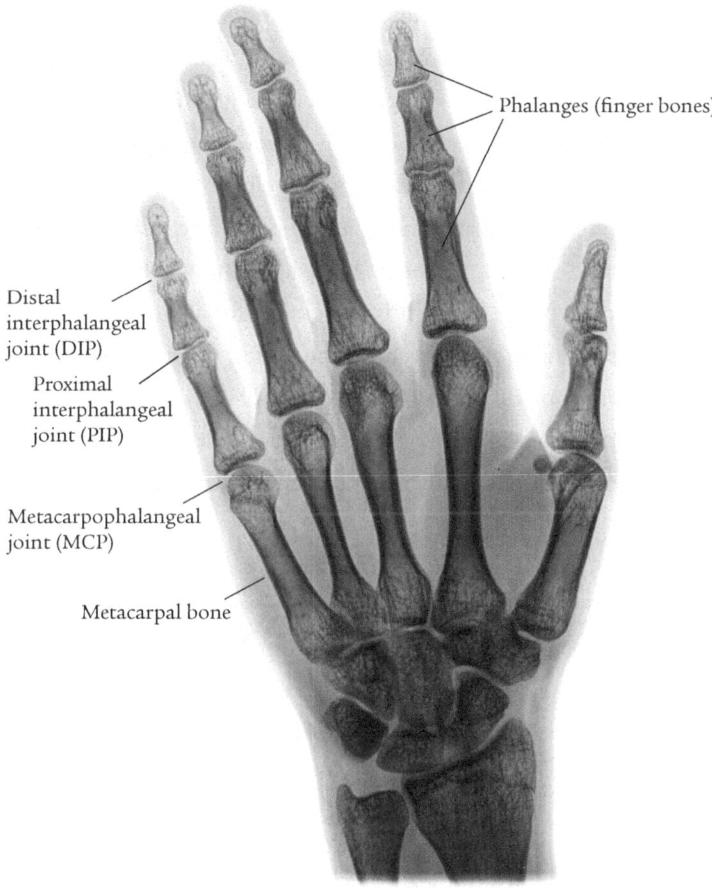

FIGURE 2.13. *Bones and joints of the fingers. Credit: wonderisland/Shutterstock.com*

energetically applied to subsequent keystrokes, this reciprocal motion results in what is often called hand staccato. Like other motions used at the piano, hand staccato can be controlled with regard to time, amplitude, and speed.

Fingers are adept at producing light, fast movements and are the smallest levers in the playing apparatus.[36] The primary muscles that control them lie mostly within the forearm. The fingers usually work in conjunction with the rest of the arm as they fatigue quickly if used to activate keys independently without assistance from larger levers higher up the chain. Often, one impetus from the arm will be translated by the fingers into numerous key attacks, forming a larger grouping or gesture.

As in most other segments of the body, the tendons controlling the fingers function like pulleys. There are opposing tendons on each side of the finger allowing for flexion and extension. These tendons are attached to the individual bones by sheaths that hold the tendons close. The muscles primarily responsible for finger movement actually originate at the elbow

[36] For practical purposes, the finger will be considered as one lever in this book, rather than three separate tiny levers.

and run the full length of the forearm before crossing the wrist and attaching as tendons to the ends of the farthest finger phalanges. The flexors allow the fingers to be brought in toward the palm while the extensors allow the fingers to be straightened back out or lifted above the key surface. Since these tendons originate at the elbow, they function at their maximum effectiveness when they form a straight line from the elbow to the fingertip, a principle of piano technique often referred to as alignment. A finger may be considered optimally aligned when a straight line can be drawn from the fingertip all the way to the elbow, as observed in the transverse (horizontal) plane. While it is impossible for all fingers to remain optimally aligned at all times—certainly not when chords are being played—this point of anatomic design can greatly inform choices at the piano, particularly with regard to fingering.

When examining the configuration of the extensor tendons (Figure 2.14), it becomes apparent that the fingers are not equally independent. As the extensor digitorum muscle

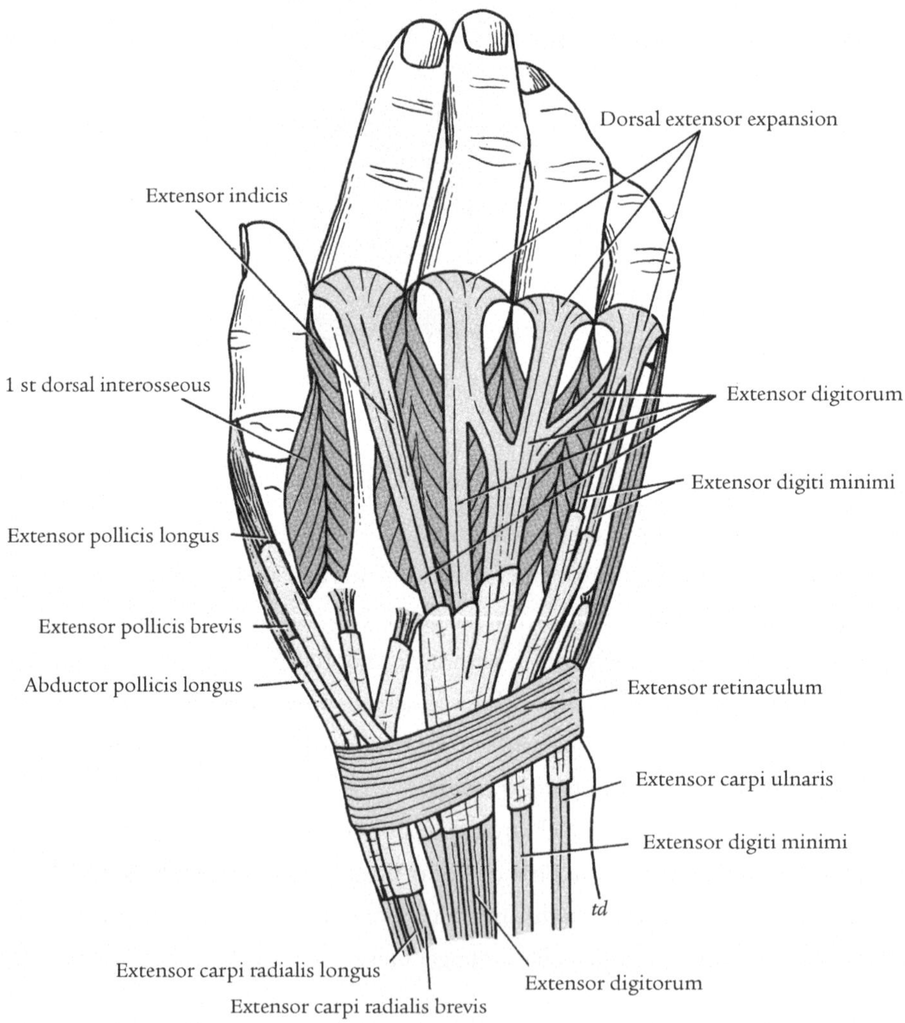

FIGURE 2.14. *Finger extensors. Credit: Richard S. Snell,* Clinical Anatomy for Medical Students, *5th ed. (1995), 450, © Little, Brown and Company. Reproduced by permission.*

passes from the elbow through the wrist, it divides into several tendons. Both the fifth finger and the thumb are strong because they have multiple extensors and because their power can be supplemented by forearm rotation. The second and third fingers have their own extensors and are thus relatively independent. However, the extensor of the fourth finger is bound to the extensors of the third and fifth fingers on either side, limiting its independence. The inherent differences between the fingers can be appreciated by making a loose fist and raising each finger independently from the other fingers. The structure of the hand and fingers should be respected; attempting to artificially strengthen the small muscles and tendons that control them may contribute to injury.

While most of the musculature that allows pianists to move the fingers originates in the forearm, there are small muscles—the intrinsic muscles—that are contained entirely within the hand. They open and close the palm to position the fingers over keys to play various chord shapes. Interossei muscles move fingers two through five away from and back toward the middle finger. The lumbricals work along with the interossei muscles to flex the metacarpophalangeal (MCP) joints while simultaneously extending the interphalangeal (PIP and DIP) joints of fingers two through five, as illustrated in Figure 2.15. Both the thumb and fifth finger have their own intrinsic muscles that flex/extend and abduct/adduct.

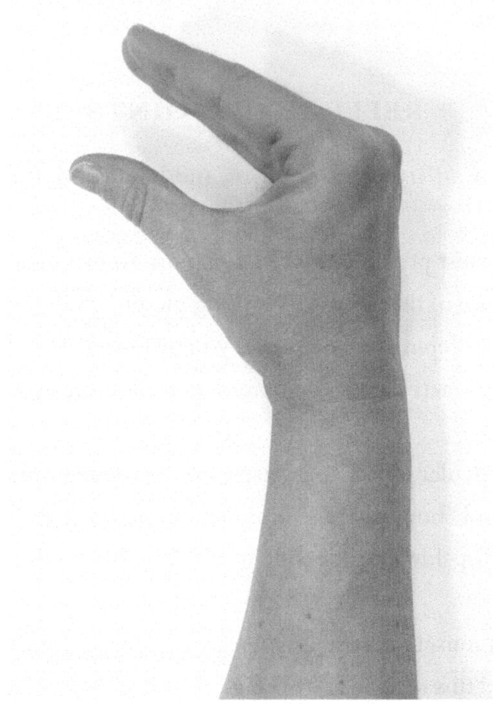

FIGURE 2.15. *The hand with the lumbricals activated. Credit: Justin Mohling, photographer, and Ella Durham, photo editor.*

As discussed in the section on leverage, the pianist chooses the playing lever to use depending on the task. The fingers, hand, forearm, and upper arm are the primary levers in piano playing. They function collaboratively, combining through joint fixation to form levers of different lengths. Because these levers are useful for different purposes, pianists must constantly negotiate the balance of work between smaller and longer levers. Different combinations of the whole arm, forearm, hand, and finger can be used to produce a wide range of available touches from quick to slow and quiet to powerful according to need. The fingers, the smallest levers, are most suited to producing fast and light touches where relatively little power is needed. As the need for power or speed grows, larger parts of the body must support and supplement the actions of the smaller levers.

CONCLUSION

Piano playing is a complex sequence of motions requiring the cooperation of many muscle groups working together in a finely balanced ballet. Intricate patterns of movement are activated to execute different pianistic tasks. Because the body is an integrated system, action in one part always affects another. Axioms and laws from physics, biomechanics, and ergonomics shed light on piano playing and help clarify how the body moves. These disciplines provide valuable tools for understanding, evaluating, and comparing the efficacy and appropriateness of technical strategies, especially those intended for the small-handed.

REFLECTION POINTS

1. With the help of a partner or a mirror, model ideal neutral playing position at the piano.
2. What is the shortest possible lever you can use to produce a sound at the piano? What is the nature of that sound? Working incrementally, add one lever at a time, focusing on the difference in the sound you produce.
3. Do the following activity to help train muscle contraction and instantaneous release:
   - Choose a particular muscle or muscle group, for example, the trapezius muscle of the neck and shoulders, the muscles on the front or back of the upper arm, those on the topside or underside of the forearm, or those controlling an individual finger.
   - Contract that muscle as hard as you can and hold it for a count of four seconds. Feel how hard this muscle is working.
   - Say or think the word "Release!"

- As you release this muscle, focus on how it feels as the muscle contraction dissipates. Does the contraction release gradually, or all at once? Do this activity again, focusing on releasing the muscle contraction instantaneously and completely.
- Next, contract the selected muscle only half as much as you did previously. Direct your awareness to feeling both the contraction and the release. Focus on releasing the muscle contraction immediately and completely. Repeat this activity several times, halving the degree of muscle contraction each time so that you become aware of minute levels of muscular effort.

4. What strategies can you think of to encourage frequent return to neutral position at the wrist?
5. Practice configuring various chord positions to find the ideal alignment of the hand and forearm. This will change according to register and the arrangement of white/black keys within a chord or shape.

THREE

Technique and the Small-Handed Pianist

BUILDING A RELIABLE AND EFFICIENT TECHNIQUE IS A PRIORITY FOR ANY pianist, but even more so for someone with small hands. Research strongly suggests that small hand size is a risk factor for injury and that psychological stress can complicate the already challenging work of piano playing. For example, performers might subconsciously contract their muscles continuously in anticipation of a passage they perceive as taxing, a phenomenon known as muscular bracing. Over time, many small-handed pianists may come to accept sensations of discomfort or fatigue as a natural state of affairs, ignoring signs of distress until after an injury occurs.

Pianists with small hands should consequently monitor how hard they are working at any given moment. To do this, they must develop an awareness of how neutral position feels and seek to return to this state as soon and as often as possible. If the activity of piano playing is viewed as water in a glass with the glass as a metaphorical measure of an individual's healthy capacity for work, a pianist must experience how it feels when the glass holds barely a drop. If the player's awareness falters and the average level in the glass rises too high, it is more likely that water will overflow the rim and that the player will suffer pain or injury.

Once small-handed pianists build this kinesthetic awareness, they can begin to identify inefficiencies in their playing. Using a less efficient gesture to communicate expressive intent may not be a poor choice in and of itself, as when a performer chooses to express the musical tension of a passage through extravagant arm gestures accompanied by raised shoulders. However, such gestures should be limited in scope and used with an understanding of their effect on the development of fatigue. They should be the exception rather than the rule.

MALADAPTIVE TENDENCIES

The technical and musical challenges faced by small-handed pianists usually fall into three basic categories: fatigue, power, and/or reach. A complex passage might impose obstacles from two or even all three areas simultaneously, and the necessity to play fast and/or loud is likely to compound these problems. Unfortunately, pianists with small hands often resort to approaches that lead to further difficulty and fail to produce the desired outcomes. This section explores the most common of these dysfunctional tendencies. It is followed by a lexicon of fundamental facilitative technique that serves as the theoretical foundation for the adaptive strategies explored in this book.

Fatigue

Small-handed pianists may try to conquer difficult passages by practicing them repetitively to the point of exhaustion without ever achieving reliable mastery. Merely increasing the quantity of practice, however, does not address a problem's root causes; rather, it quickly depletes available physical and mental reserves. As the popular colloquialism variously attributed to Mark Twain, Benjamin Franklin, Albert Einstein, or Rita Mae Brown goes, "Insanity is doing the same thing over and over and expecting different results."

A primary means of avoiding fatigue is to use the appropriate levers for the task.[1] Unfortunately, small-handed pianists tend to involve large levers far more than is necessary from either a technical or musical standpoint. As explained in Chapter 2, longer levers provide power but take more time to move through space due to their greater inertia. Therefore, overusing long levers in fast passages is a sure path to fatigue. Conversely, small-handed pianists may tire by relying excessively upon the fingers for power when a larger gesture would be a better choice. At the point where the limit of a particular muscle is reached or when more power is required, larger levers such as the forearm and upper arm must be recruited to supplement the fingers. However, these larger levers should relinquish activity once less power is required. Every key on the piano can be effectively reached in a variety of ways.[2] Since piano playing involves collaboration from all the levers of the upper limb, the same end result can be achieved from different combinations of smaller and larger levers and vertical and horizontal motions. No lever should be working if it is not necessary or appropriate.

---

[1] Refer to Chapter 2.

[2] Otto Ortmann, *The Physiological Mechanics of Piano Technique* (1929; reprint, Toronto: E. P. Dutton, 1962), 33.

Exhaustion also sets in if muscles are continuously contracted and are not rested on a frequent basis.[3] Overworked muscles take much longer to recover their contractility, and muscles that are completely depleted lose their ability to contract at all.[4] Common observable symptoms of excess or inappropriate muscle tension include jaw clenching, fingers that are habitually rigidified into a claw-like shape, stiffness at the wrist joint, or shoulders that are raised or locked in place so that the collarbones and shoulder blades cannot move independently.

Power

Producing sufficient power is a significant challenge for small-handed pianists, a predicament Ortmann noticed among children. He was concerned that teachers often expected young students to play as loudly as adults and that students typically responded by "stiffening," landing on one side of the hand in the manner of a karate chop in order to produce the necessary force.[5] A small hand using the same degree of muscular contraction as a larger hand will produce a sound of lesser intensity because of its smaller mass. When the hand is fully stretched to reach the notes of a widely spaced chord, sound projection is compromised even further. As the fingers spread apart, flexing them becomes more and more difficult, and flexing the fingers when they are fully abducted is physiologically impossible. Since small-handed pianists must frequently play with the hand stretched to its maximum, they often struggle to provide the finger joint support that is needed to deliver the power of the arm into the key.

To produce more sound, many small-handed pianists may unwittingly fall into the trap of attacking the keys from high above the keyboard. In addition to being taxing in its own right, crashing down on the fully extended hand may cause the wrist or the arch formed by the hand knuckles (MCP joints) to buckle, reducing the skeletal support of the hand and affecting the focus of the tone.[6] This approach also tends to overwork the trapezius muscle that crosses the upper back, shoulders, and neck. Small-handed players may alternatively resort to using individual high finger strokes. Unfortunately, excessive and constant finger extension overtaxes the extensor and flexor muscles of the forearm.[7] Aside from the disproportionate physical toll, vertical playing sounds distinctly harsh due to the noisy impact of the fingers against the keys; accuracy may also suffer. Furthermore, even though the larger and longer levers of the arm are designed for power, small-handed

[3] Refer to the discussion of static loading in Chapter 2.

[4] Ortmann, *Physiological Mechanics of Piano Technique*, 56–57.

[5] *Ibid.*, 303.

[6] Metacarpalphalangeal joints (MCPs). Refer to Figure 2.13.

[7] Ortmann, *Physiological Mechanics of Piano Technique*, 235.

pianists may use excessive amounts of arm weight or muscular contraction when they could channel energy through the strong bony structures of the fingers, hand, and wrist instead.

Finally, small-handed pianists may develop a subconscious habit of pressing into the immovable keybed (also known as "keybedding") well after the hammer has struck the string, extra work that statically loads the muscles of the forearm, upper arm, and shoulders. This is wasted effort, since pressing into the keybed cannot alter the tone and will make subsequent movements more taxing.

Reach

Having insufficient reach to play large chords or arpeggios and navigate through large distances at the keyboard is a pervasive problem for small-handed pianists. As discussed in Chapter 1, different anatomic dimensions can affect hand span. When fingers are short and the hand is narrow, it can be physiologically impossible to grasp wide intervals. Small-handed pianists may also be forced to adopt a higher wrist position than normal to deliver power more directly to the keys, which further reduces the reach of the hand.

Unfortunately, pianists with small hands often instinctively respond to large shapes by actively reaching for notes with the fingers. Actively abducting the hand to its maximum, however, gives rise to more internal forces than does passively opening the hand. Playing this way for long periods of time is tiring in its own right, and the misaligned tendons of the fingers, hand, and/or forearm are more vulnerable to key impact force. Flexion at the wrist becomes more limited. Active stretching limits accuracy and speed, and producing fine dynamic gradations with the fingers stretched to the extreme is very difficult.[8]

In addition, active reaching with the fingers may inadvertently cause the hand to work against the arm. For instance, when playing the broken-octave pattern in Example 3.1, a small-handed pianist might instinctively stretch the hand out to the size of an octave, extending the thumb outward to the left even as the arm moves to the right to execute the pattern. This would be particularly problematic at high tempos.

EXAMPLE 3.1. *Maladaptive extension of the hand in opposing directions*

[8] Ortmann, *Physiological Mechanics of Piano Technique*, 320.

If large distances need to be covered, pianists may unthinkingly resort to twisting the forearm to connect distant keys with the fingers, a temptation that may be greater for small-handed players because they can cover less distance with a single hand placement. In Example 3.2, unnecessary twisting gives rise to dysfunctional swings of the elbow when the forearm rotates to move the thumb under the hand in the ascending part of the pattern or when the forearm rotates to move the hand over the thumb in the descending portion. Asterisks mark places where this maladaptive rotation might be observed.

EXAMPLE 3.2. *Maladaptive twisting of the forearm*

Even though aligning each finger with the elbow maximizes the effectiveness of each tendon, successively aligning each and every finger with the forearm can result in excessive side-to-side swiveling of the wrist at faster tempos. The arrows in Example 3.3 show these maladaptive lateral wrist motions. Performed repeatedly, swiveling motions can impede speed and contribute to fatigue.

EXAMPLE 3.3. *Maladaptive swiveling visible at the wrist*

Other maladaptive approaches frequently adopted to deal with problems of reach include suspending unused fingers high above the keys, maintaining extreme wrist positions for prolonged periods of time, maximally spreading the fingers apart while simultaneously trying to flex them toward the keys, and insisting on producing a physical legato with the fingers instead of using other means to aurally connect the notes.

LEXICON OF FUNDAMENTAL FACILITATIVE TECHNIQUES FOR THE SMALL-HANDED PIANIST

To avoid falling back upon maladaptive tendencies, small-handed pianists need to develop a large and versatile toolkit of facilitative technique. Building upon earlier discussions of the principles of movement and anatomy in Chapter 2, this section delineates some important fundamental strategies.

Avoiding Fatigue

Movements must be accurately mapped in the brain to be effective.[9] Incorrect kinesthesia—for example, an inaccurate sense of how the shoulder girdle suspends the weight of the arm—can engender unnecessary muscular work. This issue is critical because the technique of small-handed pianists necessarily involves more frequent movements and a greater range of motion caused by extra shifts, unusual fingerings, and unconventional hand and arm positions. The player must vigilantly monitor the overall level of effort relative to muscular baseline.

Timing Muscular Releases

"Release" means several things within the context of music. The term is frequently used in a musical sense to describe the discharge of tension following the climax of a phrase. In this book, however, release refers either to the cessation or reduction of muscular contraction or to the physical release of a piano key. In some cases, the two types are used together—a key is released, resulting in the stopping or lessening of muscular effort. Afterward, the finger may lift from the key surface, or it may passively rest in the keybed.

Releasing muscular activity is as important as initiating motion. If muscles are forcefully or continuously contracted and are not rested adequately, lactic acid buildup occurs, causing rigidity and exhaustion. When playing large chords or other strenuous shapes successively, it is helpful to build in quick releases after as many of them as possible. Even if there is no time to return all parts of the playing apparatus back to baseline, small moments of rest will help the player to fight off muscular exhaustion and prepare for the next action. Longer series of chords should be organized into smaller subgroups, with releases inserted between the subgroups. Those with small hands should also avoid keybedding because it causes pianists to tire quickly and easily. Instead, they should focus on releasing muscles at regular intervals between strong contractions so that this becomes a natural reflex.

Using Levers Appropriately

Small-handed pianists must use levers wisely to maximize the effect of muscular effort. How the player moves in the backward/forward dimension relative to the fallboard is worth examining in this regard. Backward and forward movements that are used to navigate between the black and white keys require involvement from the full arm moving

[9] F. Matthias Alexander, *The Use of the Self* (London: Methuen, 1932); Barbara and William Conable, *What Every Musician Needs to Know about the Body* (Portland: Andover Press, 2000); Thomas Mark, *What Every Pianist Needs to Know about the Body* (Chicago: Gia, 2004).

from the shoulder joint. Since the arm is a large lever, it has greater mass and therefore greater inertia. Even though these shifts take time to execute, pianists with smaller hands may need to use them more frequently than those with larger hands. In a comparison of children and adults playing chromatic octave scales, for instance, Ortmann demonstrated that children used more backward/forward motion than larger-handed adults.[10] Similarly, with regard to the first two measures from the last movement of Beethoven's "Appassionata" sonata, Ortmann observed that small-handed players favored a compensatory fingering that used backward/forward arm gestures as opposed to one that stretched the fingers apart (see Figure 5.1).[11] Thus, when these movements are called for, performers with small hands should make intentional choices about how they incorporate this dimension into their playing, especially at fast tempos. They should try to minimize the size of the motions by playing closer to or between the black keys or enact these motions as gradually and smoothly as possible.

As explained earlier, inertia renders large levers ineffective in passages with fast tempos and changes of direction. Instead of dropping the arm vertically many times in succession, pianists should play fast notes or chords with smaller levers and incorporate them within larger gestures, a point that will be explored later in this chapter. On the other hand, pianists should not depend on the fingers for strength but instead should explore efficient means of involving larger levers. It is imperative, however, to coordinate and control the spread of muscular involvement, recruiting larger levers and muscles only to the degree necessary and avoiding unnecessary involvement that negatively impacts speed.

Generally speaking, then, the small-handed pianist should use the smallest lever that has sufficient power to accomplish the job. Staying within the mid-range of motion for all levers and joints whenever possible is another cardinal rule. While humans are designed to move even in extreme ranges, movements outside the mid-range are inherently more fatiguing. Additionally, to minimize difficulties with inertia, smooth curvilinear movements that avoid sudden changes of direction (jerk) should be employed.

Cultivating Finger-to-Elbow Alignment

The muscle/tendon units primarily responsible for finger activity actually lie in the forearm, originating at the elbow.[12] Misalignment between the tendon at the fingertip and its point of origin at the elbow forces the muscle and tendon to work against the resistance introduced at the lateral bend in the wrist, contributing to fatigue. Whenever the tempo

---

[10] Figure 5.1 in Chapter 5 reproduces Ortmann's photographic record. Ortmann, *Physiological Mechanics of Piano Technique*, 287.

[11] *Ibid.*, 279.

[12] Refer to Chapter 2.

ACTIVITY 3.1
CULTIVATING FINGER-TO-ELBOW ALIGNMENT

As explained in the Preface, activities included in this book are not intended as technical exercises to be endlessly repeated for the purpose of mastery but are rather a means for readers to directly experience principles explained in the text. Play through each activity a few times to understand the movements described.

Begin by playing broken chord inversions at a slow to moderate tempo, one hand at a time. Play the right hand as notated; play the left hand two octaves lower. Adjust the alignment at the wrist for every pitch, forming a straight line from the tip of the playing finger to the elbow. Rather than leaving the arm in place and reaching for successive notes with the fingers, keep the hand compact and allow the arm to position the finger over the appropriate key. The elbow should hang neutrally and should not be lifted away from the torso. Drop onto each key, keeping all the finger joints supported—do not let them buckle or collapse. Allow the fingers to freely release notes to accommodate the movement of the arm. Transpose the activity to different keys.

permits, small-handed pianists should thus focus on keeping the hand as compact as possible to permit the most optimal alignment from fingertip to elbow and should move the arm successively behind each playing finger.[13] Arpeggios at slow to moderate tempos are excellent for fostering a good sense of finger-to-arm alignment (Activity 3.1).

Combining Basic Movements into Larger Gestures

Movements that work at slow tempos or in isolated contexts may be dysfunctional at faster speeds. Thus, it is important to be able to integrate small individual movements into larger gestures. Each lever of the body moves around a joint. As explained in Chapter 2, it follows that efficient individual movements describe arcs instead of straight lines and that efficient larger gestures delineate curvilinear shapes formed by combinations of arcs. Curvilinear motions are inherently preferable to angular ones because they allow the pianist to seamlessly change the direction of travel without first stopping movement in one direction and then restarting it along a completely different trajectory. Small-handed pianists need to think deliberately about the shapes, motions,

[13] However, as shown in Example 3.3, note-by-note alignment can become maladaptive at faster tempos.

and gestures they use to play challenging passages and to reimagine those passages in alternative ways.

The primary vertical action of the arm at the keyboard is a down-up or drop-lift motion observable in the sagittal plane. In preparation for this movement, the weight of the forearm is lifted. This weight is then dropped through a stabilized finger into the key to produce sound. The forearm is then lifted again and the next keystroke is played with the weight of the arm still suspended. Given that hyperextension of the wrist is well-documented as a factor leading to fatigue, the wrist should not be allowed to drop lower than the level of the keybed. Often initially taught in conjunction with two-note melodic slurs, the drop-lift motion can be extended to encompass larger numbers of notes within a single forearm gesture. Activities 3.2 and 3.3 can be used to develop an awareness of straightforward applications of this movement.[14]

When playing down-up groupings as illustrated in Activities 3.2 and 3.3, the forearm describes small shallow arcs that rise slightly above and dip slightly below the key surfaces. These arcs, most visible at the wrist, are the fundamental building blocks for larger and more extended gestures that emerge naturally as tempo increases (Activity 3.4). Arc-shaped movements may proceed in either a clockwise or counterclockwise direction, and the lateral range of the arcs can be comfortably enlarged to encompass vast expanses of the keyboard when supplemented with side-to-side movements of the upper arm or torso.

As suggested in Activity 3.4, underarcs can combine with overarcs to create ovals for facilitating changes of direction (Figure 3.1). These arm gestures are visible in the frontal plane, from behind the player.

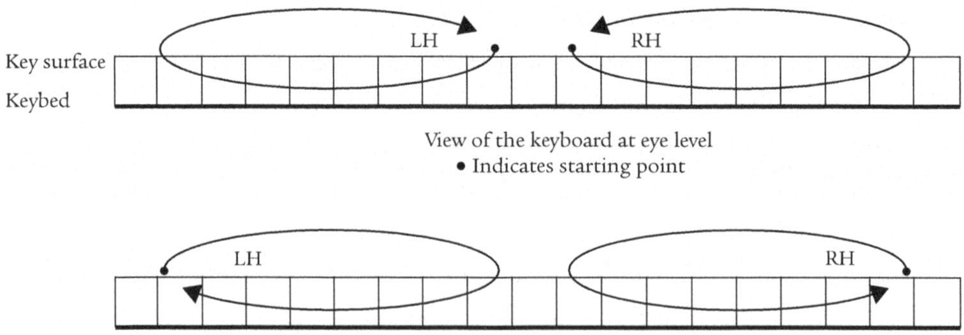

FIGURE 3.1. *Basic oval gestures viewed from behind (frontal plane). Credit: Matthew Bailey, graphic artist. © University of Nebraska-Lincoln. Used by permission. All rights reserved.*

[14] The system used in this book for marking drops, lifts, and arcs is adapted from Victoria von Arx, *Piano Lessons with Claudio Arrau* (New York: Oxford University Press, 2014), 168–171.

ACTIVITY 3.2
USING DOWN-UP (DROP-LIFT) MOTIONS TO ORGANIZE SERIES OF NOTES

This simple activity will acclimate the hand, wrist, and arm to the feeling of playing two-, three-, four-, and five-note slurs with different finger combinations and on different combinations of black and white notes using a basic down-up motion. Play the right hand as notated; play the left hand two octaves lower.

- For a two-note slur, start from a neutral position on the keys and lift the hand off the keys using the forearm. Keeping the finger stabilized, drop the weight of the forearm down on the first note of the slur, taking care to support the wrist so it does not collapse at the moment of impact. The wrist will not drop lower than the level of the keybed; the wrist and forearm will merely return to neutral starting position. Experiment with controlling the speed of the drop, noticing the different sounds produced. Quickly dropping, or even throwing the weight, increases volume, but may result in a more percussive tone quality.
- Next, suspend the weight of the arm and move the forearm very slightly in the direction of the second note until it lines up with the next playing finger.
- With the arm weight continuing to move upward, play the second note and release the key. As the forearm rises away from the key, the wrist may gently flex

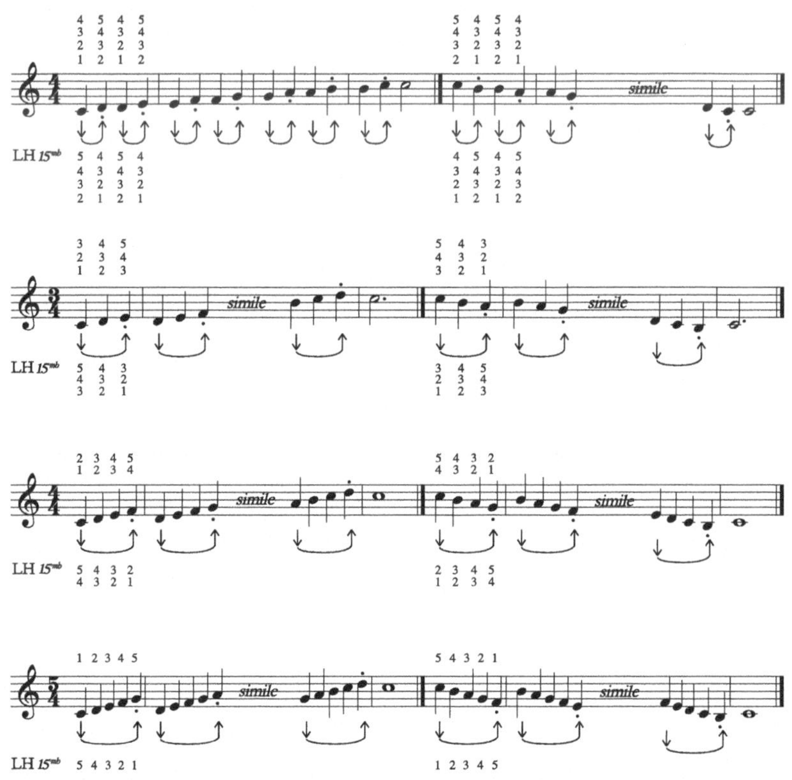

so the finger can stay in contact with the key longer, thus allowing for a controlled release of the tone. Note that if this upward movement of weight is very quick, an accent may result. Arm weight is now suspended in preparation for the drop on the first note of the next slur.

- Transpose the activity to different keys.
- Gradually expand the motion to encompass three-, four-, and five-note slurs.
- When playing black notes or combinations of black and white notes, keep forward and backward movements to a minimum by playing nearer the fallboard or between the black keys.

ACTIVITY 3.3
USING DOWN-UP (DROP-LIFT) MOTIONS TO ORGANIZE SERIES OF DOUBLE NOTES

The same movement used in Activity 3.2 can be used in Activity 3.3 to join together strings of consecutive double notes. The overall down-up gesture can accommodate any fingering of the performer's choice. There is no need to maintain finger legato, especially in the lower voice. The activity is shown in thirds but can be gradually expanded to address fourths, fifths, sixths, sevenths, and octaves. Play the right hand as notated; play the left hand two octaves lower. Transpose to different keys.

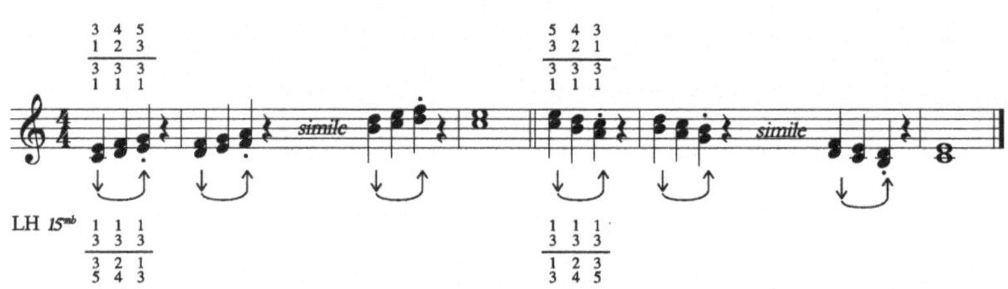

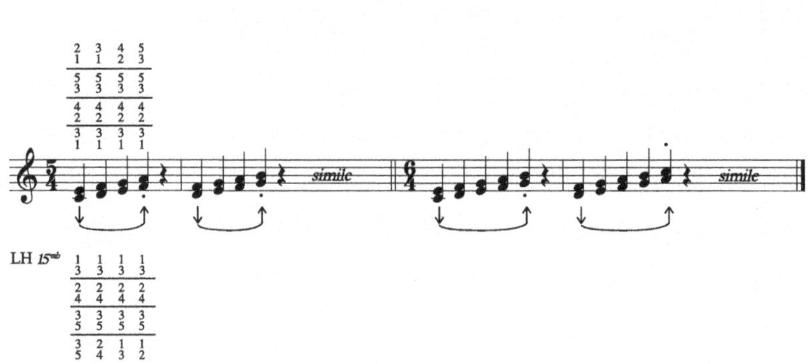

ACTIVITY 3.4
USING OVERARCS AND UNDERARCS

The same series of pitches found in Activity 3.1 is used in Activity 3.4, now played at a faster tempo. Practice each arpeggio one hand at a time. In Activities 3.4 and 3.5, the abbreviation UA stands for underarc, a curvilinear gesture of the forearm which starts with a downward drop toward the keybed that is extended horizontally, ending with a lift. OA stands for overarc, a curvilinear gesture of the forearm commencing from the key surface that starts with an upward lift of the forearm that is extended horizontally, ending with a drop back toward the keybed.

Let the overall gesture of the forearm and arm govern the shaping of each arpeggio. Allow the fingers to release the keys one by one so that the hand gradually closes up into a more compact shape as the notes ascend. The movement of the forearm should be smooth and continuous when transitioning from underarcs to overarcs. This continuity of forearm movement functions as a musical "glue" that binds the individual notes together even in the absence of finger legato.

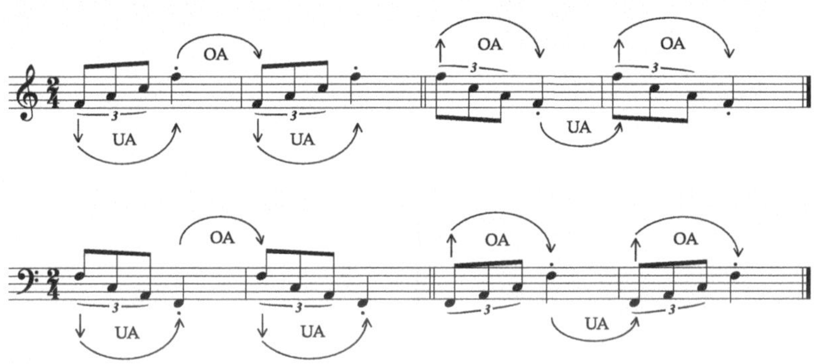

When using ovals, a visible elliptical movement of the wrist and elbow echoes the movement of the forearm and upper arm. The tempo must be slow enough to permit this level of involvement. To accommodate tempo increases, however, the amplitude of the motion can be flattened to the point where the forearm barely rises above or falls below the surface of the keys and where circularity is internally felt but is scarcely visible to the eye. The movement of the arm encourages the fingers to sequentially release notes, reducing the discomfort associated with grasping large intervals. Oval gestures used in conjunction with the pedal allow pianists to project a fluid line in the absence of an actual physical legato. Activity 3.5 can be used to learn basic interactions between the underarcs and overarcs frequently encountered in piano music.

Small-handed players, especially those with shorter arms or narrower shoulders, often benefit from conscious exploration of wrist angles necessitated by a particular passage of music. For instance, they may be tempted to radically swivel the wrist in order to maintain legato connections between distant notes or accommodate changes in direction.

ACTIVITY 3.5
COORDINATING OVAL GESTURES

Coordinating under- and overarcs can be complicated because the hands are shaped as mirror images of each other. The simple activities shown below illustrate the most basic interactions between the hands. In the first example, both hands start with underarcs and move to overarcs. In the second example, both hands start with overarcs and move to underarcs. In the third and fourth examples, over-and underarcs are juxtaposed against each other.

Start out by playing the activity slowly. Then, expand the range so that it spans several octaves instead of just one and gradually increase the tempo. Transpose to several different keys.

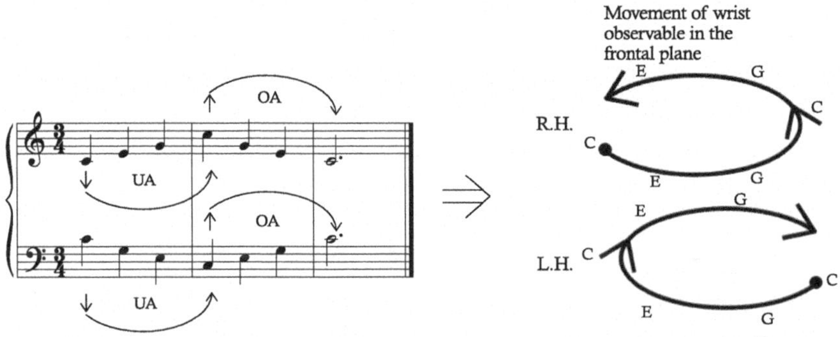

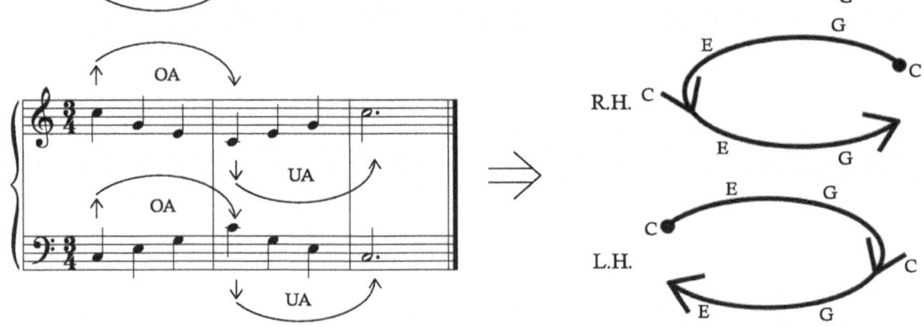

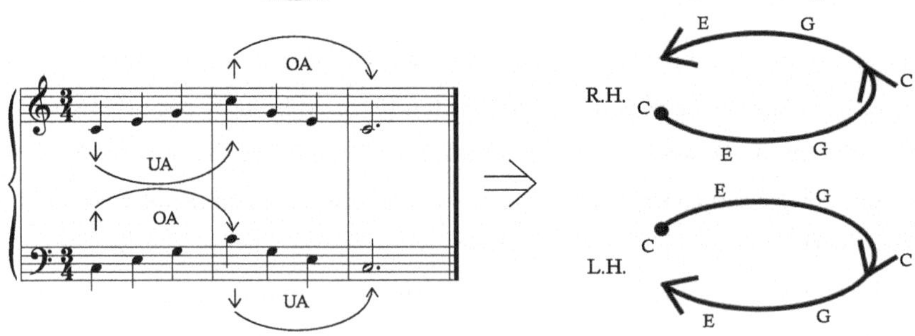

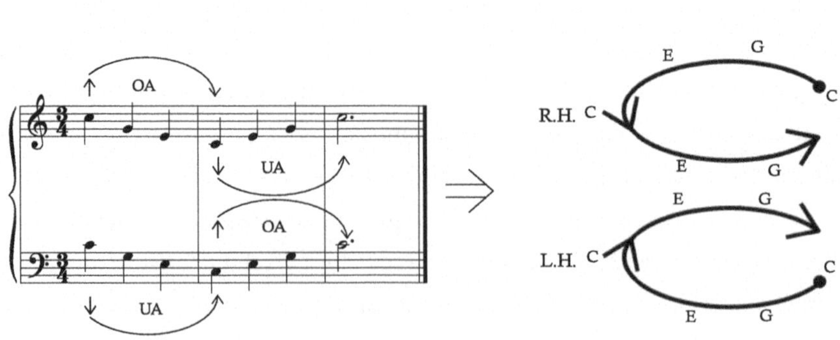

However, as previously discussed and illustrated by Example 3.3, it is important to avoid or minimize rapid lateral movements at the wrist joint. Instead, it may be necessary to adopt a compromise position of the wrist that encompasses a group of pitches rather than individual pitches, particularly at the extremes of the keyboard or when the hand is positioned across the torso. Using smaller arcs and ovals in combination with moderate wrist deviations can help avoid swiveling in fast passagework. Similarly, playing with the wrist lower or higher than normal, or gently alternating between these positions, may be a reasonable compromise where resetting to neutral between keystrokes would compromise accuracy. For example, it may be advisable to position the wrist relatively high when traversing various combinations of black and white keys at a fast tempo in order to be able to play in the black-key area and to eliminate backward/forward motions that would otherwise be required. Playing with the wrist slightly lowered below the level of the key surface may also be useful for brief periods of time. Because using this position necessitates that the fingers extend higher to reach black keys, lowered wrist positions are most useful when playing large shapes on white keys. Out-of-neutral wrist positions must be strategic responses to the demands of a particular combination of keys, and the wrist should be returned to its neutral position as soon as conditions permit.

Reducing the Amplitude of Movements and Gestures

Small-handed pianists use a greater number of positioning motions and arm gestures than larger-handed players. Care must thus be taken to minimize the amplitude of these gestures— the magnitude of vertical distance above or below the level of the piano keys—in proportion to the tempo. A large dramatic gesture may work well in a stately march but not in a rapid octave etude.

A common mistake is to conflate horizontal motion (left/right) with vertical motion (down/up). When shifting or reaching for distant notes, players may arc the fingers, hand, or arm much higher over the keyboard than necessary. Because this further reduces the horizontal distance that can be comfortably reached by the fingers and/or the speed of covering these distances, the amplitude of these strokes should be minimized to keep the fingertips as close to the keys as possible. If the joints of the fingers are relaxed, the palm and unused fingers can often graze or rest on the surface of the keys without making unintended sounds. This is especially crucial at fast tempos. When playing large leaps, reducing amplitude helps maximize the speed of horizontal positioning movements so that the pianist arrives in time to make the next keystroke in an accurate and controlled way.

Producing Power

Finding effective means of producing either a sufficient volume of sound or adequate power in the form of stamina is one of the primary challenges for small-handed pianists.

It may be tempting for players to approach this problem solely by increasing muscular effort, but as noted above, there will be negative physical and musical consequences. Successful power production depends on getting the most for the effort expended.

Maximizing Key Speed

Tonal intensity is a product of both mass and key speed, as derived from Newton's Second Law.[15] For small-handed pianists, the mass of the hand and arm are often limited to begin with. More to the point, increasing mass by involving the large lever of the arm increases inertia, which is problematic at faster tempos. Therefore, key speed is the more easily manipulated variable. Rather than using the full weight of the arm in consecutive vertical attacks to produce a *forte* sound, the small-handed pianist should activate the key in a way that sends the hammer toward the string at maximum speed.

Forearm rotation—pronating toward the thumb or supinating toward the fifth finger—is a highly effective means of delivering power by maximizing key speed. Rotation involves negligible amounts of weight but is capable of generating the high key speeds necessary for louder sound production even when the hand is extended to its maximum. Instead of pressing or crashing vertically down on the keys with the dead weight of the arm, small-handed pianists can substitute the speed and strength of rotational movements of the forearm. In Example 3.4, a quick and powerful forearm pronation can be used to play each chord, as indicated by arrows showing the direction of rotation.

EXAMPLE 3.4. *Using key speed to produce volume*

Depending on the tempo as well as the shape of the chord and how comfortably it fits within the hand, an alternative method of increasing key speed in Example 3.4 is by resting the fingers on the notes of the chords, stabilizing the finger joints, and using a whip-like thrusting gesture of the upper arm with an upward follow-through of the wrist to produce a sound that is resonant rather than harsh. However, since the upper arm is a larger lever, ample time between subsequent keystrokes is required.

In order to manipulate the speed of key attack most effectively, small-handed players should cultivate an awareness of where the point of mechanical escapement lies in the

---

[15] Refer to discussion on Newton's Laws in Chapter 2.

piano action. Playing to this point, which is located fractionally above the keybed, and then immediately releasing the muscular effort used to depress the key will help pianists to avoid wasted effort in the form of keybedding.

Transmitting Force through Skeletal Structures

To maximize sound, small-handed pianists should take advantage of skeletal structures to reduce the need for extraneous muscular contraction. When the fingers of the hand are naturally curved and the joints of the fingers are fixated at the moment of key impact, the individual bones of the fingers—the phalanges—line up and brace against the stabilized joints, allowing for the forces generated by the arm to be transmitted more directly into the key.[16] When playing with looser joints, some of the energy of the keystroke is dissipated before it is imparted to the key, resulting in a softer tone. Thus, the degree to which the joints of the fingers are stabilized at the moment of impact greatly influences the resultant sound.

As previously noted, the small-handed pianist has limited choice in terms of skeletal support of the fingers when playing with the hand at maximum extension. It is impossible to fully open the hand and fingers while simultaneously maintaining a naturally curved and supported shape. Therefore, small-handed pianists should search for ways to keep the hand as compact as possible so that finger joint support can be maintained. A compact hand shape can frequently be maintained through redistribution, refingering, early release of notes, well-considered note omissions, judicious use of the pedal, or other strategies that will be explored in detail in the chapters that follow.

Using Arm Impulses

Using arm impulses requires coordinating activity along the entire upper limb. With the hand resting on or near the keys and the weight of the forearm suspended by the upper arm and shoulder, the upper arm fires impulses through a loose elbow down through the forearm and flexible wrist, to which the hand responds with a bounce.

The motion of the hand in an arm impulse may bear visual resemblance to that produced by a light hand staccato. However, appearances are quite misleading, as the two movements feel radically different. In a hand staccato, the upper arm is not actively involved. Rather, muscles in the forearm actively flex and hyperextend the hand up and down, the MCP joints of the hand and all the joints of the fingers are firmly stabilized, and the impact of the fingers upon the keys is very noticeable.[17] Hand movement in an arm impulse, however, results from muscle activity higher up the arm, and the resulting touch

---

[16] It should be noted that the muscular work to stabilize the joints should be released *immediately* after key impact in keeping with the fundamental principle of the release of muscular work.

[17] Refer to Figure 2.13 showing the bones and joints of the hand and fingers.

on the key feels lighter and less percussive. Force is applied at a more oblique angle to the key than would be the case in a hand staccato, and the hand is not actively lifted prior to the next keystroke but remains on or very close to the key surfaces.

Impulse movements of the arm are particularly valuable for playing long series of rapidly repeated chords. In the chordal sequences contained in Activity 3.6, using hand

ACTIVITY 3.6
USING IMPULSES

The notes from the following passage are taken from the opening of Grieg's "Puck." Play it at different dynamic levels and tempos using just the fingers. Notice that it is quite difficult to keep all the notes even and controlled.

Now, play the same pattern again at the lower tempo indicated, this time using a small arm impulse on each pitch. The fingers never leave the keys entirely. Next, play the example several more times at gradually increasing speeds. As the tempo increases, reduce the size of the keystroke until the impulses feel like very small vibrations; gradually find the upper tempo limit. At that point, transition to using more selective arm impulses on strong beats, taking over with the fingers for the notes in between. Notice how clarity and control are improved at each speed by adding selective arm impulses.

Next, play the following pattern of repeated chords. Use small arm impulses for each eighth-note chord, adding a slightly larger impulse on the accented downbeat. Start at the lowest metronome marking provided. After resting for a moment, play the excerpt markedly faster. Notice that the degree of movement decreases as the tempo increases until the hand appears to almost vibrate at the end of the forearm. Once the notated pattern feels comfortable, try playing continuous eighth-note chords for two to four measures at a time, searching for the perfect balance of impulse degree and timing while maintaining awareness of the fatigue level.

Finally, use arm impulses to play chords that move up and down the keys. At higher tempos and dynamic levels, notice how much less fatiguing this technique is in comparison to playing the same pattern using either forearm or hand staccato.

staccatos alone is fatiguing. Using longer levers connected by stabilized joints—for instance, the forearm plus the hand or the full arm—would generate the necessary power, but at the expense of speed. Instead of relying on one longer lever, arm impulses use the various levers of the arm separately but in a finely coordinated sequence, borrowing power from larger muscles while minimizing inertia. Thus, arm impulses can provide more sustained power than hand staccatos, forearm staccatos, or the full arm without sacrificing speed. The degree of contraction of the muscles of the upper arm is very small. Depending upon the speed and context, arm impulses can be activated for each note or chord in a series or used more selectively on the major beats, alternating with hand staccatos.

Power can be enhanced further by situating the wrist slightly higher than normal, while still keeping it pliable enough to allow the impulses from the arm to translate to the hand. Similar to using the fingers in a position of skeletal support, this alignment sets the bones of the fingers and hand against the wrist, which helps transmit the impulse to the keys. This coordination is inherently less tiring than maladaptive attempts to use long levers for every keystroke. Even so, as with all muscular activations, it is important to return the forearm and arm to baseline as soon as possible.

Addressing Large Reaches and Distances

Because actively abducting the fingers apart to play widespread notes has so many negative consequences, pianists with small hands should constantly search for ways to eliminate or alleviate reach. Remembering that neutral position for a small hand might span a fifth or sixth instead of an octave or more, a good general rule is to keep the hand as compact as possible and to frequently return to muscular baseline by using strategies proposed in this book.

Using the Arm to Navigate Distances

For the small-handed, learning to bridge large expanses with the arm instead of the fingers is pivotal to developing confidence at the keyboard. When executing a wide leap, pianists should refrain from aggressively abducting the fingers apart and instead move through the gap using the arm. Widely spaced melodies or accompanimental patterns that cannot be covered within a single hand placement can be refingered into smaller groups that are then subsumed within larger arm gestures such as overarcs, underarcs, or compound ovals.

The forearm, linked to the hand by a stabilized wrist, can also be used to negotiate wide leaps. With the elbow remaining at roughly the same point in space, the forearm swings back and forth, facilitated by rotation of the humerus in the shoulder joint.[18] This produces a fan-shaped motion visible in the transverse plane from directly above the

[18] The humerus is the bone of the upper arm.

FIGURE 3.2. *Forearm fan gestures viewed from above (transverse plane). Credit: Justin Mohling, photographer, and Ella Durham, photo editor.*

pianist (Figure 3.2). The pianist should focus primarily upon the movement of the forearm rather than the upper arm in producing this gesture to avoid over-involving larger muscles that will cause the fan movement to slow down and fatigue to ensue.

The distance at the keyboard that can be covered this way is largely determined by the length of the forearm, but it can be further increased by using moderate lateral deviations at the wrist or by shifting the torso from side to side. Particularly at faster speeds, the momentum generated by a fan shape can facilitate its continuation because the arm will rebound from one side of the gesture to the other.

Forearm fans are often used to play a series of leaps where there are constant changes of direction, but they are also effective when playing large one-way leaps. Forearm rotations may also be used in conjunction with fan gestures to provide additional power for key attack or to extend the reach of the fan. As with ovals and arc-shaped motions, it is important to minimize the amplitude of fans at faster tempos to avoid fatigue. Activity 3.7 allows the pianist to learn this motion in incremental steps.

Using Passive Spread

Of course, small-handed pianists must often open the hand to cover large chords or groups of pitches. One facilitative strategy to consider in this regard is passive spread. As previously established, actively reaching for notes should be avoided. As an alternative, the pianist can

ACTIVITY 3.7
USING FAN-SHAPED FOREARM GESTURES

Small-handed pianists can practice this movement by starting with smaller intervals, progressing incrementally to very large ones. Play the right hand where notated, and the left hand two octaves lower. To play the simple exercise below,

- Position the elbow approximately in the middle of the interval.
- Swing the forearm and hand as a single unit quickly to the left, landing squarely on the first note.
- Then swing the forearm quickly in the opposite direction to play the upper note of the interval, keeping the elbow in place. Repeat a few times until the motion feels balanced and strong.
- Then expand the size of the interval until it reaches out to three octaves. For each interval, shift the elbow a little to the right so that it is positioned between the two notes. With the largest intervals, it will feel as if the forearm is being thrown from side to side. The vertical amplitude of the stroke will be very low, just barely clearing the tops of the keys.
- Transpose the activity to include different combinations of black and white notes and gradually increase the tempo.
- Then, practice the motion in the opposite direction, beginning with the upper note and swinging down to the lower one.

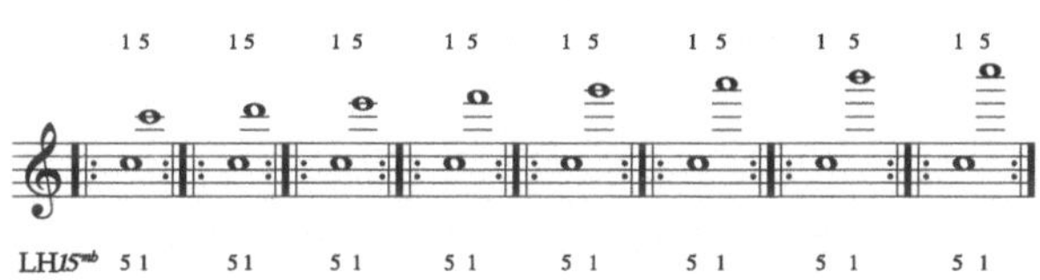

The basic fan shape can accommodate more than two notes as well as many different fingerings. It can also combine with forearm rotation. In the next pattern, the second finger is pronated over the thumb, which is used as a pivot, to extend the range of the fan. As the range of forearm pronation approaches its limit, the upper arm rotates slightly at the shoulder joint to supplement the forearm rotation. The elbow stays close to the torso as the upper arm rotates, leaving intact the overall feeling of the fan shape.

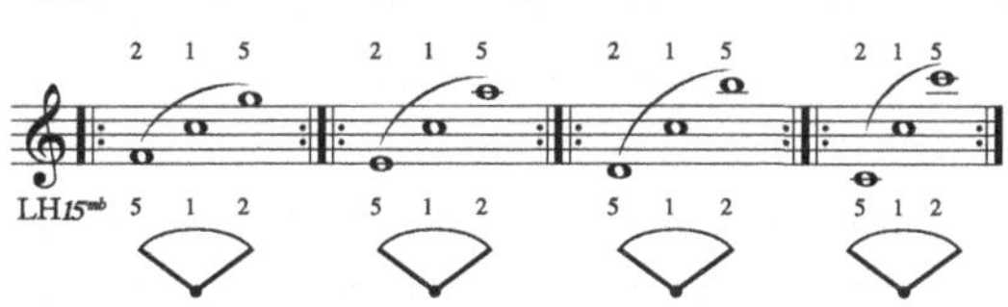

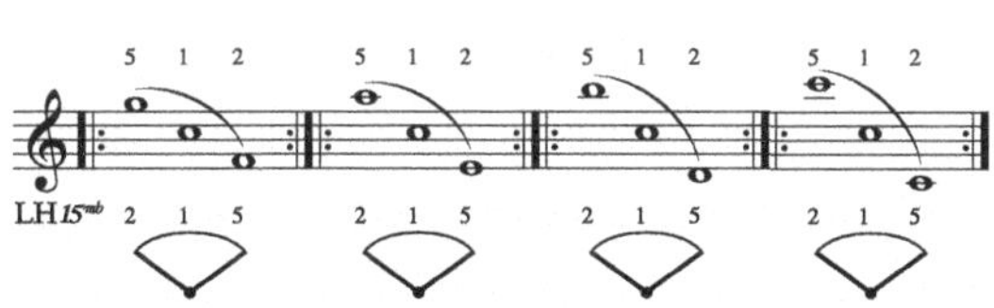

Finally, practice the fan shape in conjunction with the traditional and alternate fingerings noted in the following exercise. Instead of initiating a new shape for every octave, extend the fan shape to cover the full two-octave range. Notice that the basic fan shape remains the same regardless of whether the traditional or the alternate fingering is used. Start slowly and gradually increase the tempo. Transpose the activity to black-key octaves.

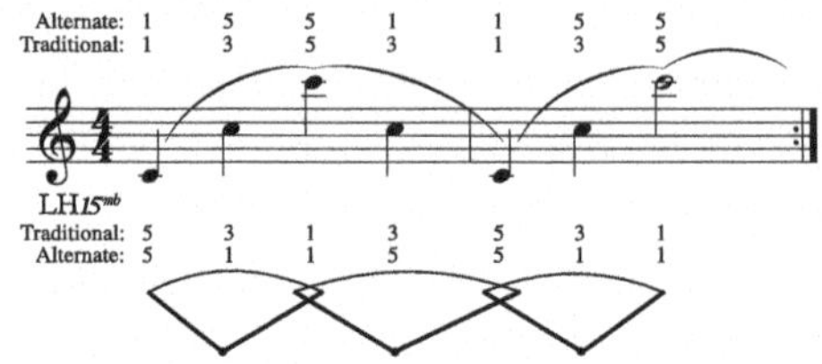

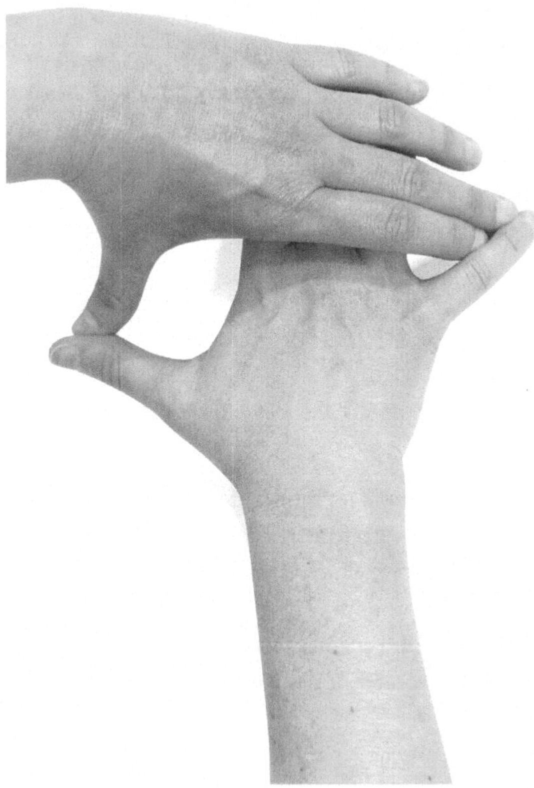

FIGURE 3.3. *Passive spread. Credit: Justin Mohling, photographer, and Ella Durham, photo editor. © University of Nebraska-Lincoln. Used by permission. All rights reserved.*

use the piano keys to gently spread the fingers apart. The difference can be experienced by comparing the feel of actively stretching open the hand to the shape of an octave, against the feel of using the other hand to passively spread the fingers open to encompass that same octave (Figure 3.3). The degree of muscular tension and the sensation of stretch are radically different. The actively stretched hand creates internal tensions within the hand, but the relaxed hand that is opened up by the other exerts no muscular activity of its own. Even so, it is important to release passive spread as soon as conditions permit.

It is often possible to open the hand to a larger degree using passive spread as opposed to active spread. The trade-off is that the fingers do not have much skeletal support from the structure of the hand and are thus limited in power. It should be noted that when the hand is passively spread open, the palm will lie flat. It is important, however, not to suspend any unused fingers above the key surfaces. As long as the finger joints are not stabilized, unused fingers will not usually produce sound even when they are touching the keys.

Using Vertical Spanning

Another useful strategy for reaching large arpeggios and chord shapes is vertical spanning. When pianists think about positioning the fingers over the keys, they typically think of abducting the fingers apart in the horizontal dimension. Vertical spanning exploits the vertical rather than the horizontal distances between fingers (Figure 3.4). If the hand is spread out on a flat surface such as a table or a piano keyboard, the horizontal spans between any two adjacent fingertips ranges from approximately 1 to 4 inches. However, the vertical span between any two fingers, adjacent or non-adjacent, can be significantly larger. These spans can be accessed by flexing one finger in a downward direction while

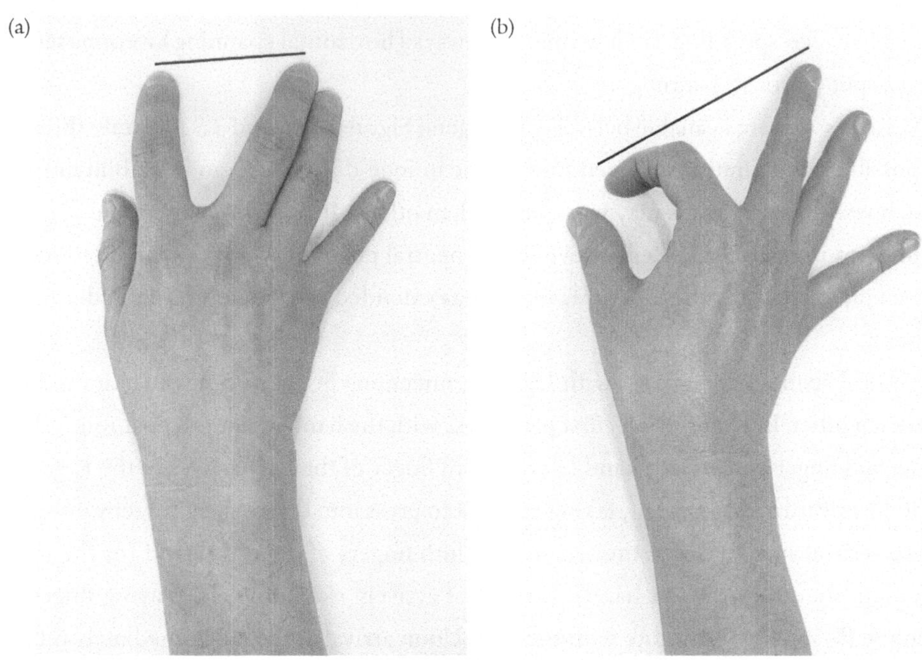

FIGURE 3.4. *Horizontal versus vertical spanning between the second and third fingers. Credit: Justin Mohling, photographer, and Ella Durham, photo editor. © University of Nebraska-Lincoln. Used by permission. All rights reserved.*

(a) (b)

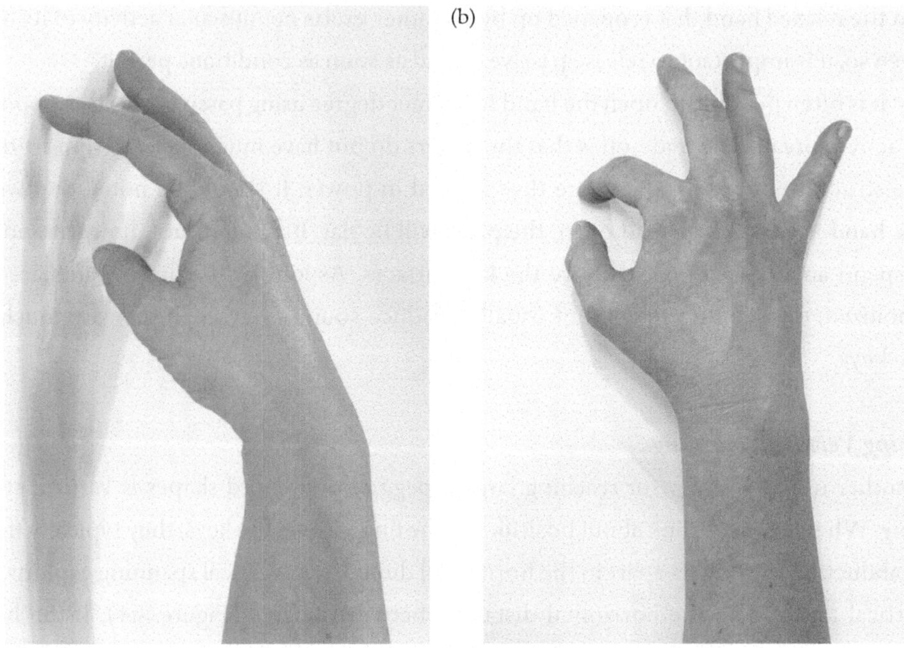

FIGURE 3.5. *Selected vertical spans. Credit: Justin Mohling, photographer, and Ella Durham, photo editor. © University of Nebraska-Lincoln. Used by permission. All rights reserved.*
a. Vertical span between thumb and second finger
b. Vertical span between second and fifth fingers

extending the other upward in the opposite direction. An analogy can be made to the difference between spreading the legs apart sideways (horizontal spanning) as opposed to doing a split (vertical spanning).

Vertical spans are available between all fingers; Figures 3.4b and 3.5 illustrate three of the possible configurations. Depending on the unique dimensions and flexibility of the hand, however, some spans may work better than others. In addition, for certain players, the forearm may need to briefly move out of neutral position. Vertical spanning should be used judiciously for short periods of time, as extended use would undoubtedly stress the hand.

Vertical spanning can assist with legato connections between notes that are distant from each other. In Example 3.5, first play the C with the hand in neutral position. When the second finger of the right hand (or the third finger of the left) arrives at the F, gently flex it downward and to the left, taking care not to press into the keybed, thereby opening up the vertical span between the second and fifth fingers of the right hand (or the third finger and thumb of the left hand). Instead of actively extending the playing finger to reach the E♭, move to that note with the arm. Upon arrival at the E♭, immediately return the hand to its preferred compact shape. A significant advantage of using vertical spanning is that it allows the hand to reach through large intervals without having to swivel left or right with the wrist.

EXAMPLE 3.5. *Vertical spanning in ascending pitches*

For a downward moving series of notes, the third finger of the right hand (or the second finger of the left) gently flexes downward and to the right in order to allow the hand to continue moving in the leftward direction of travel toward the E♭ (Example 3.6).

EXAMPLE 3.6. *Vertical spanning in descending pitches*

Taking advantage of vertical spans in arpeggiated figures and in melodic lines containing large intervals thus allows fingers to access widely spaced pitches, helps position the hand at an angle that assists the direction of travel for the arm, and prevents rapid adjusting motions at the wrist. This promotes a physical sense of connection, even where fingers must release individual pitches. Activity 3.8 explores vertical spanning between various finger combinations.

Vertical spanning might also be activated to play large chords. Most adult pianists are able to play the first chord in Example 3.7 while maintaining a straightforward orientation to the keyboard and using horizontal spans between the fingers. However, in the second and third chords, vertical spanning could be used to maximize the spread between fingers two and four (refer to Figure 3.5) and to avoid depressing unwanted keys.

EXAMPLE 3.7. *Vertical spanning used to play large chords*

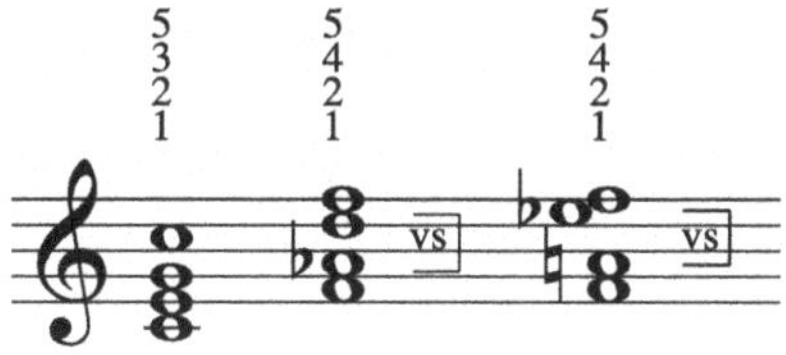

BOUNDARIES

The overriding purpose of this book is to empower pianists with small hands to devise and use strategies that enable them to play with confidence and ease. However, these adaptive measures will not suffice if the repertoire drastically exceeds a player's

ACTIVITY 3.8
USING VERTICAL SPANNING

Practice vertically spanning notes between the thumb and the other fingers of the hand by trying the following activity.

Instructions for the right hand:

- Play the first note, C, with the thumb and at the same time quickly open the hand outward to the right in preparation for playing the second finger.
- Now play the G securely with the second finger, flexing it downward and to the left without pushing into the keybed. Using the arm to move horizontally across the keyboard, allow the hand to close naturally and move the thumb under the second finger to play the next note, A.
- Continue this scissoring motion up the keyboard.
- Practice the exercise backward, going down the keyboard using the same fingering. Play the high C with the second finger. Then play the F with the thumb and flex it down and to the right without pushing the finger into the keybed. Using the arm to move laterally across the keyboard, allow the hand to close and move the second finger over the thumb to play its note, E.
- Continue this scissoring motion down the keyboard.
- Practice the exercise with 1-3, 1-4, and 1-5, gradually increasing the size of the interval.

Instructions for the left hand:

- The direction of flexion will be reversed. For instance in the first example, play the first note, C, with the thumb, quickly opening up the hand to the left in preparation for playing the second finger.
- Play the F securely with the second finger, flexing it downward and to the right. Avoiding keybedding and using the arm to move laterally across the keyboard, allow the hand to close naturally and move the thumb under the second finger to play the next note, E.
- Continue this scissoring motion down the keyboard.
- Practice the exercise backward, going up the keyboard using the same fingering. Play the low C with the second finger. Then play the G with the thumb, flexing it downward and to the left. Using the arm to move horizontally across the keyboard, allow the hand to close and move the second finger over the thumb to play its note, A.
- Continue this scissoring motion up the keyboard.
- Practice the exercise with 1-3, 1-4, and 1-5, gradually increasing the size of the interval.

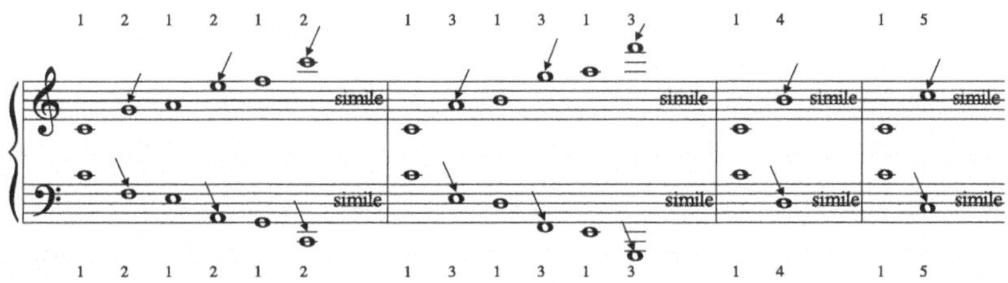

Practice vertically spanning notes between other finger combinations, such as 2-3, 3-4, 4-5, 2-4, 3-5, and 2-5. Watch for the scissor-like motion produced as the fingers "walk" up or down the keyboard. Remember that the arm is the primary actor in moving the hand and fingers horizontally across the keys between vertically spanned pitches; there is no need to press the finger into the keybed.

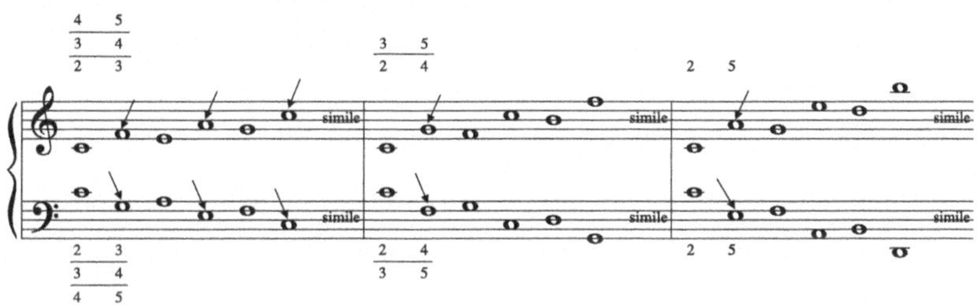

physical and musical limitations. The legendary pianist Josef Hofmann, who traveled with a reduced-size keyboard, stated that "whether the hands are too small or the stretch between the fingers too narrow—if you attempt a piece which for these or other physical reasons you cannot fully master, you always run the serious risk of overstraining. . . . If you cannot play a certain piece without undue physical strain, leave it alone."[19] Pianists should exercise caution when they find themselves playing in compromised situations where

- The wrist is constantly held in an exaggeratedly high or low position.
- The hand is constantly stretched to its maximum due to large numbers of massive chords or extended arpeggios.
- There is excessive swiveling or twisting.
- Facilitative movements that could render the piece more playable cannot be used because of the tempo or because they significantly detract from the musical impression.

[19] Josef Hofmann, *Piano Playing with Piano Questions Answered* (1920; reprint, New York: Dover Publications, Inc., 1976), 131.

Small-handed pianists must have ample opportunities to return the hand, wrist, and arms to neutral positions, even when using adaptive strategies. If these opportunities are limited or non-existent and if a pianist cannot find a musically satisfactory means of addressing the physical challenges embedded in a piece, then it is probably best to select different repertoire.

Even if effective strategies can be devised, the cognitive demands of implementing them may be onerous. This must be considered when working with less experienced pianists, especially children. Playing the piano should be about making music, not fabricating more and more convoluted physical adaptations. If a small-handed player has a burning desire to play a particular piece when either the physical demands cannot be met in a musically satisfactory way or when doing so results in an inordinate cognitive load, then performance on an ergonomically scaled piano keyboard (ESPK) may be the only viable option.

Pianists should never feel compelled to play pieces that exceed their unique abilities. Among singers, for example, voice types are carefully matched to repertoire. No one would expect a lyric soprano to have the range and power for a Wagnerian role. Likewise, small-handed players should select music that is a good fit for their hands. Fortunately, the vastness of the piano repertory ensures that pianists will never be at a loss for exciting new works to learn and explore.

CONCLUSION

Problems related to small-handedness become more apparent as a player progresses from the beginning levels of the pedagogical repertoire to works containing contrapuntal textures, large chords, extended arpeggios, widespread melodies, broken octaves, double notes, and situations demanding both speed and power. At that point, specialized approaches more specifically tailored to meet the needs of the small-handed may be needed. When devising or applying facilitative strategies, the following questions can help the pianist evaluate whether they are in accord with the fundamental principles of movement overviewed in Chapter 2:

- What phase of a movement feels most active?
- Which part of the anatomy is primarily responsible for the work?
- What is the minimum amount of muscular activity necessary to initiate and control movement?
- How does the work spread from one part of the anatomy to the next?
- How soon can this muscular activity be released?
- How soon can each part of the playing apparatus be returned to its neutral position?

Tables 3.1 and 3.2 serve as quick reference tools, summarizing common maladaptive tendencies and basic facilitative techniques. The application of these techniques to specific musical contexts will be explored in subsequent chapters.

TABLE 3.1 Maladaptive tendencies for small-handed pianists to avoid

Problem	Tendency
Fatigue	• Focusing exclusively on the quantity of practice • Practicing beyond physical limitations • Using the wrong lever for the task • Failing to return to neutral positions and muscular baselines • Using external devices or exercises in an attempt to make the hand bigger
Power	• Using successive vertical attacks ("pounding" on the keys) • Using repetitive individual high finger attacks • Using the wrong lever for the task • Keybedding
Reach	• Actively abducting the fingers to their extremes to reach for notes, particularly for extended periods • "Twisting" the forearm and trying to connect distant pitches with the fingers • "Swiveling" the wrist quickly and/or repetitively • Locking the wrist into contorted positions and/or suspending unused fingers over keys • Relying excessively on physical legato to connect pitches

TABLE 3.2 Facilitative strategies for small-handed pianists to cultivate

Problem	Strategy

Avoiding fatigue	• Timing muscular release
	    • Frequently releasing muscular work to baseline and returning to anatomically neutral positions
	    • Avoiding keybedding
	• Using levers appropriately
	    • Using fingers for small movements that are fast or repetitive; involving larger levers for power
	    • Minimizing backward and forward movements
	    • Using the smallest effective lever to accomplish the task
	    • Staying in the mid-range of motion for all levers and joints
	    • Adopting smooth curvilinear movements that avoid jerk
	• Cultivating finger-to-elbow alignment
	• Combining basic movements into larger gestures
	    • Down/up (drop/lift)
	    • Underarcs and overarcs
	    • Combining underarcs and overarcs into ovals
	    • Reducing the amplitude of gestures and movements

(*Continued*)

TABLE 3.2 *(Continued)*

Problem	Strategy
Producing power	• Maximizing key speed • Transmitting force through skeletal structures • Using larger levers for power • Using arm impulses
Addressing large reaches and distances	• Moving through distances using the arm • Using passive spread instead of active spread when appropriate • Using vertical spanning instead of finger abduction when appropriate • Using the forearm to navigate distances

REFLECTION POINTS

1. To produce a powerful sound, play large chords, or navigate distances at the keyboard, what strategies have you used to avoid fatigue? Did these methods produce desirable results? Were there any undesirable musical or physical consequences? Would these methods be more suitable for a pianist with larger or smaller hands?
2. Have you tried to play a piece that you eventually set aside because you felt it was too big? Were the problems primarily related to fatigue, power, or reach? Do you think that one of the fundamental techniques introduced in this chapter might have made it possible for you to play the piece? Why or why not?
3. How might a typical plan for developing piano technique be adjusted to account for hand size? Which principles of movement should be taught, and in what sequence?

FOUR

Redistribution

REDISTRIBUTION—MOVING NOTES DESIGNATED FOR ONE HAND INTO the other—can resolve or even eliminate many obstacles to small-handedness. It is, however, a skill that does not come naturally to many pianists. Reading habits ingrained over years of study are difficult to change, and playing notes in arrangements that deviate from the visual appearance of the score can tax the cognitive networks of the brain. Redistribution is thus frequently overlooked as an adaptive strategy.

Redistribution is the optimal answer to a legion of commonplace problems. Notes in large chords that are out of the reach of one hand can be diverted to the other. Passages of double notes can be divided between the hands to improve shaping or voicing. Leaps can be eliminated, legato connections restored, or harmonies rendered complete. In many cases, then, redistribution allows pianists to exercise greater control over the tonal and rhythmic fabric of the music.

All the same, redistributions should be viewed against the backdrop of their musical contexts. The pianist Claudio Arrau, for instance, felt that physical ease and musical expression sometimes worked at cross-purposes. He specifically advised against redistributing passages in certain late Beethoven sonatas because he believed that when composers "[want] to express something difficult and hard, . . . to simplify is . . . a fundamental mistake because the difficulty written in has an expressive purpose."[1]

The examples provided here are necessarily limited in number but nevertheless typify problems found throughout the literature. As always, pianists are encouraged to modify all solutions to address individual needs.

[1] Claudio Arrau quoted in Victoria A. von Arx, *Piano Lessons with Claudio Arrau* (New York: Oxford University Press, 2014), 32.

General categories of redistribution are grouped under the following subheadings:

- Uncrossing parts
- Eliminating stretches in chords and arpeggios
- Facilitating leaps or hand shifts
- Increasing accuracy, power, and control
- Maintaining more neutral hand and wrist positions
- Facilitating trills and tremolos
- Maintaining legato and line
- Projecting harmony

UNCROSSING PARTS

Romantic and Post-Romantic composers often preferred to notate large chords with multiple doublings as interlocked octaves. When the bracketed notes in Example 4.1a are redistributed and the thumbs are unlocked, however, the passage is transformed into the series of identical parallel chords shown in Example 4.1b. Only the chord marked with an asterisk deviates from this pattern. Hand spans are reduced to sixths or sevenths instead of octaves, facilitating both voicing and note accuracy for those with smaller hands.

EXAMPLE 4.1 *Rachmaninoff, "Prélude in C# Minor," from* Morceaux de fantasie, *op. 3 no. 2, mm. 1–3*
a. Printed version
b. Redistribution

(a)

(b)

Uncrossing parts has its advantages even when the intervals are relatively compact. The solutions in Examples 4.2 and 4.3 bring the thumbs into a more comfortable position for playing in front of the body.

EXAMPLE 4.2. *Grieg, "Wedding Day at Troldhaugen," from* Lyric Pieces, *op. 65 no. 6, m. 6*

EXAMPLE 4.3. *Beethoven, Sonata in A♭ Major, op. 110, ii, mm. 1–4*

Pianists can also uncross parts to maximize speed and power. In measures 41–42 of Example 4.4, the right hand is compromised by having to reach a sixth between the first and second fingers. Exchanging the thumbs enables both hands to contract into more compact shapes: the right hand spans a fourth with 1-2 and the left hand a third with 5-1.

EXAMPLE 4.4. *Debussy, "Prélude," from* Pour le Piano, *mm. 40–42*

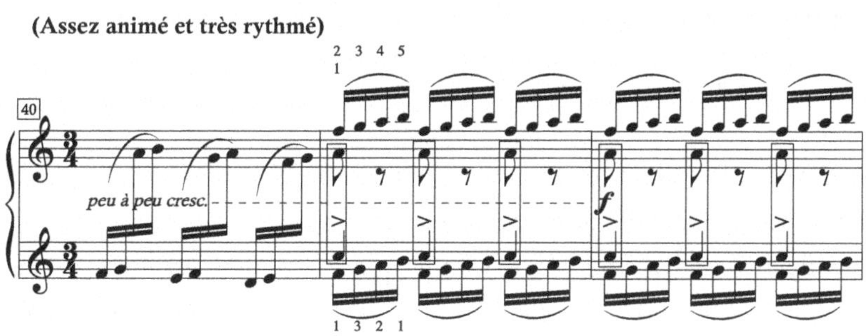

ELIMINATING STRETCHES IN CHORDS AND ARPEGGIOS

Many small-handed pianists instinctively react to widespread chords by rolling them. However, redistributions often work better than rolls. If the notes fall outside the easy reach of one hand, pianists can often redistribute selected notes into the other. In measure 8 of Example 4.5, the left hand can move up to eliminate the ninth in the right hand, thereby maintaining the uninterrupted flow of the triplet eighth notes.

EXAMPLE 4.5. *Beethoven, Sonata in C♯ Minor, op. 27 no. 2 ("Moonlight"), i, mm. 8–9*

Full first-inversion chords (for instance, the right-hand chord in measure 1 of Example 4.6 as originally notated) are a challenge for many small-handed players because of the troublesome spread between the second and fourth fingers. Instead of rolling the chords, players can redistribute notes into an underused hand wherever possible.

EXAMPLE 4.6. *Heller, 25 Etudes for Piano op. 45 no. 15 ("Warrior's Song"), mm. 1–4*

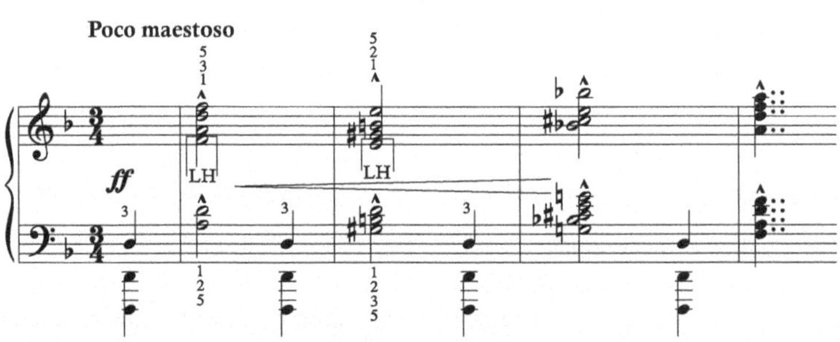

Redistribution also allows pianists to minimize the discomfort of large accompanimental chords. In typical waltz or stride-bass patterns, for instance, the left hand is forced to cross directly in front of the body to reach the upper harmonies. Example 4.7 juxtaposes Chopin's original score against an adaptation that moves some of the left-hand notes into the right hand to reduce shifting between registers and to eliminate wide reaches. The alternate version, however, results in right-hand chords with wider extensions; in addition, both the fingering and voicing are more complex than in the original version.

EXAMPLE 4.7. *Chopin, Polonaise in C♯ Minor, op. 26 no. 1, mm. 5–12*

In Example 4.8a, a player would need to extend the right hand to its maximum to hold the high D♯ through the measure while playing the accompanimental chords, or else rely upon the pedal alone to prolong the D♯. On today's more resonant pianos, the staccato articulations of the chords might be lost. The redistribution shown in Example 4.8b enables the pianist to hold the D♯ with ease and to articulate the underlying chords crisply and cleanly with the help of partial pedal changes.

EXAMPLE 4.8. *Chopin, Polonaise in C♯ Minor, op. 26 no. 1, mm. 32–33*
a. Printed version
b. Redistribution

When one hand is at rest, it can assist the active hand. In Example 4.9, taking the bottom note of the octaves with the left hand improves speed, power, accuracy, and comfort.

EXAMPLE 4.9. *Liszt,* Étude en douze exercices, *S. 136 no. 4, mm. 25–27*

Sometimes, redistributing just one note of an inner voice to the other hand can make a noticeable difference in continuity and comfort. In Example 4.10, eliminating the skip of a tenth in the right hand at measure 16 by taking the low F♯ with the left hand can improve continuity and physical ease (see Chapter 6 for an alternate solution).

EXAMPLE 4.10. *Rachmaninoff, Prelude in C♯ Minor, op. 3 no. 2, mm. 14–16*

Awkward combinations of double notes of the type frequently encountered in Beethoven's piano sonatas and chamber music can sometimes be made less taxing by radically redistributing the workload. In Example 4.11, the passage is converted into a series of contrary motion rotations to divide the task more evenly between the hands.

EXAMPLE 4.11. *Beethoven, Sonata in B♭ Major, op. 22, i, mm. 147–148*

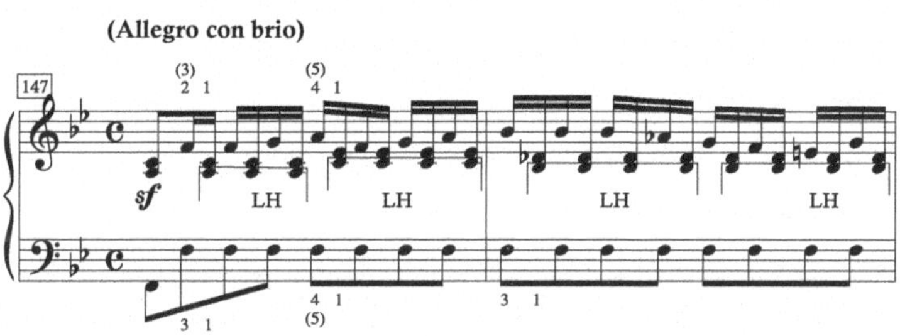

Small-handed pianists may need to extensively redistribute three-part textures where an inner accompanimental part is shared between the hands. In Examples 4.12 and 4.13, some of the notes are exchanged to eliminate large reaches.

EXAMPLE 4.12. *Mendelssohn,* Songs without Words, *op. 19 no. 1, mm. 16–19*

EXAMPLE 4.13. *Brahms, Intermezzo in A Major, op. 118 no. 2, mm. 25–30*

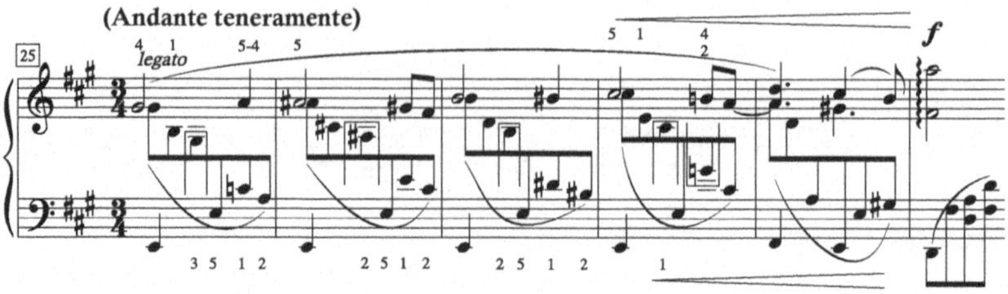

Conversely, redistributing the notes of a widespread accompanimental pattern into a three-part texture shared between the hands may eliminate technical problems related to stretch. With the fingers outstretched to play the extended left-hand arpeggios as written in Example 4.14, it may be difficult to achieve the desired qualities of flow and luminosity. Because the redistribution creates problems of its own with fingering and voicing, pianists will need to experiment to reach an acceptable compromise.

EXAMPLE 4.14. *Clara Schumann, "Notturno," from* Soireés musicales, *op. 6, mm. 1–6*

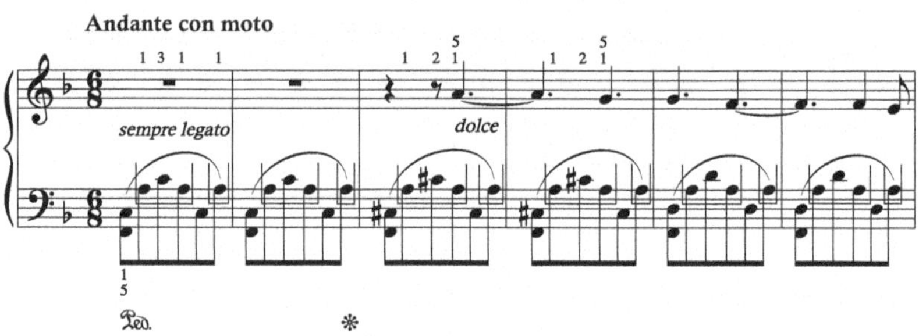

Even when a small-handed pianist can physically reach the notes, there still may be musical reasons to redistribute them. Moving accompanimental notes out of the right hand and into the left may make it easier to project the melody in Example 4.15.

EXAMPLE 4.15. *MacDowell, "To a Wild Rose," from* Woodland Sketches, *op. 51, mm. 17–25*

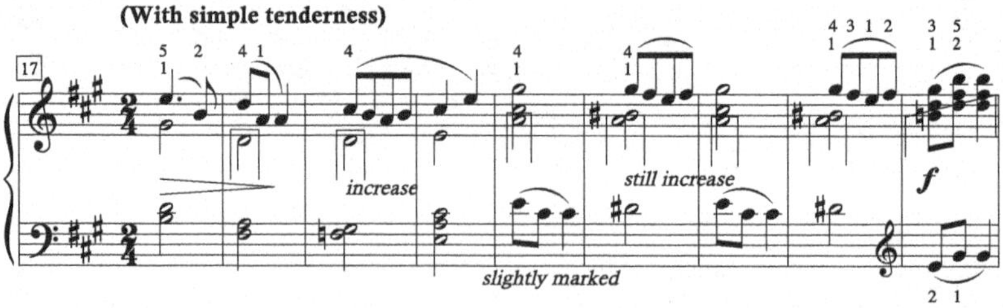

Similar strategies may apply in Neo-Romantic works by modern composers. The entire third movement of Lowell Liebermann's *Gargoyles,* especially the final section where the theme is presented in canon in the top and bottom with a shared accompaniment in the middle, would benefit from various redistributions (Example 4.16). Small-handed pianists would undoubtedly make very different choices from those with larger hand spans.

EXAMPLE 4.16. *Lowell Liebermann,* Gargoyles, *iii, mm. 50–53*

Passages where the hands play in parallel thirds, sixths, diminished sevenths, or octaves can often be facilitated by redistribution. As shown in Example 4.17, however, once this decision has been made, large sections may need to be reconfigured. In this

EXAMPLE 4.17. *Liszt,* Un sospiro, *mm. 46–51*

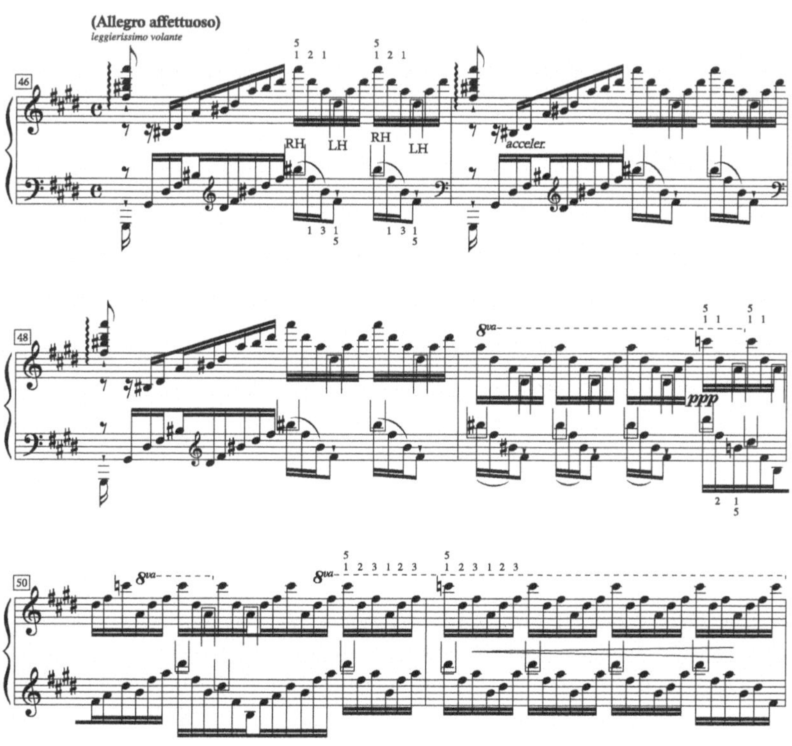

excerpt, accuracy and fluency are essential because the same transparent chord occupies six entire measures.

Redistributions can also be used in conjunction with rolling to play chords that fall outside the reach of the hand. In Example 4.18, the top notes of the left-hand chords are moved into the right to improve the projection of the tenor melody.[2]

EXAMPLE 4.18. *Brahms, Rhapsody in B Minor, op. 79 no. 1, mm. 81–82*

Similarly, in measures 52 and 54 of Example 4.19, a small-handed pianist might find it difficult to sustain the rhythmic flow of the eighth notes and to catch the grace notes cleanly in the pedal unless the upper note of the left-hand rolled tenths is redistributed into the right hand.

EXAMPLE 4.19. *Schumann,* Kreisleriana, *v, mm. 52–55*

The potential benefits of redistribution to reduce stretch are apparent in excerpts from Chopin's first scherzo shown in Examples 4.20–4.22. Beginning in measure 69, the top notes of large left-hand chords can be transferred into the right hand (Example 4.20). In Examples 4.21 and 4.22, selected notes may also be reassigned.[3] However, any technical gain must be weighed against the substantial challenges of learning these redistributions, a choice that will be different for every player.

---

[2] An alternate solution for this excerpt is discussed in Chapter 6.

[3] In Example 4.21, redistributing the first note in the bass clef in m. 85 into the right hand will necessitate depressing the pedal early in order to catch the right-hand C♯ in the preceding measure.

EXAMPLE 4.20. *Chopin, Scherzo no. 1 in B Minor, op. 20, mm. 69–76*

EXAMPLE 4.21. *Chopin, Scherzo no. 1 in B Minor, op. 20, mm. 85–88*

EXAMPLE 4.22. *Chopin, Scherzo no. 1 in B Minor, op. 20, mm. 110–113*

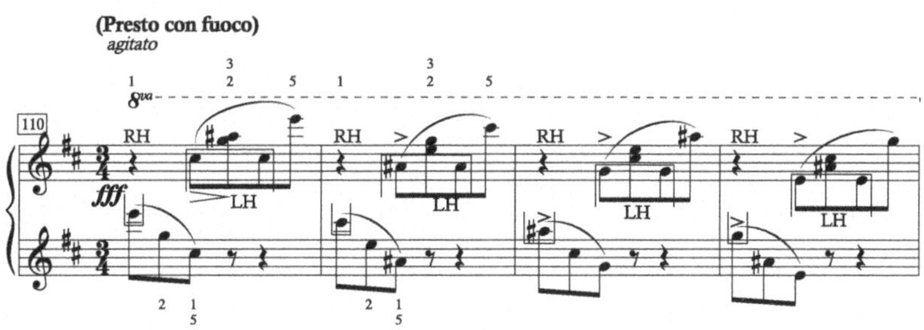

Small-handed pianists may take heart from knowing that Chopin endorsed redistribution as a performance strategy. Rather than insisting on playing extended single-line arpeggios in one hand as he had written them, Chopin advised his students to consider redistributing the notes.[4] This principle could apply in other nocturnes, scherzos, and

[4] Fryderyk Chopin, *Nocturnes*, ed. Jan Ekier (Krakow: PWM, 2012), Performance Commentary, 4.

ballades. Examples 4.23a and 4.23b show Chopin's solutions for a passage from the Nocturne, op. 27 no. 1.

EXAMPLE 4.23. *Chopin, Nocturne in C♯ Minor, op. 27 no. 1, mm. 27–28*
a. First Chopin solution
b. Second Chopin solution

(a)

(b)

In Example 4.24, the redistribution for a difficult arpeggio passage from Chopin's second scherzo takes the arpeggiated tenths and systematically reallocates them into smaller chords and melodic intervals played cooperatively by the hands.

EXAMPLE 4.24. *Chopin, Scherzo no. 2 in B♭ Minor, op. 31, mm. 118–132*

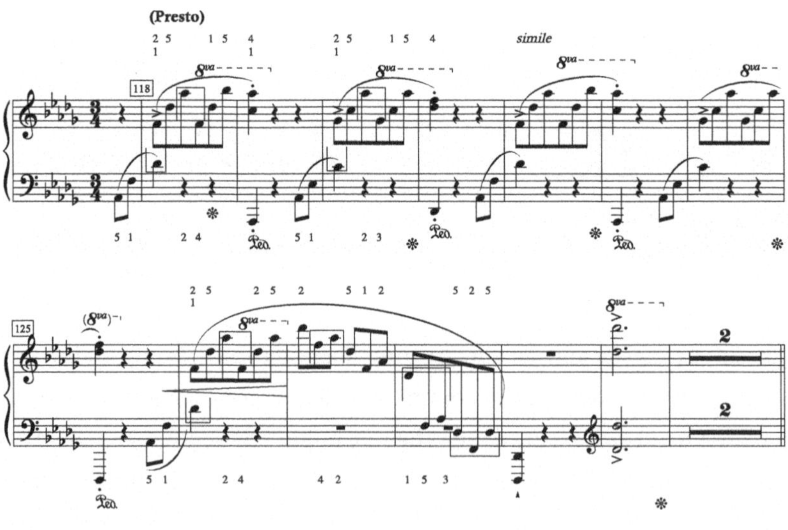

FACILITATING LEAPS OR HAND SHIFTS

Leaps at fast tempos can be treacherous for small-handed pianists, causing problems with accuracy or continuity, or provoking anxiety that can lead to muscular bracing. In Example 4.25, moving the first note of measure 24 from the right hand to the left hand allows the delicate flow of the sixteenths to continue undisturbed. The redistribution allows the pianist additional time for rounding off the cadence with a sense of repose by allocating the widely separated notes between the hands. The dynamic level of this F♯ should be sensitively tapered to match the rest of the descending scale.

EXAMPLE 4.25. *Mozart, Sonata in C Major, K. 279, i, mm. 22–24*

Awkward leaps between phrases often occur at moments when pianists can ill afford to disrupt the rhythmic flow. Examples 4.26, 4.27,, and 4.28 show leaps that would be unremarkable for larger-handed players. For those with smaller hands, however, redistribution of just a few notes provides crucial time to reposition the hand.

EXAMPLE 4.26. *Beethoven, Sonata in E♭ Major, op. 31 no. 3, ii, mm. 43–45*

EXAMPLE 4.27. *Beethoven, Sonata in D Minor, op. 31 no. 2 ("Tempest"), i, mm. 117–120*

EXAMPLE 4.28. *Beethoven, Sonata in C Major, op. 53 ("Waldstein"), i, mm. 272–274*

Nevertheless, there are times when musical objectives might override concerns about technical ease. Using redistributions at the beginnings of Beethoven's Sonatas Op. 111 and Op. 106 may lead to increased power, accuracy, and comfort.[5] Some argue, however, that by eliminating the element of risk, the psychological, physical, and visual drama of the opening gestures will be lost or diminished (Examples 4.29 and 4.30).

EXAMPLE 4.29. *Beethoven, Sonata in C Minor, op. 111, i, mm. 1–2*

EXAMPLE 4.30. *Beethoven, Sonata in B♭ Major, op. 106 ("Hammerklavier"), i, m. 1–4*

Pianists often overlook leaps hidden between broken octaves. Patricia Eger pointed out that if the broken octaves in Example 4.31 are played in the right hand instead of the left hand, the intervals between the notes are transformed into sevenths instead of ninths.[6] Extending that line of reasoning, ascending broken octaves will contain smaller shifts if played by the left hand instead of the right hand.

[5] Pianists who cannot reach the ninth in the first and third measures of Op. 106 may also have to consider omitting a note or rewriting the chord, a topic discussed in Chapter 6.

[6] Patricia Eger, "A Study of the Problems Encountered by the Pianist with Small Hands and a Compendium of Practical Solutions" (DMA dissertation, Ball State University, 1982), 154.

EXAMPLE 4.31. *Beethoven, Concerto no. 5, op. 73 ("Emperor"), i, mm. 438–439*

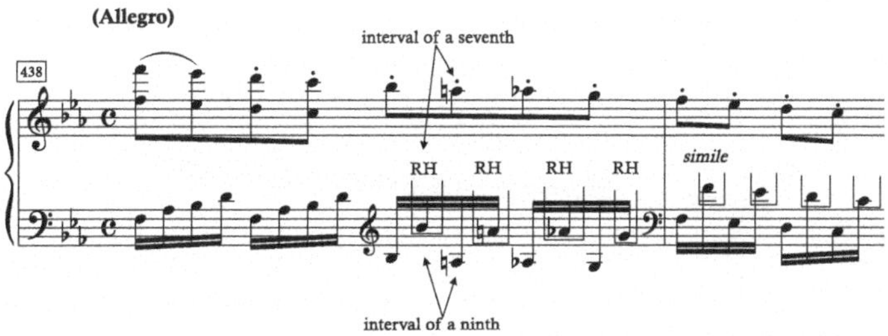

Redistribution can also help resolve the rapid octave transpositions in Example 4.32. The solution could work equally well for pianists with larger hands.

EXAMPLE 4.32. *Prokofiev, Sonata no. 2 in D Minor, i, mm. 85–90*

Pianists can redistribute to avoid awkward hand shifts in the same hand. As shown in Example 4.33, the notes can be allocated into smaller shapes that are passed from one hand to another to produce a unified sweep of sound. The damper pedal sustains the longer chords of the melody, allowing the right hand to move down to assist the left. The hands do not interfere with each other due to the layout of the black and white keys.

EXAMPLE 4.33. *Debussy, "Poissons d'or," from* Images, *Book II, mm. 14–15*

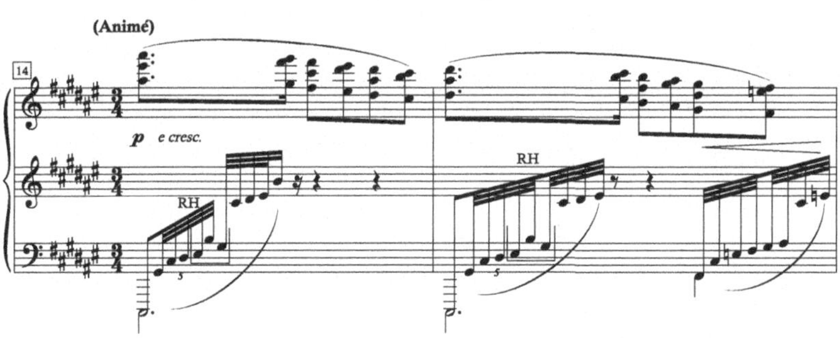

In Examples 4.34 and 4.35, small-handed pianists would be more likely than large-handed pianists to consider using multiple hand exchanges in order to increase the speed and note accuracy of extended arpeggios.

EXAMPLE 4.34. *Debussy, "Reflets dans l'eau," from* Images, *Book I, m. 67*

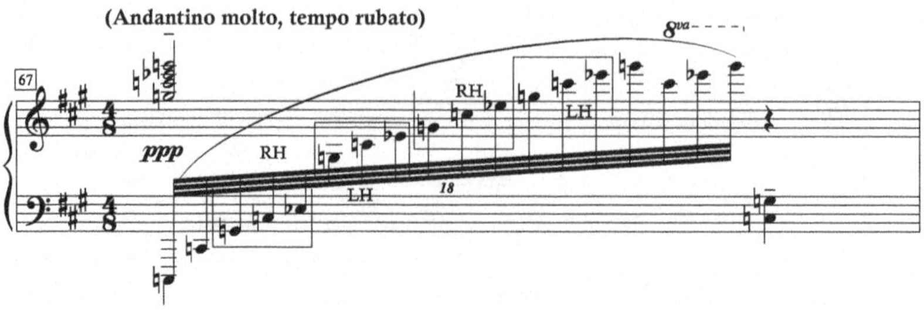

EXAMPLE 4.35. *Debussy, "Hommage à Rameau," from* Images, *Book I, mm. 51–53*

INCREASING ACCURACY, POWER, AND CONTROL

Pianists frequently need solutions that they can rely on to be error-free. In unison passages, for example, mistakes can be detected by even the most casual listeners. Provided the hand can encompass an octave, the redistribution in Example 4.36 improves accuracy by eliminating the need for shifts of any kind.[7] The hands stay in one position to execute the entire melody. Like Example 4.32, this solution would work for pianists with larger hands as well.

EXAMPLE 4.36. *Schumann, "Hunting Song," from* Album for the Young, *op. 68 no. 7, mm. 1–4*

Reducing the size of shifts between chords is the goal of the redistribution in Example 4.37a. If the hand is big enough, the lowest note in the right-hand chords could be alternately exchanged between the hands, minimizing shifting while distributing the workload. A notable bonus is that, beginning with the last chord in measure 43, the three lower notes of the right hand and the three upper notes of the left hand move chromatically downward in perfect parallel

[7] Alternative solutions are presented in Examples 5.46 and 6.21.

motion (Example 4.37b), making this passage easier to learn and memorize. Redistributions of this type are generally possible when the hands are in close proximity to each other.

EXAMPLE 4.37. *Debussy, "Prélude," from* Pour le Piano, *mm. 43–46*
a. Redistribution
b. Parallel voice leading

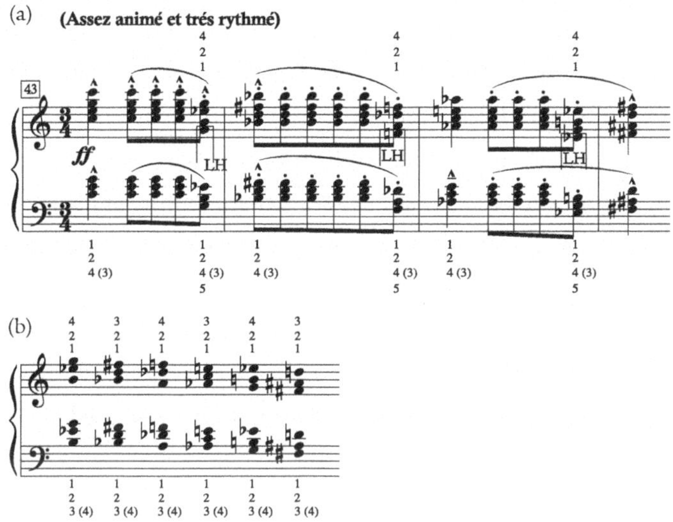

Dividing the octaves between the hands in Examples 4.38 and 4.39 results in a significant increase in power because the pianist no longer has to keep the hand outstretched.

EXAMPLE 4.38. *Liszt,* Un sospiro, *m. 29*

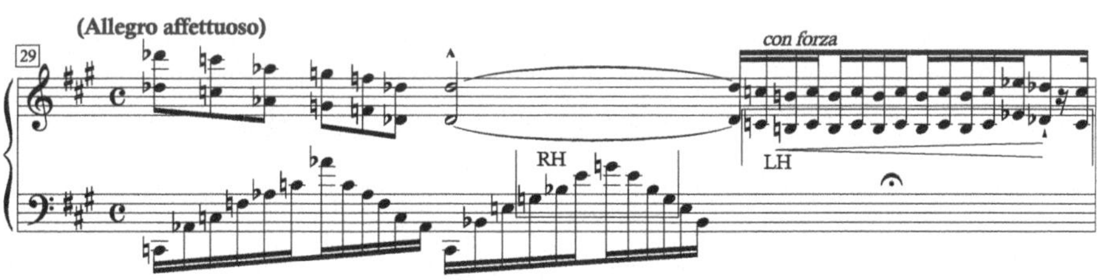

EXAMPLE 4.39. *Chopin, Nocturne in C♯ Minor, op. 27 no. 1, m. 83–84*

MAINTAINING MORE NEUTRAL HAND AND WRIST POSITIONS

A neutral hand position is the position of the hand relative to the wrist and forearm in all planes when a pianist allows the arm to hang naturally at the side. The more the hand deviates from this position and the greater the degree of deviation, the greater the physical demands on the body. Playing at the extremes of the keyboard, where the upper arms and forearms are farther away from the torso, tends to result in radial deviation at the wrist. Playing in the very middle of the keyboard with the hands directly in front of the torso or with the forearm positioned across the torso tends to result in ulnar deviation. In these situations, it is important to periodically return the thumb and wrist to a more neutral alignment as soon as possible. Sometimes even a brief return to neutral can greatly reduce fatigue.

Example 4.40a shows a series of chords moving from left to right across the body. To preserve a more neutral alignment, a pianist can distribute more notes into the right hand at the beginning of the passage and more notes into the left hand at the end (Example 4.40b). The difficulty of memorizing the fingering changes would have to be balanced against gains in comfort.

EXAMPLE 4.40. *Grieg, "Wedding Day at Troldhaugen," from* Lyric Pieces, *op. 65 no. 6, mm. 45–46*
a. Printed version
b. Redistribution

(a)

(b)

Notes located to the extreme right of the piano can be distributed into the right hand to achieve greater comfort and accuracy, as shown in Example 4.41.

EXAMPLE 4.41. *Chopin, Etude in F Major, op. 10 no. 8, mm. 93–95*

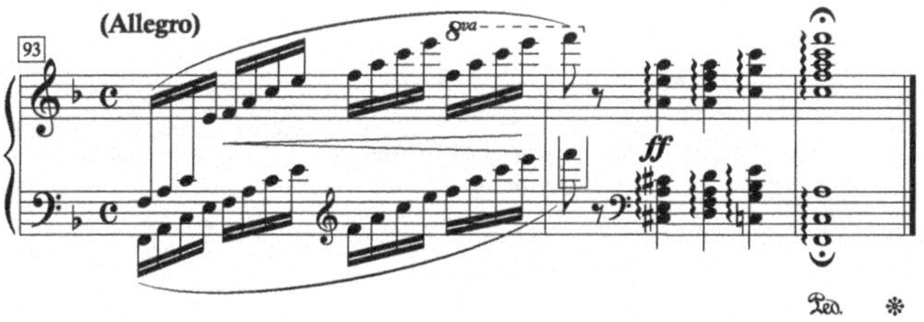

In Example 4.42, the indicated redistribution allows the left hand to keep a more neutral wrist angle when it crosses over the right hand to play notes in the highest register of the keyboard.

EXAMPLE 4.42. *Muczynski, Prelude, op. 6 no. 1, mm. 61–65*

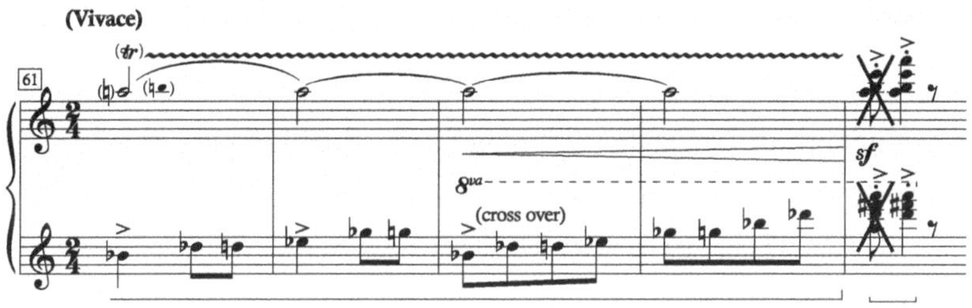

Example 4.43 reallocates notes so the right hand can maintain a more neutral wrist position.

EXAMPLE 4.43. *Tchaikovsky, "February," from* The Seasons, *op. 37a, mm. 85–88*

FACILITATING TRILLS AND TREMOLOS

Trilling with some fingers is much more difficult than with others. Trilling with the fourth and fifth fingers is particularly awkward if the hand is simultaneously extended to play a chord. Examples 4.44 and 4.45 show how redistributing chord tones into one hand releases the other to play the trill with greater ease.

EXAMPLE 4.44. *Beethoven, Sonata in E♭ Major, op. 27 no. 1, iii, mm. 21–24*

EXAMPLE 4.45. *Debussy, "Prélude," from* Pour le Piano, *mm.74–76*

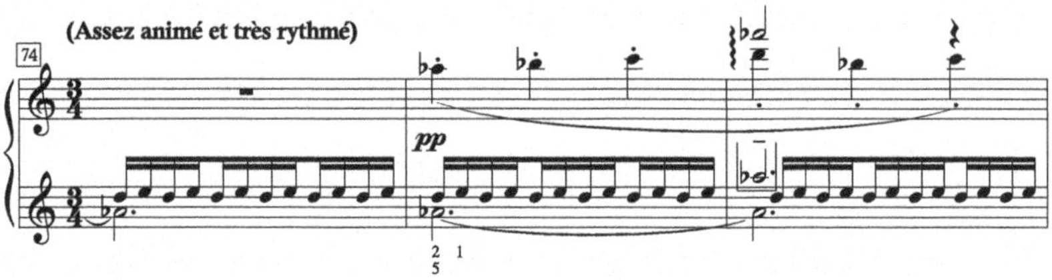

Redistributing notes of the lengthy ***fff*** tremolo shown in Example 4.46 frees pianists from the task of playing rapid ninth chords that can strain the right arm. The solution uncrosses the thumbs of the right and left hands so that the trill can be played as a series of forearm rotations.

EXAMPLE 4.46. *Debussy,* L'isle joyeuse, *mm. 252–255*

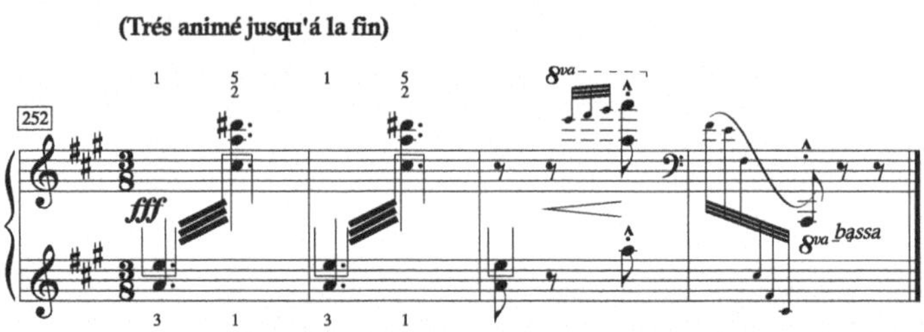

MAINTAINING LEGATO AND LINE

Redistribution is normal protocol for playing strict counterpoint. The right and left hands must work together seamlessly to preserve the integrity of each contrapuntal line. However, small-handed players often need to redistribute parts sooner and more frequently than pianists with larger reaches, as shown in Examples 4.47 and 4.48.

EXAMPLE 4.47. *J. S. Bach, Sinfonia no. 4 in D Minor, mm. 1–2*

EXAMPLE 4.48. *Händel, Suite in F Major, HWV 427, iv, m. 23, m. 26, and m. 42*

The same strategy can be applied in more homophonic music, for example, in passages of parallel sixths or octaves, where pianists may struggle to maintain the line, shaping, and tempo of a phrase (Example 4.49).

EXAMPLE 4.49. *Haydn, Sonata in D Major, L. 50/Hob. XVI: 37, i, mm. 7–8*

Three-voiced passages with doublings at the octave or the sixth can often be redistributed by alternating the notes of the middle voice between the left and right hands (Examples 4.50 and 4.51). Disruptive shifting that might disturb the shaping of the line is thereby minimized.

EXAMPLE 4.50. *Debussy, "Danseuses de Delphes," from* Préludes, *Book 1, mm. 18–20*

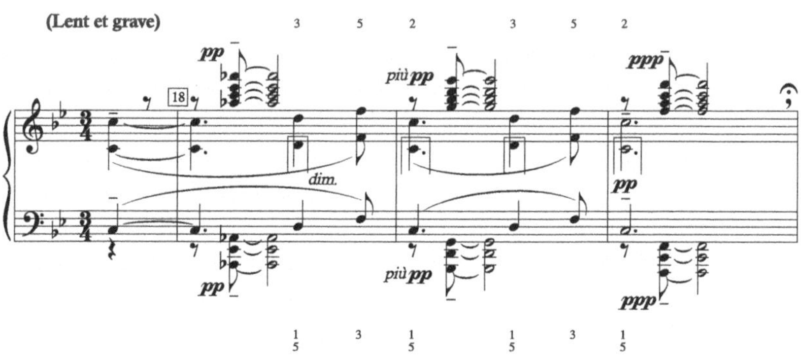

EXAMPLE 4.51. *Schumann,* Kreisleriana, *ii, mm. 1–2*

PROJECTING HARMONY

In music where the harmonic rhythm is very slow, large structural chords often must be held for long periods of time. In Example 4.52, pianists who are able to reach the ninth in the left hand from the G♯ in the tenor of beat 2 down to the F♯ a beat later may nevertheless be unable to play the E with the second finger. Taking the G♯ with the right hand enables the pianist to play and sustain the harmony through pedal changes marked by the composer.

EXAMPLE 4.52. *Bartok, "Buciumeana," from* Six Romanian Folk Dances, *mm. 17–18*

To make clean pedal changes and avoid obscuring the melody or harmony, it is necessary to physically hold low bass notes for full value at the beginnings of measures 1 and 2 in Example 4.53. Pianists unable to reach the tenth in the left hand between the grace note F and the high A can simply redistribute the A into the right hand.

EXAMPLE 4.53. *Schumann,* Kreisleriana, *vi, mm. 1–2*

CONCLUSION

Redistribution is a powerful adaptive strategy that tends to be underutilized by the small-handed. At first, performers may be reluctant to allocate notes in new ways due to a strong sense of loyalty to the score. However, scores exist for different purposes. Composers use them to transmit information about how a work should be physically executed but also to delineate aspects of voice leading, melodic line, phrasing, and harmonic content. These two domains—the performative and the descriptive—often clash. It may thus be more meaningful to see the score as a musical blueprint or guide, not an ironclad contract.[8]

Redistribution for its own sake, however, is pointless and can sometimes create more problems than it solves. Even if a passage is awkward, the printed score may already represent the best possible solution. Pianists can make wise choices if their decisions are driven by coherent values that balance physical and musical needs. If these conditions are met, there is no reason not to explore creative solutions in places where the notes just do not fit the hands.

APPLICATION PROBLEMS

The following excerpts contain problems that might be addressed through redistribution. Find possible solutions by exploring the strategies presented in this chapter or by devising your own. Be sure to follow basic principles of effective technique as presented in Chapters 2 and 3 to avoid maladaptive tendencies.

1. Grieg, "Gade," from *Lyric Pieces,* op. 57 no. 2, mm. 1–8
2. Schumann, "Am Kamin," from *Kinderszenen,* op. 15 no. 8, mm. 1–16
3. Mendelssohn, *Songs without Words,* op. 19 no. 2, mm. 1–21
4. Kuhlau, Sonatina in A Minor, op. 88 no. 3, iii, mm. 53–63
5. Beethoven, Sonata in E♭ Major, op. 31 no. 3, ii, mm. 155–157

[8] Kenneth Hamilton, Neal Peres Da Costa, Richard Taruskin, and others dispute the idea that *urtext* editions should serve as the ultimate arbiters of composers' ideations of musical works. For example, analysis of early acoustic recordings and piano rolls provides concrete proof of unwritten traditions of performance practice such as unmarked note, tempo, and dynamic modifications that were widely accepted among educated musicians of the time. In addition, a degree of translation is required to play repertoire originally written for earlier keyboard instruments on the modern piano. Kenneth Hamilton, *After the Golden Age: Romantic Pianism and Modern Performance* (New York: Oxford University Press, 2008); Neal Peres Da Costa, *Off the Record* (New York: Oxford University Press, 2012); Richard Taruskin, *Text and Act* (New York: Oxford University Press, 1995).

FIVE

Refingering

FINGERING IS A CRAFT DEVELOPED OVER AN ENTIRE LIFETIME OF EXPERIENCE. Wise choices not only expedite music learning but also provide a first line of defense against the difficulties of small-handedness. Fingering is also an art. A close study can heighten a performer's awareness of the relationship between how music feels and how it sounds. According to Chopin, "Everything is a matter of knowing good fingering."[1]

What is a "good fingering"? First and foremost, a good fingering embraces both the technical and musical goals of the performer. Ideally, it is easy to play and easy to remember. Since every hand is different, however, fingering is highly individual. A solution that works for one player may be disastrous for another. To find the perfect fingering is to embark on a journey, and not all pianists will travel by the same road.

For small-handed players, however, the path forward is sometimes difficult to find. They are often unduly influenced by editorial fingerings printed in the score, many of which are inappropriate for small hands. Less experienced players may not know how to tailor these fingerings for a better fit or how to select alternatives that will work reliably under performance conditions. Otto Ortmann also observed that players "seldom use wrong fingering through sheer carelessness. The fingering they substitute is, for them, an easier fingering, determined primarily by the physiological structure of the hand and the amount of training."[2] Pianists should therefore pay careful heed to why they use the fingerings that they do and consider whether there are better alternatives.

[1] Jean-Jacques Eigeldinger, *Chopin: Pianist and Teacher as Seen by His Pupils* (London: Cambridge University Press, 1986), 18.

[2] Otto Ortmann, *The Physiological Mechanics of Piano Technique* (1929; reprint, New York: E. P. Dutton, 1962), 320–321.

The scale and arpeggio fingerings that are widely used today were promulgated and embedded into numerous compositions and exercises by famous early nineteenth-century composer-pianists and teachers such as Clementi, Czerny, and Cramer. It is easy to forget, however, that these "standard" fingerings are not immutable but instead evolved over time. For instance, C. P. E. Bach listed not one, but three possible ways to finger a right-hand C-major scale: the now traditional 123-1234 as well as 123-34-34 and 1234-123.

Modern fingering practices can perhaps be traced most directly to Czerny, who recommended the following:[3]

- Neither the thumb nor the fifth finger should be used on the black keys.
- Unnecessary changes of position should be avoided.
- The third finger is not permitted to cross over the fourth.
- If necessary, one should silently change fingers on a single note in order to produce a *legato* transition into the next note.
- One should avoid using the same finger on consecutive keys, unless one is sliding from a black to a white key or moving from the end of one phrase to the beginning of the next.
- One should use the same or similar fingering for analogous or sequential passages.
- For repeated notes, one should use different fingers.

Such mandates reflect Czerny's preference for long legato phrases unbroken by small internal articulations, leaving the hand in the same position for the longest possible time. His approach, however, can create problems for pianists unable to reach the octaves, ninths, and tenths that are ubiquitous in literature written for the modern piano.

Czerny's maxims diverge in significant ways from the fingering principles advocated by Chopin, a composer who was revered as a teacher during his lifetime. Chopin recognized that there were inherent differences in the strengths of individual fingers and recommended innovative strategies such as these:[4]

- Using the thumb on a black key
- Changing fingers silently on a single note

---

[3] Jeffrey Swinkin, "Keyboard Fingering and Interpretation: A Comparison of Historical and Modern Approaches," Claremont Graduate University, http://scholarship.claremont.edu/cgi/viewcontent.cgi?article=1210&context=ppr (accessed February 10, 2014).

[4] *Ibid.* Czerny was a highly prolific teacher whose ideas were widely disseminated by means of the thousands of technical studies published during his lifetime. He outlived Chopin by eight years. Chopin's planned piano method, on the other hand, was never completed.

- Moving the thumb under the fifth finger
- Sliding or repeating one finger across multiple keys
- Moving 3, 4, and 5 over each other instead of shifting to the thumb
- Repeating a finger on the same note

In contrast to Czerny, Chopin believed that the aural impression of legato could be conveyed without an actual physical connection, thereby freeing the pianist to use the thumb on black notes or to play consecutive notes with the same finger. He often relied on the damper pedal or fluid motions of the arm to "glue" the notes together. Chopin's path is one that can help lead small-handed pianists out of a metaphorical forest of pianistic difficulties.

FINGERING AND HAND SIZE

As explained in Chapters 2 and 3, those with small hands risk fatigue or injury when they use fully extended positions that repeatedly force the hand far outside its preferred neutral shape (see Figures 2.9 and 2.10). Small-handed pianists use larger angles of abduction between their fingers than pianists with larger hands. When fingers are at full extension, it is impossible to properly stabilize the joints. A good fingering reduces strain by decreasing the size and angles of finger spread.

A well-chosen fingering can further benefit small-handed players by improving wrist alignment and allowing the arm to position itself more directly behind the playing fingers. Force can then be directed and absorbed through the bony structures, lightening the workload of the muscles. While it is impossible to be in perfect alignment all the time, especially when playing large chords, this point of anatomic design should inform choices about fingering.

In addition to fostering proper alignment, a well-constructed fingering can accommodate the inherent inequalities of the fingers.[5] It harnesses the greater or lesser strengths and tendencies of individual fingers and uses them in ways that are appropriate to specific musical contexts.

Pianists often choose fingerings because of idiosyncrasies of hand size. Otto Ortmann tested a passage from Beethoven's "Appassionata" sonata that was fingered in two different ways. The photographs he obtained showed that "pronounced forward and backward arm-shifts" were required for the second option, making it less efficient than the first option in terms of speed (Figure 5.1).[6] Nevertheless, Ortmann found that small-handed

---

[5] For a more detailed explanation of the inequalities of the fingers, please refer to Chapter 2.

[6] Backward and forward motions are discussed in Chapter 3 of this book.

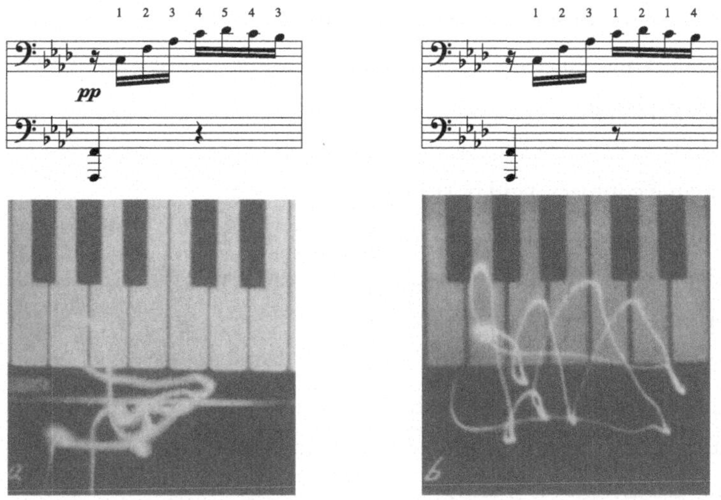

FIGURE 5.1. *Ortmann's photographs illustrating the effects of fingering upon hand movements. Credit: Justin Houser. Adapted from Otto Ortmann,* The Physiological Mechanics of Piano Technique *(1929; reprint New York: E. P. Dutton, 1962), 278.*

players preferred the second option because it contained smaller hand spans. In other words, fingerings that work for some pianists may prove inadequate for others, especially since pianists vary greatly not only with regard to hand size but also hand shape.

With these factors in mind, pianists should spend time getting to know the specific attributes of their own hands. They can start by playing three- or four-note chords and chord inversions in many keys, eventually progressing to five-note seventh chords and chord inversions. Traditional scale, arpeggio, and chord routines are also useful in this regard. Players should ask themselves:

- What is the maximum span between the thumb and fifth finger in each hand?
- What is the reach between pairs of adjacent and non-adjacent fingers? Are some chords more uncomfortable to reach than others? Why?
- How long are the fingers in relation to each other? Are any fingers unusually long or short? Is it difficult to reach chords containing black notes?
- How long or short are the fingers as compared to the palm?
- How much cushioning is present at the fingertips? Does this cushioning extend beyond the nails or not?
- Are there any other idiosyncrasies, such as high webbing between certain fingers, limited ability to move the wrist horizontally to the left or right or vertically up and down, thick fingers that depress several keys at once, difficulty spreading certain fingers apart from each other, or joints that tend to collapse?

Pianists who thoughtfully engage in this process of self-discovery may notice several hallmark problems that this chapter is designed to address. Seven alternative approaches to fingering are explored:

- Encouraging compact hand positions
- Repeating the same finger to recover space in the hand
- Optimizing sound production and voicing
- Moving longer fingers over shorter fingers
- Substituting the thumb for inner fingers
- Using positional fingerings to achieve maximum speed
- Substituting one finger for another on the same key
- Retaking released notes silently

Some sections include activities that explore more unconventional aspects of alternative fingering strategies. These are intended to foster tactile awareness of new motor patterns, not to build endurance. Therefore, activities should not be repeated endlessly and ought to be discontinued before any fatigue ensues.

It is useful for all pianists to develop a deep resource of automated fingering sequences for commonly encountered keyboard patterns. Learning traditional fingerings in tandem with alternatives suggested here can further broaden small-handed pianists' musical and technical options. Extensive experimentation is often needed before an appropriate fingering can be determined. Practicing multiple fingerings, however, is not wasted effort; rather, each fingering provides valuable kinesthetic information to the performer. Even if a pianist ultimately decides against adopting alternative fingerings, execution will often improve as a result of working through several options. However, the complexity of changing an ingrained fingering should not be underestimated. Even a small fingering adjustment can profoundly affect the amount of involvement of playing levers, the angles of alignment, and the shape and timing of the movement. Slow and thoughtful practice is needed before physical and mental readjustments can take hold.

The fingerings in this chapter lean toward solutions for the smallest of small hands, and minor modifications may be needed to address individual needs. However, pianists may also find that certain strategies work equally well for hands of many sizes, both large and small.

ENCOURAGING COMPACT HAND POSITIONS

The primary principle of good fingering is to maintain the hand in or as close to its preferred neutral shape as much as possible. Traditional methods of fingering chords

ACTIVITY 5.1
BASIC TRIADS AND CHORD INVERSIONS

Practice Activity 5.1 slowly, one hand at a time, up and down for several octaves. Retain the pliability of the hand when moving from one chord to the next. Transpose the activity to keys with different configurations of black and white notes, such as C♯ major and D major. Try altering the chords to minor, diminished, or augmented triads.

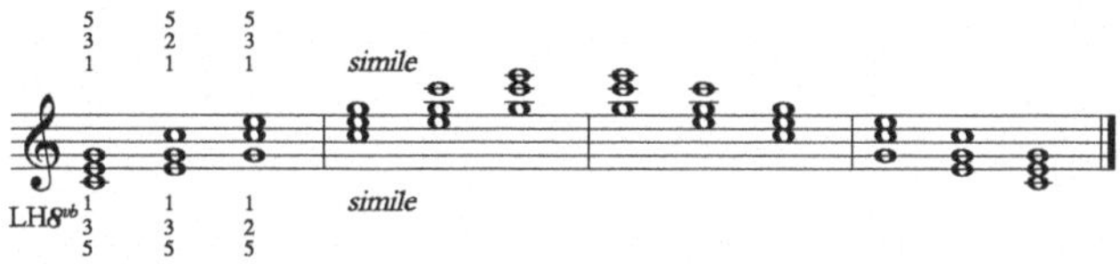

and arpeggios, however, favor pianists with large hand spans. It is therefore often necessary for small-handed pianists to scale these fingerings down to fit a hand that is more compact.

Chords

With typical four- or five-note triadic chords, a player has few fingering options. With two- or three-note chords, however, more alternatives may be available. In sequences involving these smaller chords, pianists may automatically default to fingerings that spread the hand to the octave, playing several elements of a pattern in a single hand position instead of choosing the most compact fingering for each chord. Activity 5.1 shows traditional fingerings for triads and their inversions. Activity 5.2 shows how these fingerings can be implemented within typical waltz- or stride-bass accompaniments.

In Example 5.1, pianists with small hands can substitute a fingering of 1-5 for the more traditional 2-5 to bring the hand into a stronger, more compact shape.

EXAMPLE 5.1. *Grieg, "Wedding Day at Troldhaugen," from* Lyric Pieces, *op. 65 no. 6, mm. 1–2*

ACTIVITY 5.2
WALTZ- OR STRIDE-BASS TEXTURES

Transfer the chord fingerings of Activity 5.1 to the waltz-bass accompaniments in Activity 5.2. The thumb of the right hand and the fifth finger of the left hand are repeated so that compact hand positions can be used for chords on the offbeats. Practice the activity one hand at a time, at several different tempi, and in many different keys.

Almost all pieces using waltz- or stride-bass textures can be refingered according to the principles developed in Activities 5.1 and 5.2 to avoid stretching the hand beyond neutral (Example 5.2).

EXAMPLE 5.2. *Chopin, Waltz in A Minor, op. posth., mm. 1–4*

Pianists with small hands should also choose fingerings that deliver the most compact hand shape possible to the largest chords. In Example 5.3, the two augmented fourths on beats 2 and 4 of measure 10 should be played with fingers that are widely separated.

EXAMPLE 5.3. *Clementi, Sonatina in C Major, op. 36 no. 3, i, mm. 10–12*

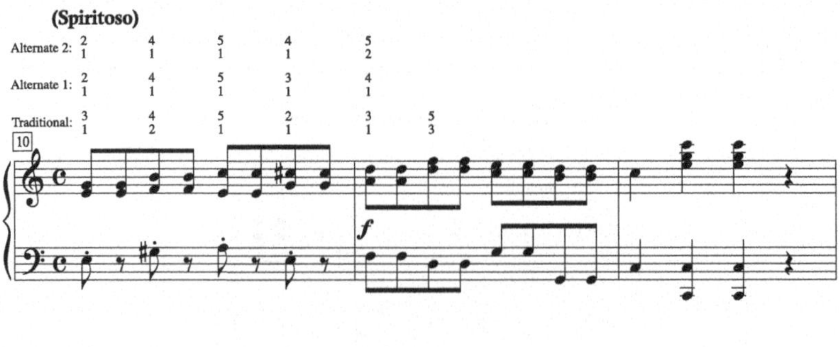

Arpeggios, Broken Chords, and Melodic Intervals

Many editors, even editors of pedagogical literature, use the octave as a basic fingering referent not only for chords but also for linear elements of music such as arpeggios, broken chords, and melodies. For most small-handed pianists, however, the size of the neutral hand is significantly smaller than the octave. Thus, as a first step, players should seek opportunities to refinger these shapes into smaller intervals of fifths or sixths.

Fingering in smaller spans inevitably leads to the relocation or addition of shifts within an arpeggio or melody. The pianist must ask whether these shifts affect the integrity and shape of the line. In most literature written for the modern piano, however, legato does not necessarily depend on physical connections between the fingers. Often, this task can be relegated to the pedal, and at other times, minor gaps can be subsumed within larger arm gestures. Therefore, while keeping musical goals firmly in mind, small-handed pianists should routinely consider relocating shifts in extended arpeggio passages, searching for repeating patterns with the smallest spans. Casting the widest possible net, they should not summarily reject reasonable alternatives, including those that break from traditional fingering practices. Activities 5.3, 5.4, and 5.5 explore several arpeggio fingerings that include added shifts or that keep the hand in a more compact shape.

ACTIVITY 5.3
THREE-NOTE BROKEN ARPEGGIOS

Practice Activity 5.3 one hand at a time in many different major and minor keys while keeping the same fingering throughout. For arpeggios containing sharps or flats, keep the backward and forward motions needed to accommodate the thumb on the black keys as small as possible. This activity may be used to practice diminished or augmented triads as well.

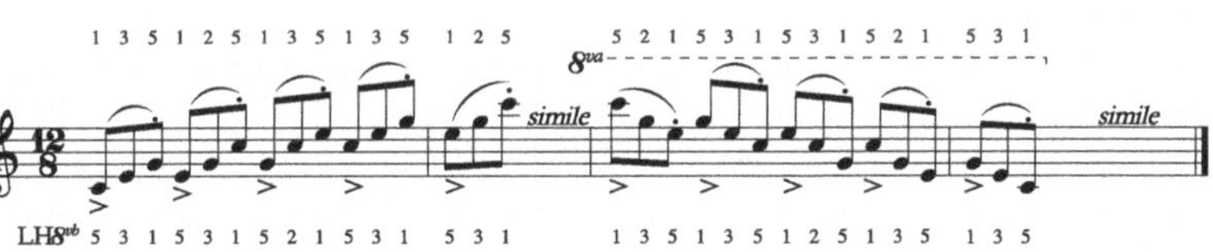

ACTIVITY 5.4
FINGERING ARPEGGIOS IN TWO-NOTE GROUPS

Activity 5.4 reinforces the idea that notes in arpeggios can be fingered in groups of twos instead of the more standard threes or fours. Transpose the activity into a variety of keys while using the same fingering so that the thumb becomes accustomed to playing on the black keys.

ACTIVITY 5.5
VARIATIONS IN ARPEGGIO FINGERINGS

Ascending and descending movements at the piano are different, asymmetrical activities. The motion required to shift the thumb under the fingers for an upward moving right-hand arpeggio is more difficult than the motion to shift the hand over the thumb for a right-hand descending arpeggio. It follows that the ideal fingering for the ascent might differ from that for the descent.

Activity 5.5 reinforces the idea that an arpeggio fingering does not have to be the same going up as it is coming down. Be sure to transpose this activity to arpeggios starting on black keys, retaining the fingering indicated. Using the thumb on black keys can provide a sense of security and power that is difficult to achieve with a traditional fingering. Practice Activity 5.5 alongside traditional arpeggio fingerings and make note of the differences and similarities between them.

In Example 5.4, a melody with a span of a ninth is divided into smaller arpeggiated shapes spanning fifths and sixths instead of being played in a single hand position.

EXAMPLE 5.4. *Schumann, "Wild Rider," from* Album for the Young, *mm. 1–4*

In Example 5.5, the melody is refingered to ease problems of stretch between inner fingers. Extra shifts are added at the points marked by the vertical arrows. The notes

EXAMPLE 5.5. *Mendelssohn,* Rondo Capriccioso, *op. 14, mm. 204–205*

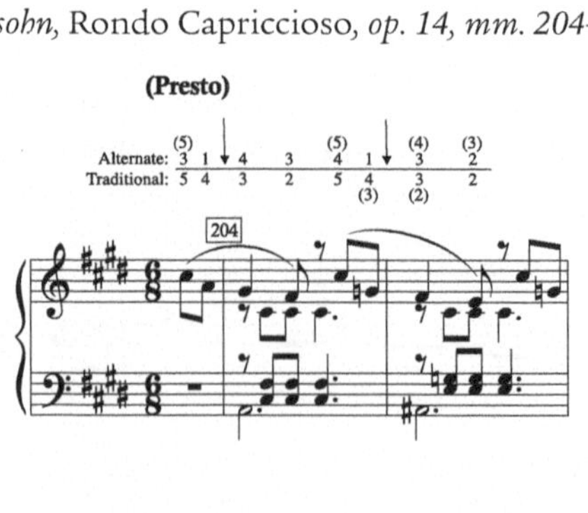

immediately preceding the shifts (A and then G) are quickly released so the hand can close up in the direction of the lateral downward shift of the forearm. The right hand remains compact, facilitating proper voicing of the melody.

A common characteristic of small hands is a limited spread between fingers 2 and 5. Examples 5.6 and 5.7 show points where the thumb could be momentarily inserted to relieve these stretches (marked by brackets), departing from more traditional arpeggio fingerings. Factors such as hand reach and flexibility will determine whether the adjustments work for individual players.

EXAMPLE 5.6. *Debussy, "Reflets dans l'eau," from* Images, *Book I, mm. 43–44*

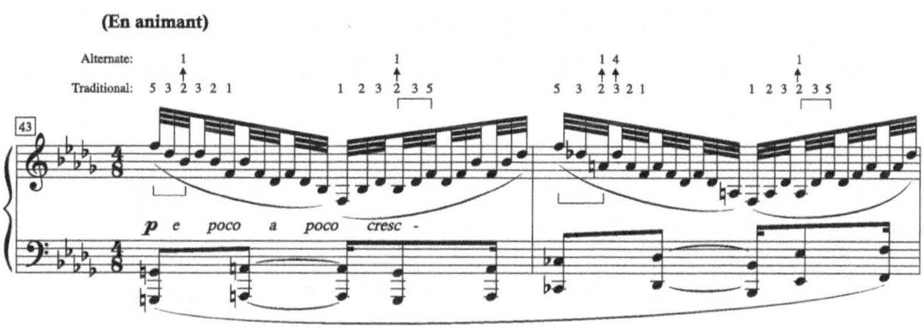

EXAMPLE 5.7. *Chopin, Etude in G♭ Major, op. 10 no. 5 ("Black Key"), mm. 74–76*

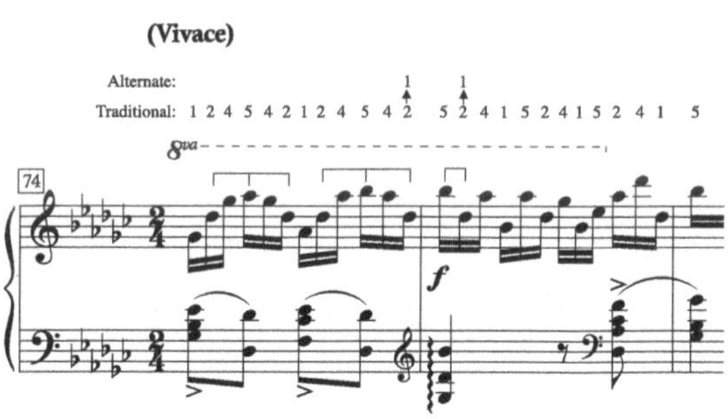

Chopin's first etude, op. 10 no. 1 serves as a classic application of adaptive refingering. The composer's fingering divides the notes into a series of interlocked arpeggios spanning ninths, tenths, and elevenths moving across the keyboard at a tempo of 176 to the quarter. Some arpeggios contain intervals as large as sevenths between fingers 1 and 2, fifths between 2 and 3, and fourths between 4 and 5. However, the famous Chopin interpreter Alfred Cortot reportedly insisted upon a different fingering, feeling that it "yielded a buoyant result."[7]

[7] Thomas Manshardt, *Aspects of Cortot* (Northumberland, Great Britain: Appian Publications, 1994), 90. Cortot's 1915 edition of the op. 10 etudes, however, does not include the revised fingering and in fact begins with two pages of suggested exercises for learning the Chopin fingering. Manshardt reports that Cortot left out the altered fingering not because he wished to keep it a secret but rather because he thought it might cause confusion.

Cortot's refingering relocates the shift one note earlier, grouping the notes into smaller units (Example 5.8a).

Alternatives 1, 2, and 3 of Example 5.8b extend Cortot's concept to the descending arpeggios that start in measure 2, grouping the notes into spans of sixths instead of tenths. In Alternative 1, the fifth finger is played twice at the top of the arpeggio. In Alternative 2, the longer fourth finger extends over the fifth finger to begin the descent. Alternative 3 reverts to Chopin's fingering for the last four notes before descending in the same manner as Alternative 1. The proposed refingerings are considerably more difficult to learn and memorize but may dramatically increase technical ease, dramatic power, and accuracy at performance tempo.

EXAMPLE 5.8. *Chopin, Etude in C Major, op. 10 no. 1, mm. 1–2*
a. Chopin's and Cortot's fingerings
b. Alternate refingerings

(a)

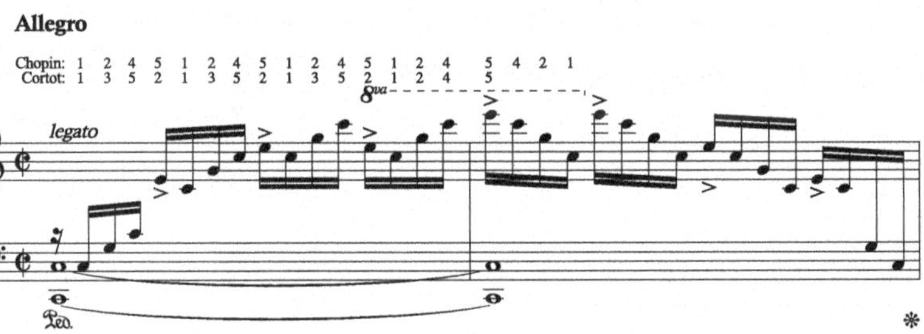

(b)

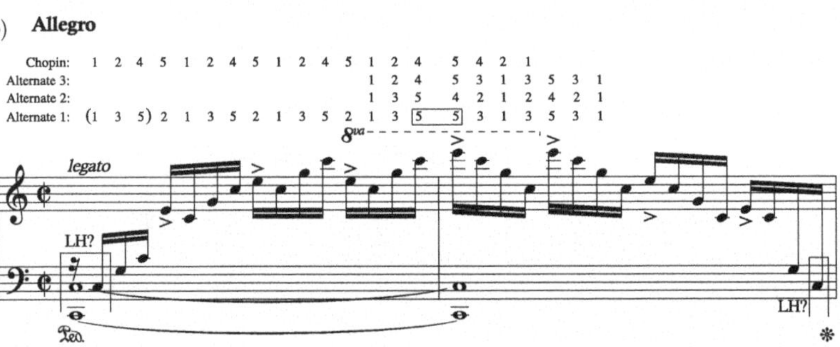

In exposed passages where security and accuracy are of paramount concern, fingerings can be customized to reflect moment-to-moment fluctuations in the comfort level of individual hands. Example 5.9 shows just one possible solution for a problematic passage in Beethoven's op. 110.

EXAMPLE 5.9 *Beethoven, Sonata in A♭ Major, op. 110, i, mm. 14–17*

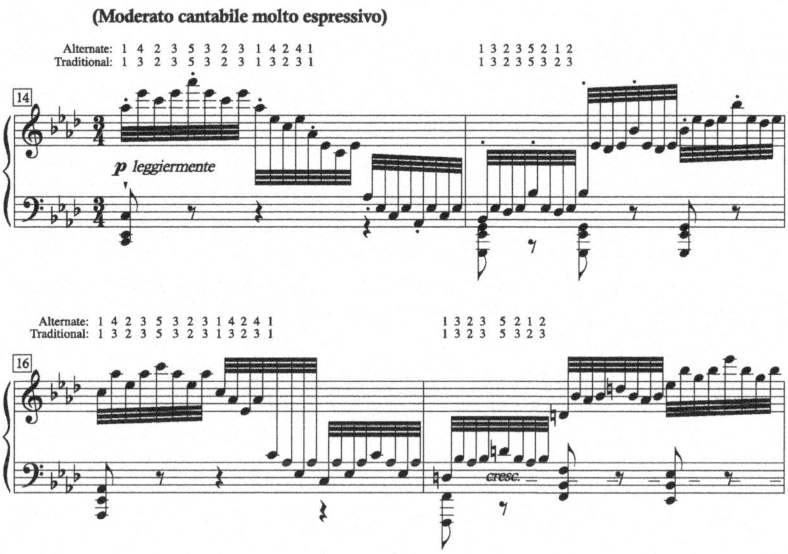

Consecutive Double Notes and Octaves

It is difficult for small-handed players to preserve the alignment of the hand and arm through passages of consecutive double intervals. In lieu of traditional fingerings, children and small-handed adults can experiment with smaller-than-normal hand positions that provide more power and endurance. In Example 5.10 from Bartok's *For Children*, the composer's suggested fingering could be slightly altered as indicated to condense the hand and improve its stability.

EXAMPLE 5.10. *Bartok, "Absent is My Sweetheart," from* For Children, *Sz. 42 no. 36, mm. 1–4*

In Example 5.11, the double fourths of measure 43 and the double thirds of measures 44–46 can be fingered with more compact hand positions than are traditional. Notice that the black note, F♯, is consistently assigned to 2 or 4, since both are longer fingers.

EXAMPLE 5.11. *Mendelssohn,* Rondo Capriccioso, *op. 14, mm. 43–46*

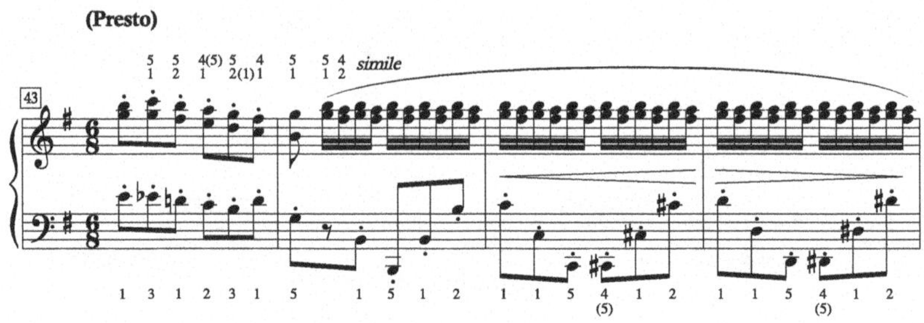

Playing long series of larger double intervals, such as parallel sixths, sevenths, and octaves, can severely test the small-handed pianist. Using fingerings that stretch the hand unnecessarily can result in discomfort or pain and a reduced ability to voice the melody. If the tempo is slow enough, the muscles should release between successive chords as often as possible. In octave passages, the default fingering for most players is almost certainly 1-5. Depending on the context, 1-4 can be considered, but virtually never 1-3.[8] Pianists may even prefer to avoid using 1-3 or 2-5 for parallel sixths, instead opting for 1-4 and 1-5, as shown in Example 5.12.

EXAMPLE 5.12. *Schubert, Waltz in B Minor, op. 18 no. 6, mm. 31–32*

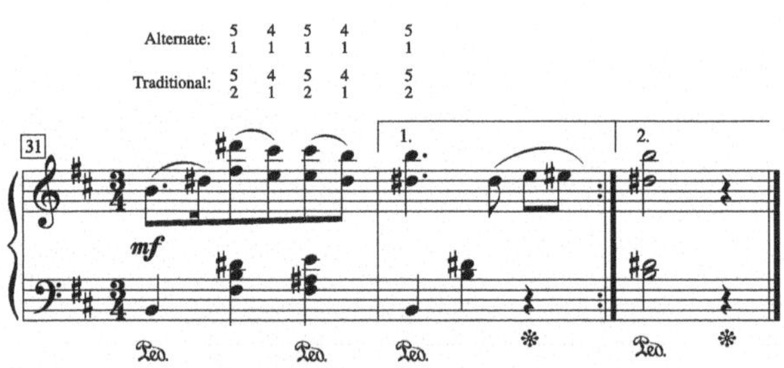

The typical editorial fingering for the double sixths shown in Example 5.13 maintains the spread of an octave between the right-hand fifth finger and the thumb. However,

[8] Octave playing is discussed in more detail in Chapter 6.

many small-handed pianists would prefer smaller shapes. In the alternate fingering, every note of the alto voice, black or white, is played by the thumb to free up space in the hand. These accompanimental notes are released early, and the role of maintaining the legato is relinquished to the pedal. Note that the left-hand waltz bass can also be refingered to keep the hand compact.

EXAMPLE 5.13. *Brahms, Waltz in A♭ Major, op. 39 no. 15, mm. 1–4*

In *forte* or *fortissimo* passages, it is imperative to search for more compact fingering solutions for double notes. In Example 5.14, the traditional fingering would force the left hand to play the harmonic sixths with 2-5 (D–B♭) and 1-2 (B♭–G). The wrist would have to swivel sharply to the right to allow the thumb to reach notes located directly in front of the body. The alternate version allows more room in the left hand for each double-note chord but requires playing the black notes with the thumb on the second chord of each measure; the longer second finger is then extended over the thumb for the third chord of the measure.

EXAMPLE 5.14. *Chopin, Sonata in B♭ Minor, op. 35, i, mm. 137–140*

Readjusting the Hand to a More Compact Shape

Small-handed pianists should take advantage of opportunities to bring the hand back into a more compact shape. In Example 5.15, the hand is reset at every long note or break in the phrasing of the melody, as indicated by the circled notes.

EXAMPLE 5.15. *Chopin, Polonaise in C♯ Minor, op. 26 no. 1, mm. 38–46, right hand*

Playing Seconds with the Thumb Alone

Besides finding ways to refinger chords and arpeggios to favor a compact hand shape, small-handed pianists should also look for seconds that can be played simultaneously with the thumb so that the rest of the hand can open up for wider extensions. In Examples 5.16 and 5.17, playing the circled contiguous black or white notes with the thumb alone minimizes uncomfortable stretches between inner fingers, enabling a small-handed player to move through the passages with speed and accuracy. In Example 5.17, because the first several measures of the piece are prolonged in a single pedal, the chords marked *m.d.* could be played with the left hand and the chords marked *m.g.* could be played with the right hand to avoid having to reach the left hand over the right hand to play an extended chord in front of the torso. If the pianist prefers to perform these measures as written, then the left thumb could extend to play both the F♯ and the G♯.

EXAMPLE 5.16. *Schumann, "Glückes genug," from* Kinderszenen, *mm. 1–6*

EXAMPLE 5.17. *Debussy, "Pagodes," from* Estampes, *mm. 1–3*

Using Non-Consecutive Fingerings

Non-consecutive fingerings condense the hand into strong, ultra-compact shapes by skipping over a finger in an ascending or descending fingering progression. They can be extraordinarily effective in combating stretch and loss of power. Frequently, the fourth finger will be left out, allowing a pianist to avoid weak 3-4 or 4-5 fingering sequences. Another common strategy is to substitute 3-1, an exceptionally strong grouping, for 2-1. Activity 5.6 can be used to acclimate a pianist to these unconventional maneuvers.

ACTIVITY 5.6
NON-CONSECUTIVE FINGERINGS

Practice these fingerings one hand at a time. Begin slowly and loudly, gradually working up to an extremely fast tempo. Transpose each activity to several different keys.

Practice the next activity in the same manner, hands separately and then hands together. In the second half, the pattern is inverted and the fingerings of the right hand and left hand are exchanged accordingly.

Non-consecutive fingerings that skip the fourth finger are useful for executing loud scalar passages. In measures 20 and 24 of Example 5.18, playing the descending scales with a compact hand position and a wrist position that is slightly higher than normal allows the player to maximize power and minimize backward and forward motions.

EXAMPLE 5.18. *Beethoven, Sonatina in F Major, Anh. 5, i, mm. 19–26*

Example 5.19 features unison five-finger staccato scales that would traditionally be played by the fingers in consecutive order. The alternate fingering exploits the inherent strength of the thumb, third, and fifth fingers, is simple to coordinate between the hands, and yields crisp articulations.

EXAMPLE 5.19. *Tchaikovsky, "February," from* The Seasons, *mm. 43–46*

In the staccato passages beginning in measure 23 of Example 5.20, traditional 1-2-3-4-5 fingerings yield awkwardly large intervals between inner fingers. Alternate 1 adds an extra shift at a strategic spot to eliminate large stretches, and Alternate 2 goes one step further, using a compact non-consecutive fingering.

EXAMPLE 5.20. *Tchaikovsky, "February," from* The Seasons, *mm. 23 –25*

Playing Trills

Pianists with smaller hands may experience more discomfort than larger-handed pianists when playing trills with adjacent fingers, especially if the trills involve both black and white keys or are a whole step instead of a half step apart. Substituting trill fingerings that use non-adjacent fingers—1-3, 2-4, and 3-5—are often the solution. Example 5.21 shows a typical cadential trill.

EXAMPLE 5.21. *Haydn, Sonata in D Major, Hob. XVII: D1, iii, mm. 17–18*

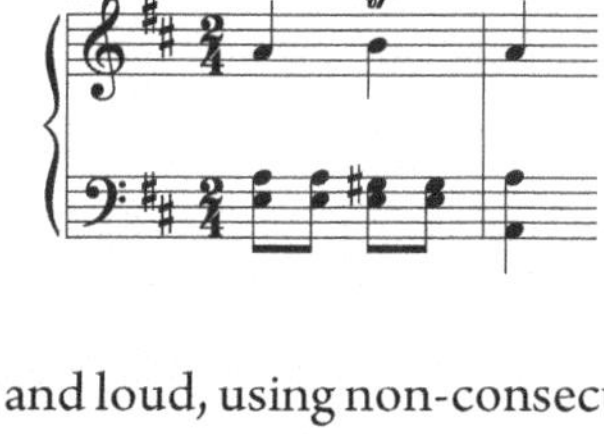

When the trill is very long and loud, using non-consecutive fingers can delay the onset of fatigue. In Example 5.22, the right-hand thumb can deliver a sharp *ff* accent on the first note, and the trill can continue at a softer dynamic with a 4-2 fingering.

EXAMPLE 5.22. *Debussy,* L'isle joyeuse, *mm. 244–245*

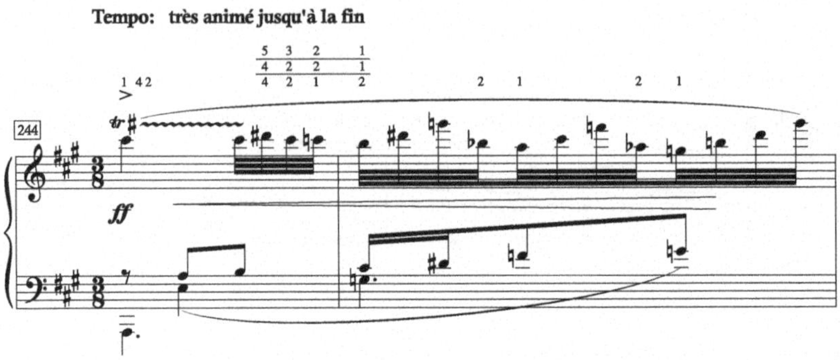

ACTIVITY 5.7
FINGER REPETITIONS AND SLIDES

Play the activity loudly and slowly and gradually increase the tempo. Transpose to many different keys. Use the held note as a pivot point for quick rotations to enact the note repetitions. In the chromatic part of the activity, minimize backward and forward motion when moving between black and white keys with the thumb or fifth finger.

Use repetitions of inner fingers to play the black- and white-note combinations shown below. Small releases between notes may be needed to get them to sound.

REPEATING THE SAME FINGER TO RECOVER SPACE IN THE HAND

Another way to maximize reach is to repeat the same finger many times or to slide a finger from a black note to an adjacent white note. Activity 5.7 can acclimate the pianist to finger repetitions and slides.

Example 5.23 shows how repetitions of the thumb in the accompaniment part and repetitions of the fifth finger in the melody line allow the hand to release uncomfortable stretches more quickly.

EXAMPLE 5.23. *Schumann, "Mignon," from* Album for the Young, *mm. 19–22*

Major ninths that span between white and black notes can be as taxing as tenths. One strategy for playing the first three notes of the right hand in Example 5.24 is to take the F♯ with a repetition of the fifth finger (Alternative 1). Otherwise, a longer finger such as the third finger can be passed over the fifth finger (Alternative 2).

EXAMPLE 5.24. *Grieg,* Norwegian Dance, *op. 35 No 2 (solo piano), mm. 5–6*

The same finger can be used to slide from a black to an adjacent white note to alleviate problems of reach. In Example 5.25, the fingering given in the score results in a wide stretch between 2 and 5 on the F♮–D♯ chord in measure 21 of the left hand. When the limit of the hand has been reached, the thumb can be repeated to play chromatically adjacent notes.

EXAMPLE 5.25. *Tcherepnin, Bagatelle, op. 5 no. 9, mm. 20–23*

Similarly, the alternate fingering in Example 5.26 uses the fifth finger of the right hand to slide from a black note to an adjacent white note on the third and fourth eighth notes of every measure. In the previously cited Example 5.20 from the same piece, the second finger of the right hand could be used to slide downward from the B♭ to the A, resulting in a fingering of 5-4-2-2-1.[9]

EXAMPLE 5.26. *Tchaikovsky, "February," from* The Seasons, *mm. 93–96*

Chopin was an early advocate of using finger repetition in chromatic sequences, and his music provides many opportunities to do so. Example 5.27 shows two typical black-to-white-note slides. In Example 5.28, Chopin recommended repeating the fifth finger to impart a speech-like articulation.

EXAMPLE 5.27. *Chopin, Nocturne in B Major, op. 9 no. 3, mm. 107–108, left hand*

EXAMPLE 5.28. *Chopin's fingering, Nocturne in E♭ Major, op. 9 no. 2, m. 16*

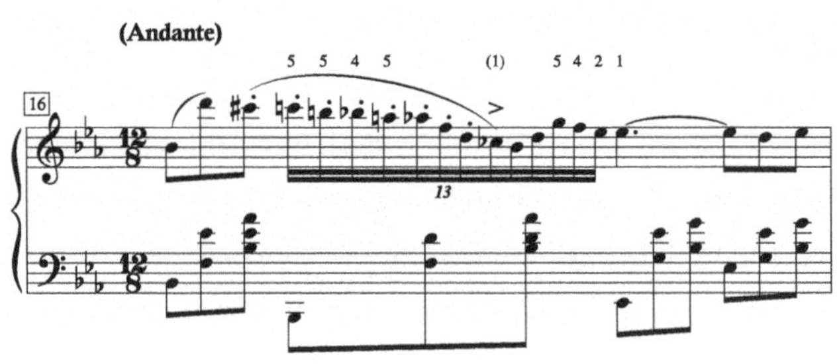

Chopin also used finger repetition to solve problems of reach. The left-hand parts in Chopin's nocturnes provide numerous occasions where this fingering strategy could be applied. In Example 5.29, he specifically recommended repeating the thumb so that the octave distance between the last two A♭s would not have to be spanned between the

[9] This fingering was suggested by Catherine Kautsky in a personal communication with the authors, April 2016.

second and fifth fingers. Example 5.30 shows three successive repetitions of the thumb, and Example 5.31 proposes using the fifth finger up to four times in a row to jump between widely separated notes.

EXAMPLE 5.29. *Chopin's fingering, Nocturne in B♭ Minor, op. 9 no. 1, m. 26, left hand*

EXAMPLE 5.30. *Chopin, Nocturne in G Major, op. 37 no. 2, mm. 1–2, left hand*

EXAMPLE 5.31. *Chopin's fingering, Nocturne in C♯ Minor, op. 27 no. 1, mm. 13–14, left hand*

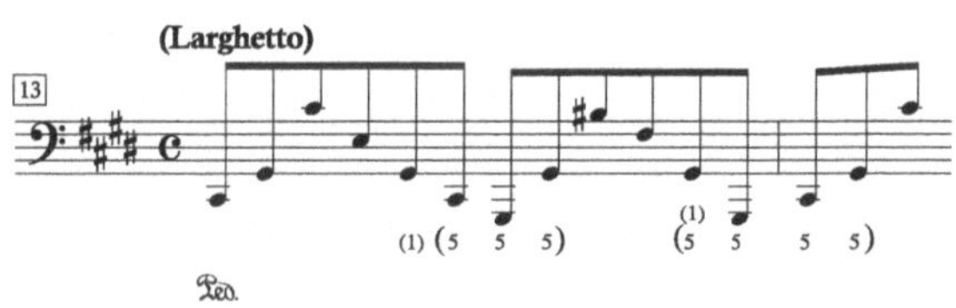

Widespread left-hand accompaniments found in music of other composers can be treated in the same fashion. Example 5.32 shows an alternate fingering that opens with a repetition of the fifth finger, causing it to leap through the interval of an octave. The chords on the downbeats act as springboards, propelling the arm through the distance. The pedal covers any gaps created by the alternate fingering.

EXAMPLE 5.32. *Grieg, "Wedding Day at Troldhaugen," from* Lyric Pieces, *op. 65 no. 6, m. 25, left hand*

In the same piece at measure 3, use the right-hand fifth finger to skip between the high F♯ and the E to avoid having to play the chord on the third beat with 1 and 4 (Example 5.33).

EXAMPLE 5.33. *Grieg, "Wedding Day at Troldhaugen," from* Lyric Pieces, *op. 65 no. 6, m. 3*

Even at high tempos, repetitions of the fifth finger may be worth considering because they eliminate the need for lateral wrist deviations (Example 5.34).

EXAMPLE 5.34. *Chopin, Etude in A♭ Major, op. 25 no. 1 ("Aeolian Harp"), m. 27, right hand*

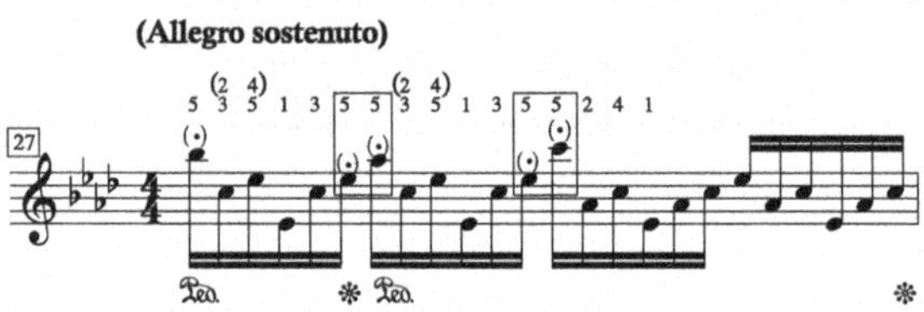

OPTIMIZING SOUND PRODUCTION AND VOICING

When the fingers are fully spread apart, it is difficult for them to produce strokes of different intensities. Thus, small-handed pianists may struggle when they must play a louder melody and softer accompaniment in the same hand or when they must voice one note in a chord.

Repeated Use of the Fifth Finger and Thumb to Project Melodies

Pianists often rely on the outside digits—the thumb and fifth finger—for melodic projection. Small degrees of rotation can be used to increase key speed and thus produce a louder tone. In Example 5.35, where the right hand must project the melody over repeated notes and chords, pianists with small hands can finger as many notes with the fifth finger as necessary to voice the melody with strength and clarity.

EXAMPLE 5.35. *Mendelssohn,* Rondo Capriccioso, *op. 14, mm. 67–74, right hand*

When the melody moves to the left hand in measure 82, players may rely on repetitions of the thumb to project the melody (Example 5.36).

EXAMPLE 5.36. *Mendelssohn,* Rondo Capriccioso, *op. 14, mm. 83–86, left hand*

The melody notes of Example 5.37 are reachable with the third, fourth, and fifth fingers of the right hand. A pianist with smaller hands might opt to take more of them with the fifth finger, however, so the hand can contract into its strongest and most compact shape.

EXAMPLE 5.37. *Beethoven, Sonata in C Minor, op. 13 ("Pathetique"), ii, mm. 1–8*

Similarly, in Example 5.38, an ultra-compact fingering allows the fifth finger to project the melody over an ostinato located in a muddy register of the piano. The hand is balanced such that power can be directed straight through the bones of the fifth finger.

EXAMPLE 5.38. *Brahms, Rhapsody in G Minor, op. 79 no. 2, m. 21, right hand*

In Example 5.39, the separation between the melody and the upper pedal point means that the task of joining the notes is largely given over to the pedal. The traditional fingering attempts to preserve as many legato connections in the melodic voice as possible. However, at certain points, small-handed players may prefer to forgo these connections in favor of repetitions of the thumb or fifth finger to ease the discomfort of the largest stretches.

EXAMPLE 5.39. *Chopin, Scherzo no. 1 in B Minor, op. 20, Trio, mm. 305–311*

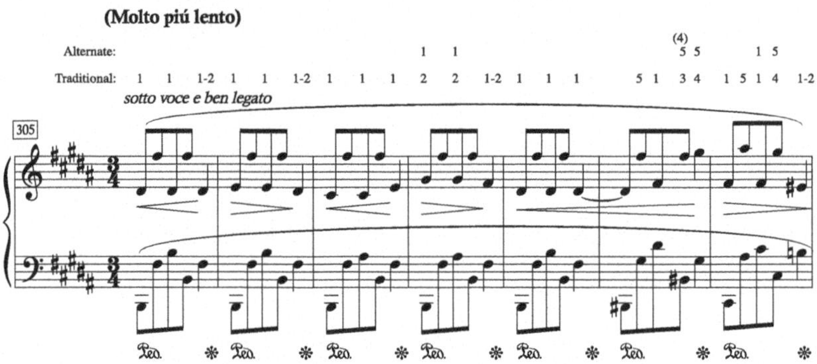

Aligning the Hand and Forearm in a Skeletal Position

Skeletal positionings that line up the hand, forearm, and upper arm in various combinations behind the playing finger can help small-handed pianists overcome inherent finger inequalities. For a note of high intensity, a finger or fingers can be extended straight downward at the knuckles (MCP joints) and dropped to the key surface by the forearm. The elbow acts as a fulcrum for this lever, and the wrist is stabilized so the force is transmitted directly through the bones with a minimum level of muscular involvement. For notes of even higher intensity, the whole arm can be dropped to the key surface from the shoulder joint with the elbow and wrist stabilized so that the force of the key strike is channeled through the bones. Fingers can be braced against each other or clustered together for greater stability. Even the weakness of the fourth finger can be overcome in this fashion.

Skeletal positions are especially useful for projecting single notes at high dynamic levels. For instance, small-handed pianists should accustom themselves to using several fingers clustered together to play single notes. Fingers 4 and 5 (or 3, 4, and 5) may be used together as a solid unit to bring out the bass notes of a waltz or stride bass, to emphasize the bass notes of important harmonies, or to play the top notes of loud triads. Specific examples can be found in Chapter 6 in the section on skeletal positioning.

MOVING LONGER FINGERS OVER SHORTER FINGERS

The Baroque practice of moving longer fingers over shorter ones, in conjunction with a quick lateral shift of the arm, can expedite the playing of large rolled chords. In Examples 5.40 and 5.41, a longer finger can move over the thumb to take the final note in a roll.

EXAMPLE 5.40. *Chopin, Etude in G♭ Major, op. 10 no. 5 ("Black Key"), m. 65, left hand*

EXAMPLE 5.41. *Chopin, Etude in F Major, op. 10 no. 8, m. 95*

The alternate fingering shown in Example 5.42 allows the left hand to move smoothly through the sextuplets by doing away with uncomfortably large stretches between inner fingers. The second finger moves over the thumb to play the top note of the arpeggio.

EXAMPLE 5.42. *Chopin, Etude in A♭ Major, op. 25 no. 1 ("Aeolian Harp"), m. 24*

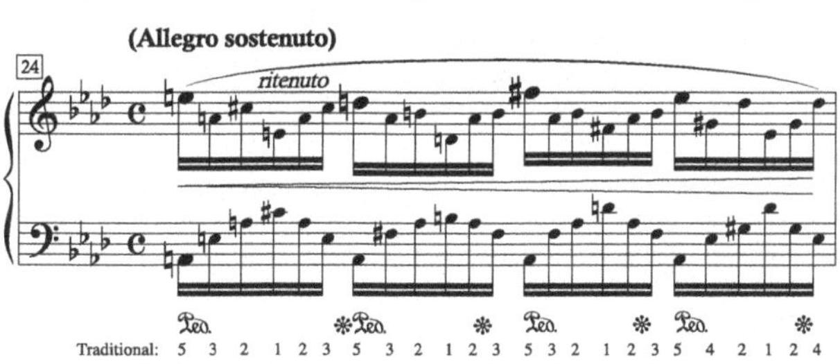

In Example 5.43, replacing the traditional 4-1-2-5 fingering with 5-2-1-5 permits the hand to move through the largest intervallic gaps with 1-5 instead of 2-5.

EXAMPLE 5.43. *Schubert, Impromptu in G♭ Major, op. 90 no. 3, mm. 82–84*

Instead of using the fingers consecutively to play the *forte* descending and ascending arpeggios in Example 5.44, small-handed pianists can add an internal shift, passing the second or third finger over the right- or left-hand thumb. The fingers can remain upright, making it easier to deliver sufficient power to the keys. The pedal sustains notes that are released early.

EXAMPLE 5.44. *Brahms, Capriccio in G Minor, op. 116 no. 3, m. 9*

The advantages of the alternate fingering for the left hand shown in Example 5.45 will be obvious to any small-handed player. Again, the second finger passes over the thumb to take the top note of the extended arpeggio.

EXAMPLE 5.45. *Liszt, "Harmonies du soir," from the* Transcendental Etudes, *mm. 38–41*

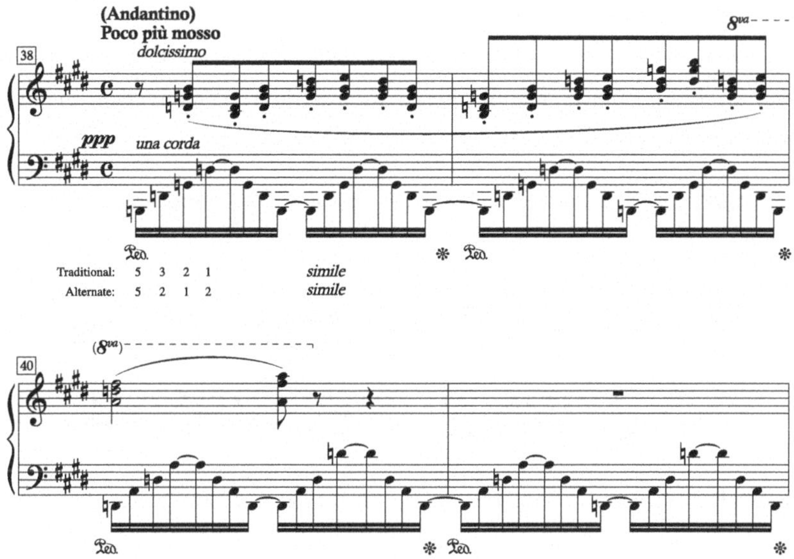

SUBSTITUTING THE THUMB FOR INNER FINGERS

Someone with small hands might occasionally find it useful to approach a passage from the viewpoint of someone with a larger reach. For instance, in Example 5.46, a pianist could start out by playing the melody using the fingers in consecutive order, but then substitute the thumb for an inner finger to artificially extend the span of the hand.[10] Players should avoid twisting the forearm, swiveling the wrist from side to side, or reaching for notes with the fingers. The fingering starts from a simple concept—1-2-3-4-5 in the right hand and 5-4-3-2-1 in the left hand—and modifies it in a way that closely replicates for the small-handed pianist

EXAMPLE 5.46. *Schumann, "Hunting Song," from the* Album for the Young, *op. 68 no. 7, mm. 3–4*

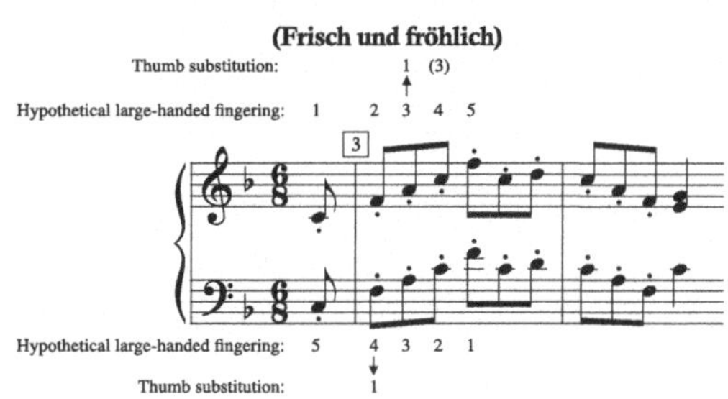

[10] Alternate solutions are presented in Examples 4.36 and 6.21.

the experience of playing with a larger hand.[11] Thinking about spreading the hand over the five notes at once places the insertion of the thumb a note earlier than where it would likely fall if a typical hand shift were employed and also encourages maximum motion economy.

The thumb could replace an inner finger to ameliorate the effect of uncomfortable stretches between fingers, as shown in Examples 5.47 and 5.48.

EXAMPLE 5.47. *J.S. Bach, Prelude in C Minor, from* Das Wohltemperierte Klavier, *Book I, BWV 847, mm. 1–2*

EXAMPLE 5.48. *C.P.E. Bach, Sonata in C Minor, Wq. 48 no. 4, iii, m. 29–30*

The alternate fingering shown in Example 5.49 encourages the hand to contract to an anatomically neutral position instead of remaining outstretched.

EXAMPLE 5.49. *Beethoven, Rondo in C Major, op. 51 no. 1, mm. 17–20*

USING POSITIONAL FINGERINGS TO ACHIEVE MAXIMUM SPEED

Small-handed pianists should acquire fluency in different fingering systems if they wish to see beyond traditional formulas that are kinesthetically entrenched through years

[11] Yet another option—using 1-2-3-5-5 in the right hand and 5-3-2-1-1 in the left—has been suggested by several pianists.

ACTIVITY 5.8
PRACTICING POSITIONAL FINGERINGS

Play these exercises from Busoni's *Klavierübung* in different keys using the same fingering.

of practice. The innovations of Liszt and his student Ferrucio Busoni may be of special interest in this regard (see Activity 5.8[12]).[13] Here, the old rule prohibiting the use of the thumb on the black keys is discarded, and fingering is instead viewed as a succession of positions. The pianist uses repeating finger combinations (for example, 1-2-3-4-5, 1-2-3-4-5, or 1-2-3, 1-2-3) to play scales or arpeggios regardless of the arrangement of the black and white notes. In editions of his music, Liszt proposed a variety of innovative positional or repetitive fingerings as solutions for many pianistic challenges:[14]

- Ascending G♯-major scales played in repeating groups of five notes in the *Rhapsodie espagnole*
- Descending chromatic scales played in four-note groups exchanged between the hands in the second *Paganini Study*

---

[12] Ferrucio Busoni, *Klavierübung in zehn Büchern*, vol. 1 (Leipzig: Breitkopf und Härtel, 1925), 1 and 27.

[13] For additional information, see Jun Young Kwak, "Awakening the Kinesthetic Senses in Piano Performance: New Approaches to Fingering and Movement Based upon Ferrucio Busoni's *Die Klavier Übung*," (DMA Paper, Texas Tech University, 2013).

[14] For illustrations of these examples, see Alan Walker, "Liszt and the Piano," Grove Music Online, s.v. "Liszt, Franz," http://www.oxfordmusiconline.com.libe2.lib.ttu.edu/subscriber/article/grove/music/48265pg9#S48265.9 (accessed May 16, 2014); Alan Walker, *Franz Liszt: The Virtuoso Years 1811–1847*, Vol. 1 (New York: Alfred Knopf, 1983), 297–301; and Kwak, "Awakening the Senses," 28–47.

- Three-finger ascending scales in the *Rigoletto Paraphrase*
- Three-finger gapped ascending scales in the *Rigoletto Paraphrase*
- Interlocked single notes exchanged between the hands in *La Campanella*
- Interlocked double notes exchanged between the hands in the *Hungarian Rhapsody,* no. 1
- Interlocked octaves in the *Paganini Etude,* no. 2
- Long series of parallel thirds played by repeating 2-4 in the *Paganini Etude,* no. 6 and the *Hungarian Rhapsody,* no. 1

On occasion, the famous pianist Alfred Cortot also advocated fingering in positions or "handfuls." For example, he recommended playing an ascending right-hand chromatic scale in Mozart's Rondo in D Major, K. 485, as a repeating five-finger scale and a descending left-hand chromatic scale in Chopin's Etude, op. 25 no. 11, as a four-finger scale.[15] A key point in favor of positional fingerings is that they often minimize the degree of backward and forward motion. This may entail playing the fingers closer to the fallboard and with a higher wrist position than customary. To determine the optimal placement of the hand and fingers, it is useful to block them out in chords.

An interesting application of this principle involves the difficult unison E-major scale from the end of Chopin's Scherzo no. 4, which traverses the entire breadth of the piano keyboard from E1 to E7.[16] The traditional fingering for this scale can be mapped as shown in Table 5.1, with vertical lines showing the finger groupings in each hand. As can be seen, the shifts in the middle of the scale do not coincide—the right hand shifts at the A but the left hand does not shift until the C♯.

Some pianists may prefer the logic of the alternate fingering noted in Table 5.2. The same fingers in both hands are grouped together—the thumbs, 2-3 with 3-2, and 2-3-4 with 4-3-2.

TABLE 5.1 Traditional E-major scale fingering

	E	F#	G#	A	B	C#	D#	E
RH	1	2	3	1	2	3	4	1
LH	5	4	3	2	1	3	2	1

TABLE 5.2 Alternate E-major scale fingering

	E	F#	G#	A	B	C#	D#	E
RH	1	2	3	1	2	3	4	1
LH	1(4)	3	2	1	4	3	2	1

[15] Manshardt, *Aspects of Cortot,* 91–92.

[16] Kwak, "Awakening the Kinesthetic Senses," 42–43.

However, Jun Young Kwak has proposed a third solution that radically tests conventional notions of scale fingering. The scale is divided into two groups, F♯–G♯ and A–B–C♯–D♯–E, played such that the hands shift at the same time (Table 5.3 and Example 5.50).

TABLE 5.3 E-major scale fingering with simultaneous hand shifts

	E	F♯	G♯	A	B	C♯	D♯	E
RH	1	2	3	1	2	3	4	5
LH	1(4)	3	2	5	4	3	2	1

EXAMPLE 5.50. *Positional scale fingering for the E-major scale in Chopin, Scherzo no. 4 in E major, op. 54, mm. 960–965*

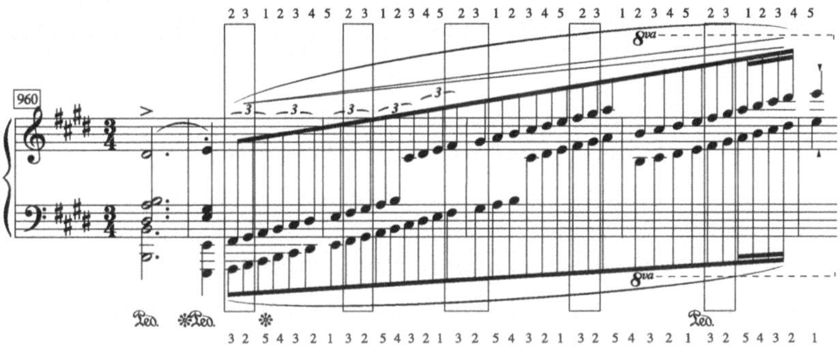

For small-handed pianists, positional fingerings have a special relevance to arpeggios. In Example 5.51, the alternate fingering allows the hands to shift simultaneously from one blocked shape to another, whereas in the traditional fingering, the shifts are staggered.

EXAMPLE 5.51. *Chopin, Etude in F Major, op. 10 no. 8, mm. 93–94*

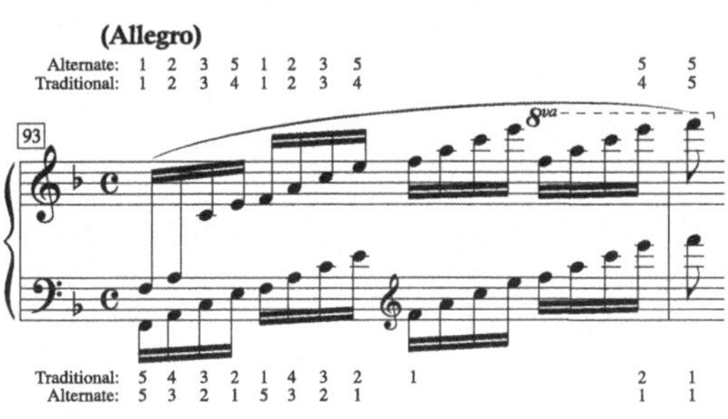

Similarly, sections of Debussy's *Reflets dans l'eau* could be subdivided into chordal groups and fingered positionally. One possibility of many is shown in Example 5.52.

EXAMPLE 5.52. *Debussy, "Reflets dans l'eau," from* Images, *Book I, mm. 56–58*

SUBSTITUTING ONE FINGER FOR ANOTHER ON THE SAME KEY

It would be remiss not to mention a common maneuver from organ playing, finger substitution. Finger substitution can be used to regain space in the hand if there is ample time. This is a necessary strategy in contrapuntal works where the pedal cannot be used, but it can also be useful in lyrical passages, where repeating the same finger might not produce a musically satisfying option (Example 5.53).[17]

EXAMPLE 5.53. *Martinu, "Columbine Remembers," from* Puppets, *Book II no. 3, H. 116, mm. 1–8*

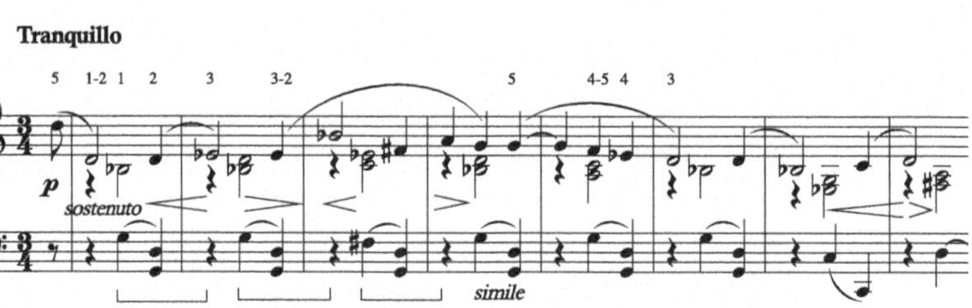

RETAKING RELEASED NOTES SILENTLY

Finally, it may occasionally be possible to silently re-depress notes that have been released. Even pianists with large hands occasionally prefer this option to half-pedaling

[17] An easier alternate solution would be to play just the notes of the melody with the right hand, redistributing the other notes into the left.

or flutter-pedaling when having to hold out a long chord or pedal point against a changing series of harmonies. In Example 5.54, the octave G♯s in the left hand in measure 72 must be released to play the upper chords, but the G♯s can be retaken silently at the beginning of measure 73 so that the pedal can be changed for the right-hand chords in the same measure. Similarly, if pianists' hands are too small to reach a tenth but large enough to reach a ninth, they may play the left-hand roll in measure 171 of Example 5.55 and then silently re-depress the F with the fifth finger before playing the chromatic quarter notes in measure 172 with the thumb. This maneuver makes it possible to change the pedal to give greater clarity to the melodious inner voice.

EXAMPLE 5.54. *Debussy, "Hommage à Rameau," from* Images, *Book I, mm. 72–74*

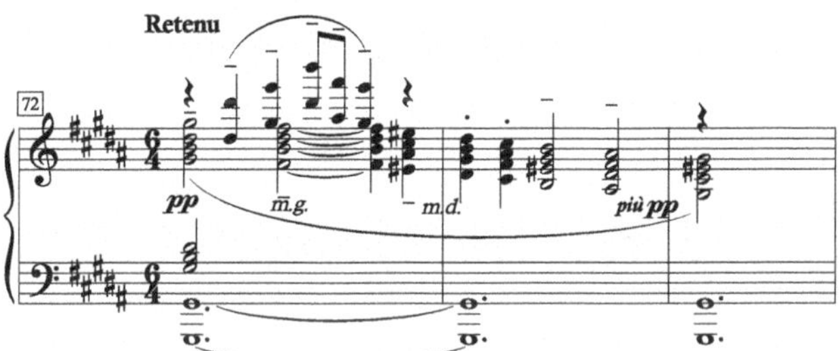

EXAMPLE 5.55. *Chopin, Sonata in B♭ Minor, op. 35, i, mm. 169–173*

CONCLUSION

In this chapter, many strategies for using refingering to address challenges posed by small-handedness have been discussed. Pianists will want to weigh the effectiveness of these alternatives against their habitual fingering preferences. The following questions may provide the basis for a more confident choice:

- Which fingering allows the hand to return to a compact relaxed state more quickly and more frequently?
- Which fingering does the most to reduce or eliminate uncomfortable stretches, including those between inner fingers?
- Which fingering optimally aligns the playing fingers behind the forearm more of the time?

- Which fingering is the easiest to execute at the required tempo?
- Which fingering is easier to learn and memorize?
- Which fingering yields accurate results more frequently?
- Most important, which fingering produces the best musical result?

Fingering is one of the primary ways by which pianists kinesthetically connect with the keyboard. While a fingering has no inherent meaning in and of itself, it acquires meaning in a musical context. The tandem goals of fingering should therefore be to facilitate the expression of the score while simultaneously maximizing physical comfort. However, fingering is highly individual, and many editorial or traditional patterns of fingering may be inappropriate or problematic for small-handed pianists. Therefore, it is important to be able to devise alternate fingerings that eliminate or minimize stretch and that maximize power.

APPLICATION PROBLEMS

The following excerpts contain problems that might be addressed through refingering. In devising solutions, follow the basic principles of effective technique presented in Chapters 2 and 3 to avoid maladaptive tendencies.

1. Beethoven, Sonata in C Minor, op. 13 ("Pathetique"), i, mm. 31–37: What fingering will enable you to play this passage with maximum accuracy and power at the required tempo? Experiment with positional fingerings for the arpeggios in mm. 33–34. Eliminate all large stretches between 1 and 2 in the right hand.
2. Schumann, "Kleine Romanze," from *Album for the Young*, op. 68, mm. 1–8: What fingering will help you project the melody clearly over the accompanimental chords?
3. Chopin, Nocturne in C♯ Minor, op. 27 no. 1, mm. 1–8: Where can repetitions of the same finger reduce the size of the stretches in the left-hand part?
4. Chopin, Waltz in A♭ Major, op. 34 no. 1, mm. 17–40: What fingering will allow you to confidently project the bass notes? Find compact fingerings that can comfortably accommodate the double sixths in the right hand, the offbeat chords in the left hand, and the trills.
5. Chopin, Polonaise in G Minor, B. 1, op. posth., mm. 23–26: What fingering eases some of the stretches between inner fingers and avoids rapid backward and forward motions caused by using the thumb on black keys?

6. Beethoven, Sonata in E Major, op. 14 no. 1, i, mm. 5–7: What fingering eliminates the need for the left hand to play black keys with the thumb when these notes are located directly in front of the body?
7. Daquin, *Le Coucou*, mm. 1–12: Will using non-consecutive fingers assist you in executing the trills? Find a fingering that relieves any large stretches between inner fingers, especially 2 and 5.

SIX

Maximizing Reach and Power

THE MOST INTRACTABLE PROBLEM FOR SMALL-HANDED PIANISTS IS DEALING with large shapes that cannot be redistributed or refingered. Otto Ortmann observed that when a player stretches the hand, "[the] freedom of playing individual keys . . . is restricted . . . [which] will explain a large number of technical difficulties that are often wrongly attributed to defects of coordination or studentship."[1] Not only are problems of reach pervasive, but they also present the potential for harm. As discussed in Chapter 1, many studies have reported a correlation between injury and playing with the hand at full extension.

Despite the clear need for facilitative strategies, pianists often shy away from them for fear of transgressing established performance traditions. For instance, concerning Chopin's Etude, op. 10 no. 1, Charles Rosen stated that "[the] opening . . . lies well for the hand, but when the tension rises, we have a passage (in measures 27–33) so difficult for pianists with only moderate sized hands that many change the fingering from the 124 5124 prescribed by the composer to the considerably simpler 125 2125—unfortunately, one can always hear the change in the phrasing of the arpeggio that this entails."[2] In another example, Artur Schnabel asserted that the trills in the last movements of Beethoven's sonatas, opp. 53, 109, and 111 should be played simultaneously with and

[1] Otto Ortmann, *The Physiological Mechanics of Piano Technique* (New York: E. P. Dutton, 1929, reprint 1962), 312–313.

[2] Charles Rosen, *The Romantic Generation* (Cambridge: Harvard University Press, 1995), 363–364. By contrast, the famous Chopin interpreter, Alfred Cortot, felt that more compact fingerings should be applied throughout this etude because they produced a superior musical result. See Chapter 5 for a fuller discussion.

not interrupted by the melody note, something that might be impossible for a small hand to maintain.[3] However, statements of opinion, even when they come from musicians of great stature, should not be deterrents to all further inquiry. The central thesis of this book is that the specific physical and musical needs of the small-handed can be addressed in a musically sensitive and appropriate way.

This chapter discusses strategies that maximize the reach and power of the hand and gives concrete examples of how these approaches work in the context of a range of piano repertoire, from standard teaching pieces to advanced concert literature. It offers numerous practical approaches for playing large chords, broken chords, double-note passages, arpeggios, and other extended shapes. Building on the discussion of movement and technique in Chapters 2 and 3, it provides a basis for the reader's continued investigation.

The chapter is organized into the following subsections, with strategies addressing problems of reach listed first, followed by strategies for maximizing power and limiting fatigue:

- Covering large distances
- Facilitating large reaches
- Dividing or rolling chords
- Releasing notes
- Using rotation for power
- Taking advantage of skeletal support
- Playing staccato chords and octaves
- Grouping notes into larger gestures
- Omitting and revising notes

One caveat should be noted: pieces filled with chords that are barely within or completely outside the reach of the hand are perhaps best avoided. The point of diminishing returns is reached when the pianist must sustain massive and unrelenting levels of sound for too long, especially at fast tempos. Small-handed performers must approach these decisions with a calm mindfulness, taking honest stock of what their hands can and cannot do at the piano.

COVERING LARGE DISTANCES

Chapter 3 presented a discussion of how different levers cooperate to cover horizontal distances at the keyboard in the most efficient manner. Two basic movements

[3] Artur Schnabel, ed., *32 Sonatas for the Pianoforte by Ludwig van Beethoven*, Vol. 2 (New York: Simon and Schuster, 1935), footnotes on pp. 494, 790, and 860.

were examined in detail: arcs and ovals (where the hand, wrist, forearm, elbow, and/or upper arm move in multiple planes around the keyboard simultaneously) and fan-shaped movements (where the hand and forearm work as a single lever and the elbow remains at roughly the same point in space). When confronted with a large leap, pianists must weigh several factors. If the tempo is leisurely and the player's goal is to project a smooth line, then arcs or ovals are usually more appropriate. However, faster tempos and changes of direction favor the use of fan shapes. Pianists may use a simple diagnostic process:

- Identify the outer notes of the leap.
- For each fingering:
  - Play the leap beginning with an underarc.
  - Play the leap beginning with an overarc.
  - Play the leap using a fan motion.
  - Try each solution in the context of the passage and decide which one is best.

For the examples in this section, the authors acknowledge the inherent difficulty of representing three-dimensional movements unfolding through time in a two-dimensional notational format: the plane of view, the left-to-right progression of the score, and the pitch layout of the keyboard are frequently in conflict. For instance, whereas a descending scale passage starts on the right side of the keyboard and progresses to the left, its notation on the score begins on the left and moves to the right. Given the variety of markings used in the examples in this chapter, the reader may wish to refer to the Key to Abbreviations and Graphic Symbols located at the front of the book.

Applications of Either Oval or Fan-Shaped Gestures

In pieces of slow to moderate tempo, oval or fan-shaped motions work equally well. Examples 6.1a, 6.1b, and 6.1c present three contrasting approaches to a passage taken from the pedagogical repertory, showing how interconnected ovals or fan-shaped movements can replace separate vertical arm drops in the left-hand parts. In the case of ovals, it often does not matter whether a pianist starts with an overarc (Example 6.1a) or an underarc (Example 6.1b). However, pianists are cautioned not to exaggerate the flexion and extension of the wrist, but to stay well within its mid-range of motion. Fan shapes are just as effective (Example 6.1c). Note that the shapes often override bar lines and phrase divisions or may even overlap (mm. 7-8 of Example 6.1c).

EXAMPLE 6.1. *Kabalevsky,* A Slow Waltz, *op. 39 no. 23, mm. 1–8*

a. Ovals in the left hand, beginning with an overarc

b. Ovals in the left hand, beginning with an underarc

c. Fan shapes in the left hand

(a)

(b)

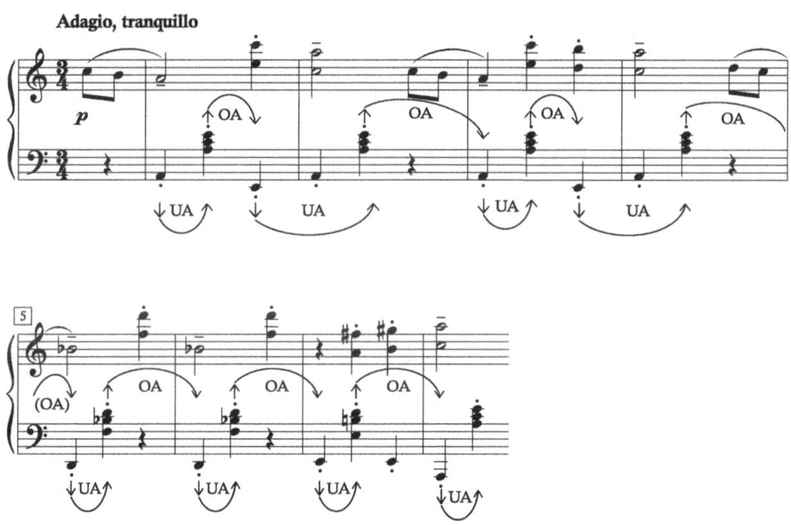

(c)

A second illustration, Example 6.2, shows contrasting approaches to performing the right-hand part of a pedagogical work by Kirnberger.

EXAMPLE 6.2. *J. Philipp Kirnberger,* Les Carillons, *mm. 13–16*
a. Ovals in the right hand, beginning with an overarc
b. Fan shapes in the right hand

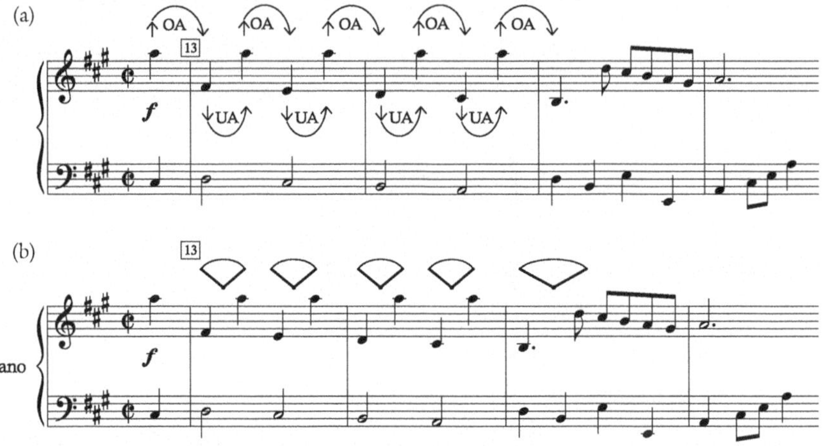

Applications of Oval Gestures

Ovals are more suitable than fans for projecting the widespread legato accompaniment parts or lyrical melodies shown in Examples 6.3 and 6.4. Notes can be connected with the pedal, not the fingers.

EXAMPLE 6.3. *Chopin, Nocturne in C♯ Minor, op. 27 no. 1, m. 1, left hand*

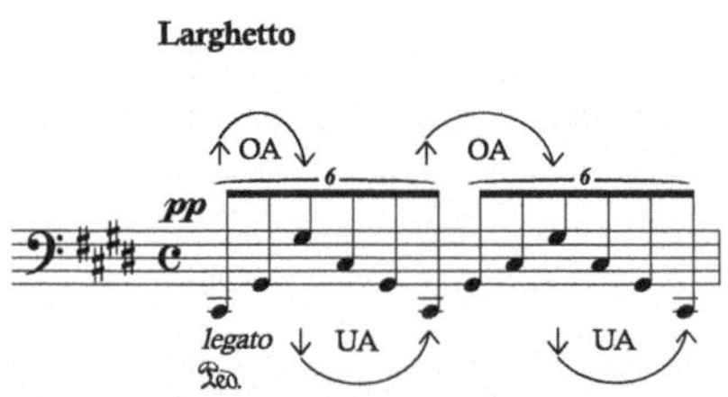

EXAMPLE 6.4. *Chopin, Nocturne in D♭ Major, op. 27 no. 2, mm. 2–6, right hand*

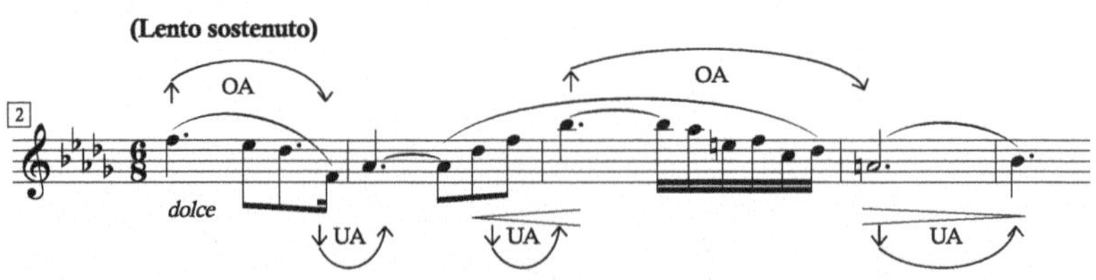

Oval movements encourage the fingers to sequentially release hold of the notes, thereby allowing the hand to close into a more compact shape. In Example 6.5, shaping the eighth-note triplets with oval motions prompts the right-hand thumb to release so that hand can periodically contract and relax.

EXAMPLE 6.5. *Beethoven, Sonata in C♯ Minor, op. 27 no. 2 ("Moonlight"), i, mm. 6–9*

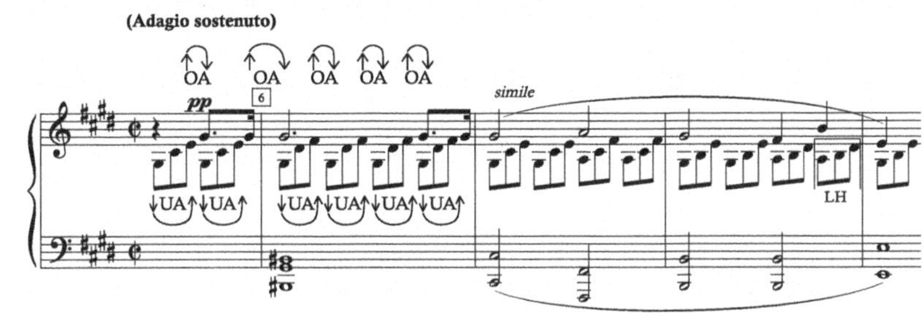

Oval motions are also more appropriate when a tapered effect is desired, such as in sequences of rolled chords. These gestures can impart a grand expression to the passage shown in Example 6.6 and mask the disruptive effect of extra shifts created by alternative fingerings for the left hand. To execute each rolled chord, underarcs can be used in both hands; subsequent overarcs will take the hands back to the starting positions for the next chord.

EXAMPLE 6.6. *Rebikov, "Une histoire triste qui finit bien," from* Tableaux pour enfants, *op. 37, mm. 21–24*

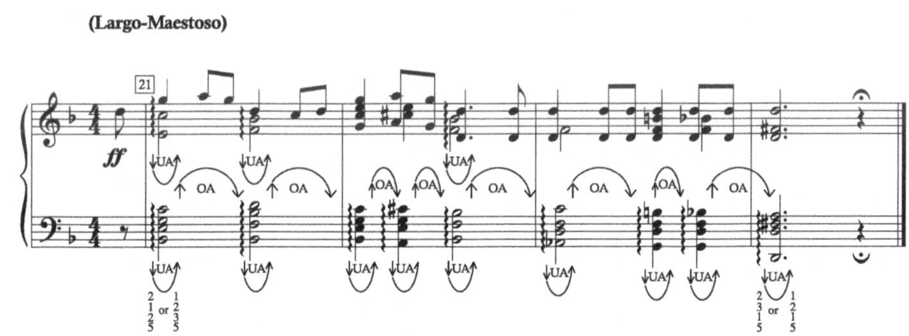

Applications of Fan-Shaped Gestures

As explained in Chapter 3, fan-shaped motions are produced by the hand and forearm acting as a unified lever with the elbow as a central pivot. Because the elbow remains relatively still, fans are more efficient than oval gestures for executing quick and accurate leaps. In Example 6.7, the hand-forearm unit can move between the highest and lowest notes of each left-hand group. Adding a quick rotation toward the fifth finger at the end of the left wing of the fan accentuates the bass note, and the rebound is used as a springboard to the upper notes.

EXAMPLE 6.7. *Scarlatti, Sonata in D Minor, K. 1, mm. 23–26*

In the famous broken-octave passages at the beginning of Liszt's *La Campanella,* the right hand can fan across the distance between the note pairs, incorporating a rotation toward the thumb at the end of the left wing of the motion to define the prominent melodic notes (Example 6.8). The boxes show which notes are grouped together in a single fan.

EXAMPLE 6.8. *Liszt,* La campanella, *mm. 5–6, right hand*

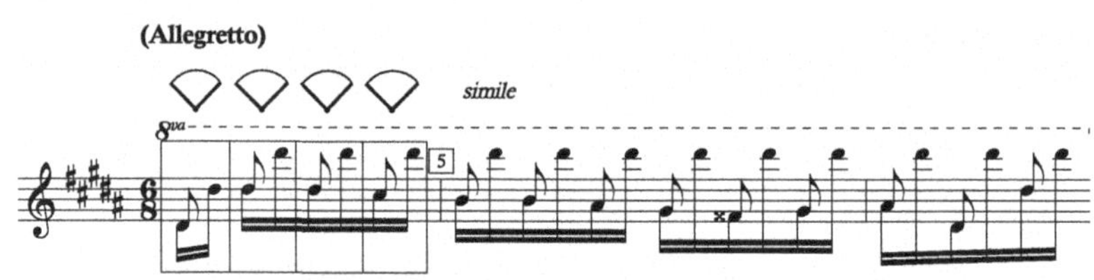

Later in the same piece, the coverage of the fans can be expanded over two octaves to fully encompass the thumb- and fifth-finger repetitions shown in the suggested alternate fingering (Example 6.9). A similar strategy can be applied to measures 165–166 in Chopin's third ballade (Example 6.10).

EXAMPLE 6.9. *Liszt,* La campanella, *mm. 41–42, right hand*

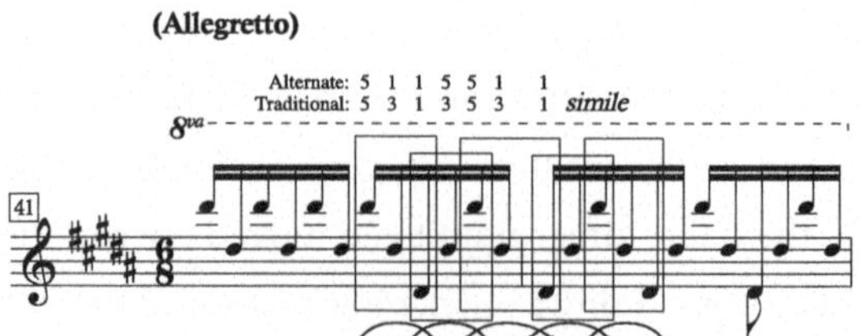

EXAMPLE 6.10. *Chopin, Ballade no. 3 in A♭ Major, mm. 165–66, right hand*

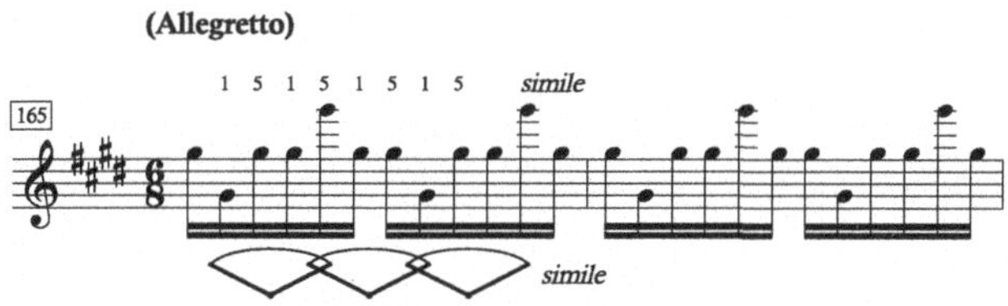

In passages containing broken intervals of different sizes, it is sometimes possible to imagine several smaller forearm rotations encompassed within a larger fan motion. In Example 6.11, this concept is represented by means of jagged lines at the top of the graphic fan symbol. The pianist pronates the hand vigorously in the direction of the thumb to execute the accents on the primary beats and plays the intervening notes of each beat with smaller rotational movements, staying as close to the keys as possible.

EXAMPLE 6.11. *Chopin, Etude in G♭ Major, op. 10 no. 5 ("Black Key"), mm. 57–58*

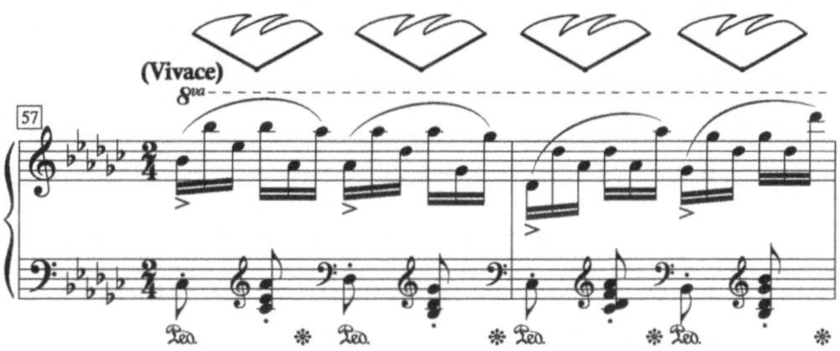

In addition, if the distance that must be covered by the fan gesture is extremely large, the hand may be passively catapulted until it reaches the end of its range of motion. In Example 6.12, the hand should be kept flexibly loose so it can be rapidly thrown by the forearm through the broken tenths marked by the brackets.

EXAMPLE 6.12. *Chopin, Concerto no. 2 in F Minor, op. 21, i, mm. 101–103*

Fan shapes are also commonly applied to waltz basses that alternate once per measure between two distant positions (Example 6.13). The elbow acts as a fulcrum for the movement and must thus be aligned approximately midway between the highest and lowest notes within each box. As mentioned in Chapter 5, many small-handed players may prefer using compact fingerings for chords on the offbeats, substituting 1-2-4 or 1-3-5 for the traditional 1-2-3.

EXAMPLE 6.13. *Schumann, Waltz, op. 124 no. 4, mm. 1–8*

Similarly, many left-hand parts in Chopin nocturnes and polonaises can be played with fan shapes. In Example 6.14, mentally map the location of the various notes and chords along the trajectory of the shape described by the forearm.

EXAMPLE 6.14. *Chopin, Nocturne in E♭ Major, op. 9 no. 2, m. 1, left hand and Nocturne in B Major, op. 9 no. 3, m. 1, left hand*

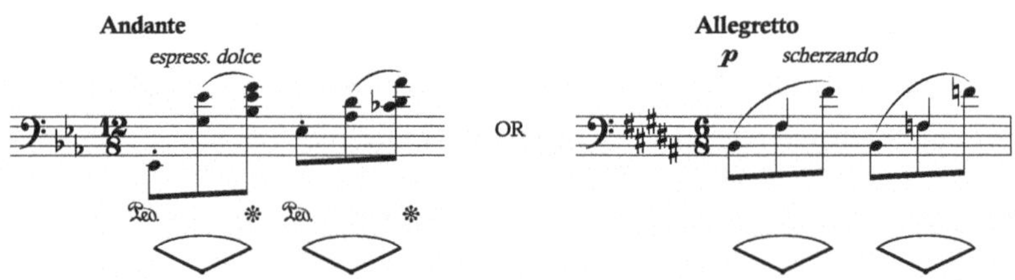

One important use of fan shapes is to expedite leaps between distant registers at extreme speeds. In Example 6.15, pianists who cannot reach a tenth must redistribute some lower notes of the alto melody into the left hand. The left hand must leap back and forth between these notes and the Cs in the low bass. To play the leaps accurately, the pianist should first position the left elbow between the top and bottom notes of the leap, near C3. After playing the B♭s and Cs of the opening measure and a half, the forearm is fanned to the left all the way down to C2 in measure 2, simultaneously rotating toward the fifth finger. Next, the forearm fans to the right through the three Cs at the beginning of measure 2, pronating the second finger over the thumb for the middle C on beat 2; the B♭ and middle C at the end of the measure can be played while still pronated. Even though the elbow will lift slightly as the second finger passes over, the general shape of the fan should be preserved. The torso will then move slightly forward and the elbow will be positioned a little to the left in order to accommodate the larger shift that is coming up. Finally, the pianist will rotate quickly down to the C1–C2 octave of measure 3. The amplitude of the fan should be kept as shallow as possible to accommodate the tempo. At the end of this sequence, the forearm will fan immediately in the opposite direction to play the C and B♭ in measure 3.

EXAMPLE 6.15. *Schumann, "Aufschwung," from* Fantasiestücke, *op. 12 no. 2, mm. 1–3*

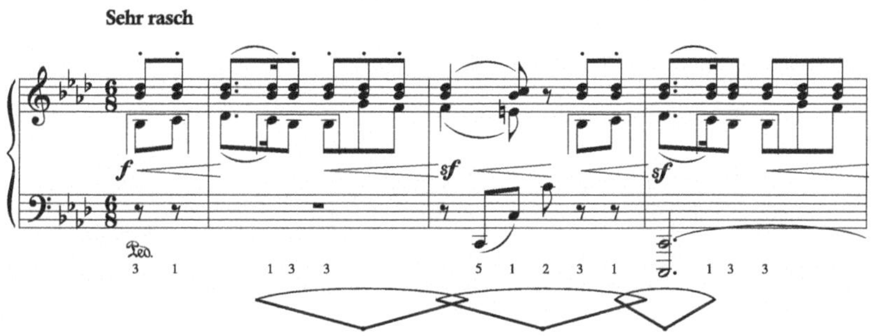

The large cross-hand shifts in the first movement of Beethoven's "Pathetique" sonata can intimidate someone whose hands, forearms, arms, and/or shoulders are smaller or shorter than average (Example 6.16). By positioning the right elbow between the lowest and highest notes of the leap and fanning the right forearm back and forth over the left hand, the notes can be executed accurately with no significant breaks in the tempo. Shifting the position of the torso to the left or right slightly in advance of and in the direction of the leap allows the hand to effortlessly "fall" onto the correct notes.

EXAMPLE 6.16. *Beethoven, Sonata in C Minor, op. 13 ("Pathetique"), i, mm. 50–59*

Fan shapes can expedite extreme leaps complicated by issues of speed and power. Modern piano rags—for instance, William Bolcom's *Raggin' Rudi* (Example 6.17) or *Rag Infernal*—contain jumps much larger than those found in classic rags of the Joplin era. The distance to be bridged is such that the upper arm must move away from the torso to extend the furthest limit of the forearm, and the rotation of the humerus will be more noticeable. Fans allow pianists to leap accurately through the left-hand jumps in measures 10, 12, and 13 in Example 6.18 as well; the shape will fully encompass the notes contained in each octave pair, extending from the thumb note of the upper octave to the fifth finger of the bottom octave.

EXAMPLE 6.17. *William Bolcom,* Raggin' Rudi, *mm. 5–8*

EXAMPLE 6.18. *Brahms, Rhapsody in G Minor, op. 79 no. 2, mm. 9–13*

When both hands must leap between registers, a pianist's spatial awareness of the keyboard can be disrupted, affecting accuracy (Example 6.19). With fan shapes, however, the elbows remain stationary, acting as clear reference points for tracking the geographical placement of the hands.

EXAMPLE 6.19. *Heller, 25 Etudes for Piano, op. 45 no. 15 ("Warrior's Song"), mm. 1–4*

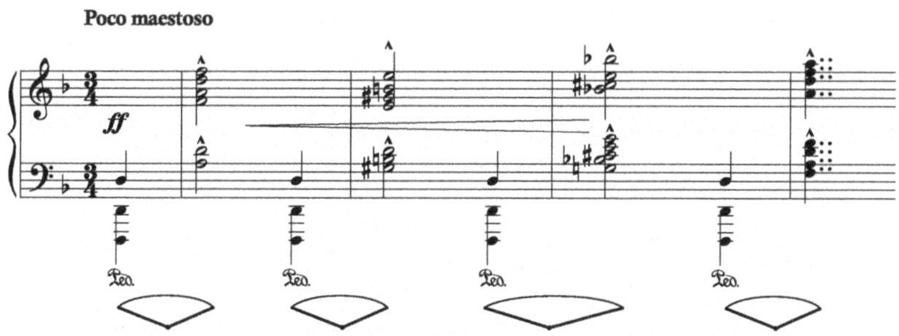

In this regard, fan shapes can help organize textural layers that must be clearly delineated in terms of color and voicing. In Example 6.20, fan gestures enable the left and right hands to comfortably swing back and forth between the octave melody, bass line, and accompanimental chords.

EXAMPLE 6.20. *Debussy, "Danseuses de Delphes," from* Préludes, *Book 1, mm. 6–9*

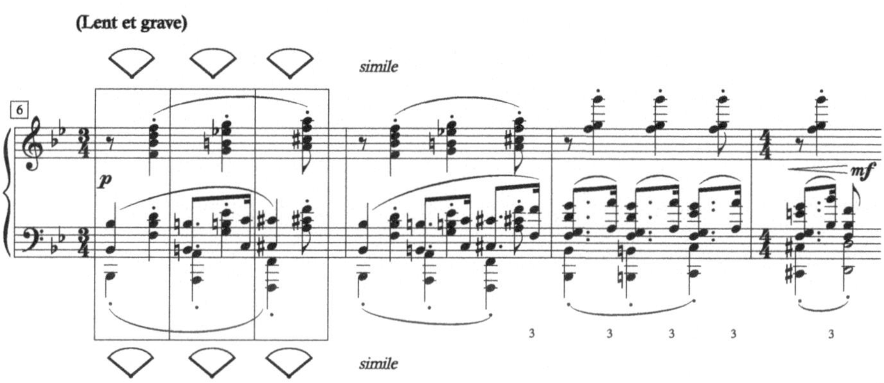

Like ovals, fans can impart a sense of shape and direction to leaping notes. The melody of Schumann's "Hunting Song" could be performed by positioning the elbow between the highest and lowest notes and swinging the forearm between the two extremes (Example 6.21).[4]

EXAMPLE 6.21. *Schumann, "Hunting Song," from* Album for the Young, *mm. 1–4*

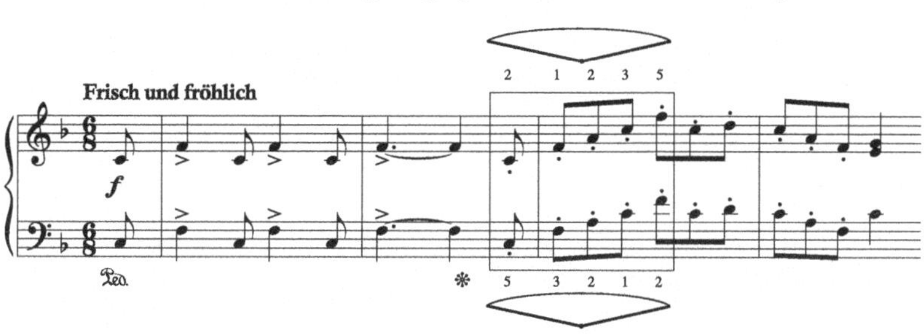

[4] Alternative solutions are proposed in Examples 4.36 and 5.46.

Finally, when a pianist is deciding between fans and ovals, speed is often a determining factor. The tempo of Example 6.22 may favor the use of fans to play the rolled chords; the traditional fingering can be modified to better conform to the curved shape of the gesture.[5]

EXAMPLE 6.22. *Brahms, Rhapsody in B Minor, op. 79 no. 1, mm. 81–82*

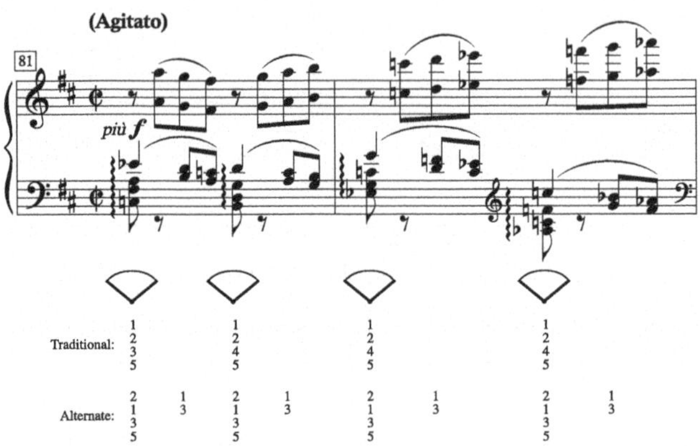

FACILITATING LARGE REACHES

The authors have repeatedly emphasized the importance of favoring the neutral shape of the hand. However, sometimes there are no alternatives to extending the hand past neutral to play large chords and arpeggios. In these instances, passive spread and vertical spanning can maximize a pianist's reach.[6]

Applications of Passive Spread

To play chords with wide stretches, pianists should avoid actively reaching for notes with the fingers or fixing the hand into rigid shapes. In the thick right-hand chords in Example 6.23, passive spread allows the hand to open into the notes of the chord by allowing the keyboard itself to spread the fingers apart. The upper arm can lift slightly without the fingers losing contact with the key surfaces; a vigorous arm impulse generates the downward force needed to depress the keys, with an upward release following immediately afterward.

EXAMPLE 6.23. *Chopin, Prelude in C Minor, op. 28 no. 20, mm. 1–4*

[5] An alternate solution involving redistribution of the notes can be found in Example 4.18.

[6] These concepts are explained in greater detail in Chapter 3.

Even tonal control of soft chords can be better achieved with passive spread. In Example 6.24, stabilizing only the joints of the finger or fingers playing the alto line will help project the inner melody. All the fingers can be released once the pedal is engaged.

EXAMPLE 6.24. *Debussy, "Danseuses de Delphes" from* Préludes, *Book I, mm. 1–2*

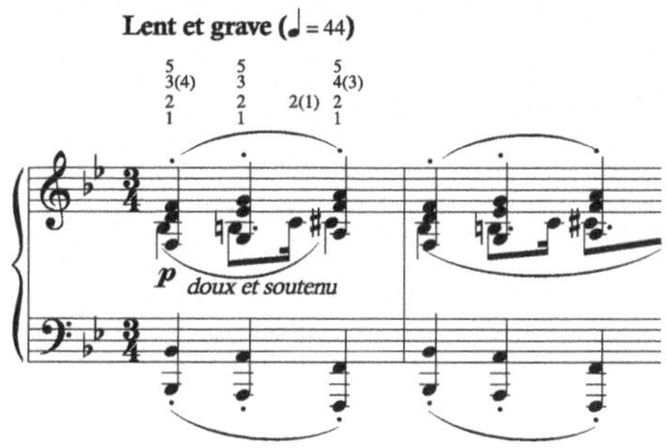

One of the most effective uses of passive spread is for playing legato contrapuntal passages where the pianist wishes to avoid using the pedal for stylistic reasons. In Example 6.25, the left hand can be passively spread to hold the tonic pedal point for the last three measures so it can play tenor notes that stretch the hand to the interval of a ninth. Similarly, in the right hand at the very end of measure 26, passive spread can be used to reach over the thumb to span a seventh between 5 and 2 to make a perfect legato connection between G4 and A4.

EXAMPLE 6.25. *J. S. Bach, Fugue in C Major from* Das Wohltemperierte Klavier, *Book I, BWV 846, mm. 24–27*

Applications of Vertical Spanning

Vertical spanning takes advantage of opposing motions in the fingers—extending one finger upward while flexing another inward toward the palm.[7] In addition to increasing

[7] Vertical spanning is explained more completely in Chapter 3.

the spread between fingers, it can connect notes in situations that limit the use of the pedal and help fingers maneuver around black keys. The flexed finger in a vertical span has better access to the end of the key, where mechanical advantage is greatest. Example 6.26, for instance, shows a wide-ranging melody set against light detached chords. Aligning the left forearm in a straight line behind the fifth finger, the pianist plays the C in the area near or slightly inside the front edge of the black keys. Then, without pushing into the keybed, he or she flexes the third finger in toward the palm to play the F. Because the third finger is curved instead of flat and activates the key lever closer to the end, delivering a slight stress on the downbeat of the first measure should be easy. Subsequently, the forearm will be realigned behind the third finger before the A and C are played at the end of the first phrase. The next phrase, which begins with the upbeat to measure 2, can be treated in a similar fashion.

EXAMPLE 6.26. *Schumann, "Happy Farmer," from* Album for the Young, *mm. 1–3*

As stated previously, limited spread between the second and fifth fingers is one marker of small-handedness. In Example 6.27, the fingering of the first sixth in the right hand (D–B) could be changed to 1-5, but a better strategy might be to implement vertical spanning. To play the chord with 2-5, the second finger flexes down toward the palm while the fifth finger extends in the opposite direction, and the hand and wrist slightly elevate to skeletally align the bones of the fingers and hand. The player can then connect the top three notes with 5-4-3, inconspicuously releasing the lower notes played by the second finger and thumb and using the arm to move from chord to chord.

EXAMPLE 6.27. *Haydn, Sonata in E Minor, L. 53/H. XVI: 34, i, mm. 30–32*

A small-handed pianist can also use vertical spanning to move rapidly up or down the keyboard. Upon reaching the third note (E) in Example 6.28, the third finger of the right hand flexes down and slightly to the left. This action will encourage the thumb to release, the fifth finger to extend out and up, and the right arm to move in the direction of the ascending arpeggio. This procedure is especially effective at measures 7 and 8, in the more difficult octave skips.

EXAMPLE 6.28. *Beethoven, Sonata in C♯ Minor, op. 27 no. 2 ("Moonlight"), iii, mm. 1–9, right hand*

In fact, almost any rapid arpeggiated passage can be facilitated by vertical spanning. The left-hand part in measures 64–65 of Example 6.29 contains only two harmonies. To minimize swiveling, pianists can use vertical spans between the first two notes of each arpeggio; notes can also be released where necessary. Return the hand to neutral as soon as the need for the vertical span has passed.

EXAMPLE 6.29. *Beethoven, Sonata in E♭ Major, op. 31 no. 3, iv, mm. 68–69, left hand*

A clear relationship can be drawn between Example 6.29 and the more difficult *doppio movimento* left-hand accompaniment in Example 6.30. The first two notes of measure 9 can be spanned vertically by extending the left-hand fifth finger out to the left and flexing the third finger in and to the right. The arm will be used to move between the bass and the harmonic third (B♭ and D♭), using the F as a pivot. The same procedure is also applicable to the extended left-hand passages in measures 69 to 72 of the same movement.

EXAMPLE 6.30. *Chopin, Sonata no. 2 in B♭ Minor, op. 35, i, mm. 9–13, left hand*

Applications of Passive Spread in Conjunction with Vertical Spanning

Passive spread used in conjunction with vertical spanning can help improve the positioning of fingers over notes of expansive chords. In Examples 6.31 and 6.32, the hand should be kept as supple as possible; at the same time, pianists can employ vertical spans to bridge uncomfortably large distances between inner fingers, to give more skeletal support to the hand, and to avoid hitting unwanted notes.

EXAMPLE 6.31. *Heller, 25 Etudes for Piano, op. 45 no. 15 ("Warrior's Song"), mm. 1–4*

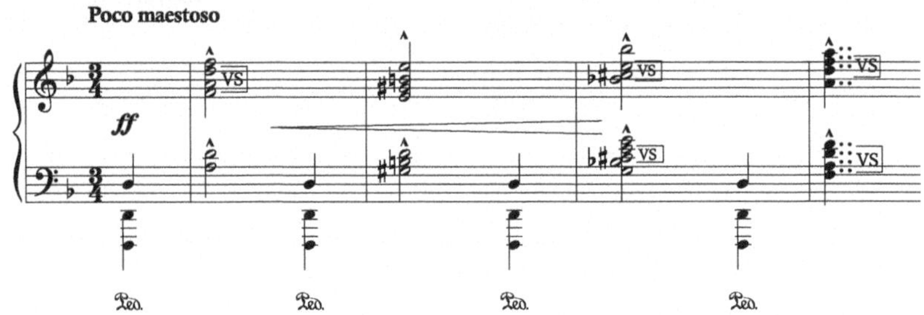

EXAMPLE 6.32. *Chopin, Prelude in D♭ Major, op. 28 no. 15, mm. 40–43*

To play the arpeggiated accompaniments in Examples 6.33 and 6.34, the fingers are spread passively into the keyboard, and vertical spanning is used between the third and fifth fingers of the left hand to bridge the gap between the lowest notes of both passages. The forearm and upper arm lead the movement, and notes are released where needed. Slight deviations of the wrist in both directions may be necessary, but excessive side-to-side swiveling should be avoided. The hand and wrist should remain supple and the player should stay close to the key surfaces, even to the point of silently touching them with the palm and fingers.

EXAMPLE 6.33. *Chopin, Nocturne in C♯ Minor, op. 27 no. 1, mm. 3–6, left hand*

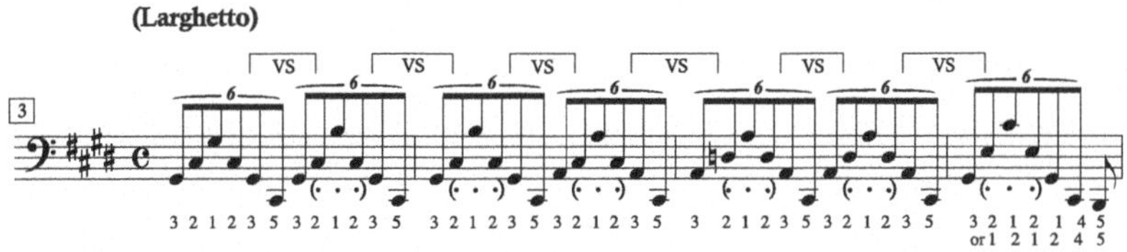

EXAMPLE 6.34. *Mendelssohn, "Venetian Boat Song," from* Songs without Words, *op. 30 no. 6, mm. 1–5*

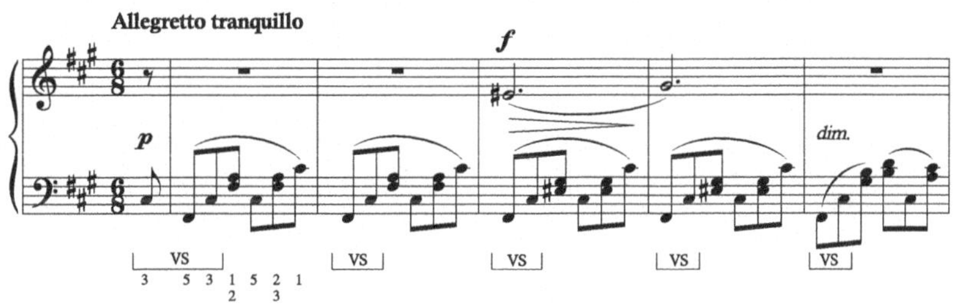

A combination of vertical spanning and passive spread might enable some pianists to play selected arpeggios in Chopin's Etude, op. 10 no. 1 using the composer's original fingerings (Example 6.35).[8] The rapid horizontal action of the arm will be the primary actor in leading the fingers to the proper notes. The goal is to eliminate or minimize the need for wrist swiveling, something that would make this etude impossible to play at a rapid tempo.

EXAMPLE 6.35. *Chopin, Etude in C Major, op. 10 no. 1, mm. 1–2*

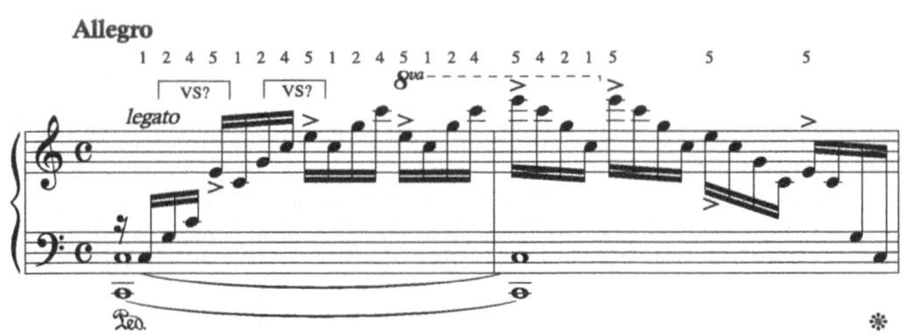

Example 6.36 demonstrates that passive spread with vertical spanning can allow pianists to play notes that once seemed far out of reach. Depending on the unique dimensions of their hands, many may still experience difficulty. Therefore, all performers should exercise discretion when attempting these excerpts, stopping at any sign of discomfort. In the seventh piece of Ligeti's *Musica ricercata,* the right hand plays two or three imitative voices against a repetitive left-hand ostinato. It is therefore best to connect as many of the right-hand notes as possible with the fingers to avoid breaks in the melodic line and muddiness in the pedal. In measures 61–62, the F can be played with the hand comfortably

[8] An alternate approach is discussed in Example 5.8.

relaxed and tilted toward the thumb. Vertical spans are activated by slightly flexing the second finger to play the G just in front of the black keys. Then, the second finger is moved silently toward the fallboard between the black keys to ready the thumb for playing the E♭. The palm flattens so it touches the keys, and the hand passively spreads open so the arm can move the thumb to its note. A similar procedure can be used to play measures 91–92 of the same piece, using the A and C at the end of measure 90 as an anchor for spreading the hand passively to reach the other notes.

EXAMPLE 6.36. *Ligeti,* Musica ricercata, *vii, mm. 61–62 and 91–92, right hand*

DIVIDING OR ROLLING CHORDS

If there is sufficient time, one option for playing chords that extend beyond the span of the hand is to divide them into smaller parts or to roll them. Different modes of execution will produce contrasting musical results:

- Consecutive—the notes are rolled one at a time from bottom to top for a transparent, semi-melodic effect. Particularly large chords can be apportioned into smaller note groups that alternate between the hands.
- Non-consecutive—the notes are played out of order. For example, when there are rolled chords in both hands, the top note of the left-hand chord can be delayed and played simultaneously with the top note of the right-hand chord.
- Simultaneous—chords in both hands are divided or rolled simultaneously so that some notes sound at the same time. They can be strictly aligned or flexibly interwoven, played in the same direction or in contrary motion. The sound is fuller and has a stronger harmonic function.

Other types of rolls common in the Baroque, such as rolls that begin at the top or rolls that go up and down several times, are not commonly encountered in music specifically written for the modern piano.

Fingering is often an issue. If the notes are so widely spaced apart that they cannot be fingered in one hand position, the pianist can add an extra hand shift or repeat a finger (especially 1-1 or 5-5).[9]

[9] For a wider discussion of finger repetitions, please refer to Chapter 5.

In addition, there are matters of harmony, melody, and pedaling to consider:

- Dividing produces a thicker and more chordal sound than rolling.
- Dividing or rolling that starts *before* the beat enables the melody to arrive on time. If the hand is too small to hold all the chord tones in the fingers, the pedal must also be engaged before the beat in order to catch the bass note. This in turn can create a gap in the melodic line that should be camouflaged either by connecting melodic notes with the fingers or by using half pedals.
- Dividing or rolling from bottom to top that starts *on* the beat results in a delay of the arrival of the melody note or inner chord tones, which can be rhythmically awkward. Normal syncopated pedaling is used to properly sustain the harmony.

Finally, even though Chopin and other Romantic composers indicated a slight preference for rolling chords in the left hand before the beat and chords in the right hand on the beat, flexibility is essential because of all the factors that must be taken into account.[10] The chord should sound as comfortable, natural, and musically appropriate as possible. Deciding to divide or to roll a chord is thus a creative choice that must balance practical with aesthetic concerns.

Consecutive and Non-Consecutive Divisions or Rolls

A standard approach for dividing a chord before the beat is shown in Example 6.37. The low G of the left hand in the second measure of Example 6.37a is played slightly before the beat and secured with the pedal. The D in the right-hand chord of measure 24 must be held through the pedal change in order to prevent a gap in the melodic line. The pianist can now play the upper three notes of the final chord, as shown in Example 6.37b.

EXAMPLE 6.37. *Schumann, "Der Dichter spricht," from* Kinderszenen, *mm. 24–25*
a. Printed version
b. Dividing the chord before the beat

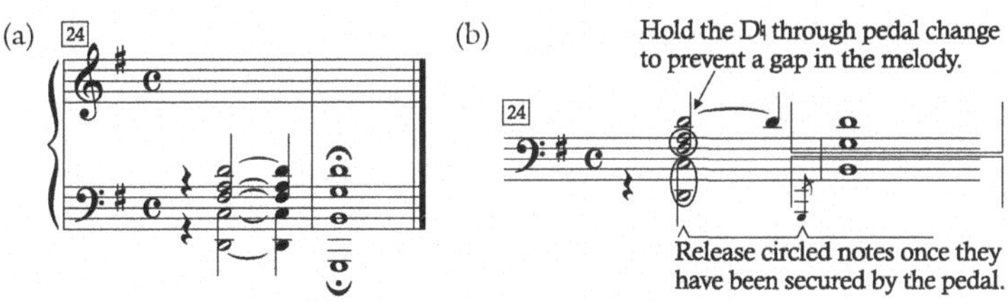

[10] Jan Ekier and Pawel Kaminsky, editors of the PWM Chopin edition, state that for rolls in the right hand, the first note of the right hand should be played simultaneously with the left-hand bass note (i.e., the chord should be rolled *on* the beat). For rolls in the left hand, however, the reverse is true: the last note of the left-hand arpeggio should be sounded simultaneously with the right (i.e., the chord should be rolled *before* the beat). Jan Ekier and Pawel Kaminski, *Performance and Source Commentary to the Sonatas of Fryderyk Chopin* (Krakow: PWM, 2010), 2.

Similarly, dividing the large chord in measure 22 of Example 6.38a before the beat, as opposed to rolling it, may communicate the harmony and eloquent simplicity of the piece more completely. Placing the two lowest notes of the bass directly against the first high A of the right hand allows more time for the leap in the left hand, makes it easier to secure the pedal, and permits the remaining notes to be comfortably redistributed between the hands, as shown in Example 6.38b. The bracketed notes in measure 21 could be redistributed to uncross the thumbs if desired.

EXAMPLE 6.38. *Schumann, "Träumerei," from* Kinderszenen, *op. 15, mm. 21–22*
a. Printed version
b. Dividing the chord before the beat

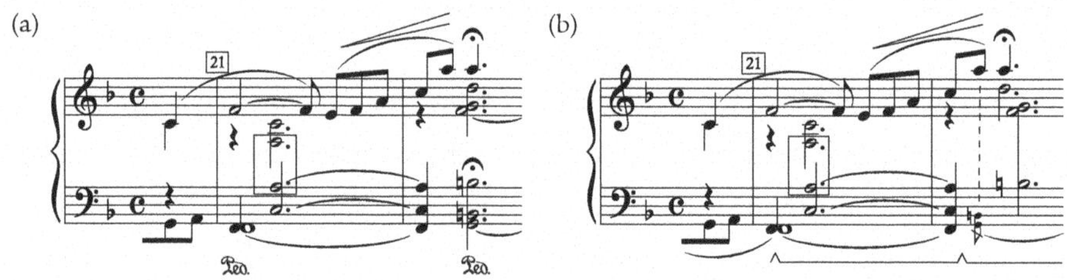

Small-handed pianists will also encounter problems when they cannot comfortably hold notated rolled chords in the fingers and must therefore rely on the pedal to prolong the harmonies. In Example 6.39, if the left-hand chords are rolled before the beat, the pedal must be activated before the beat as well in order to catch the bass. Some players may need to use alternative fingerings for the starred chords to relieve stretches in the hand. Actively reaching for widespread notes with the fingers should be avoided, even though this may seem counterintuitive. Instead, the notes can be released quickly so that the hand can close up to a more neutral position behind each note of the roll, using the pedal as a connecting device.

EXAMPLE 6.39. *Brahms, Waltz in E Minor, op. 39 no. 4 (for solo piano), mm. 23–26*

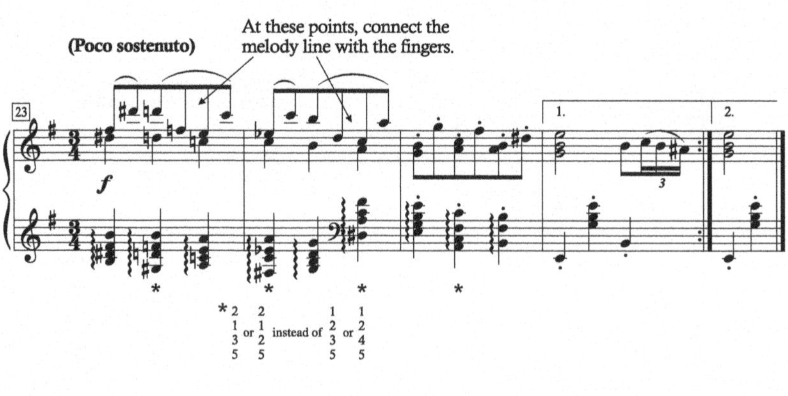

Extended chords that are beyond or just within the reach of a pianist's hand can be rolled before the beat where it is stylistically appropriate to do so, as shown in Example 6.40. Releasing lower notes as the roll unfolds permits small hands to contract to a more neutral shape and encourages an expressive lingering.

EXAMPLE 6.40. *Martinu, "Columbine Sings," from* Puppets, *Book II no. 5, H. 116, mm. 13–16*

To accentuate the brilliance of the climactic roll in Example 6.41, the left hand moves over the right to take the final note.

EXAMPLE 6.41. *Ravel, "Alborada del gracioso," from* Miroirs, *mm. 194–195*

Rolling can also be an unexpected solution to playing passages with double notes. When the fingers are maximally extended, flexing them to play the left-hand thirds in Example 6.42a is nearly impossible. Since the tempo is fast and the left hand is subordinate to the melody, imperceptibly rolling the double notes, as shown in Example 6.42b, hardly changes the sound of the passage.

EXAMPLE 6.42. *Chopin, Sonata in B♭ Minor, op. 35, i, mm. 9–16*
a. Printed version
b. Rolling the left-hand accompaniment

(a)

(b)

However, there are many cases where the choices seem less clear. Examples 6.43 and 6.44 show multiple ways of dividing up chords before or after the beat. Any notes appearing as afterbeats should be played as softly as possible to minimize the sense of rhythmic disruption.

EXAMPLE 6.43. *Chopin, Prelude in A Major, op. 28 no. 7, mm. 11–12*
a. Printed version
b. Dividing or rolling the chord before the beat
c. Playing the bass as an afterbeat

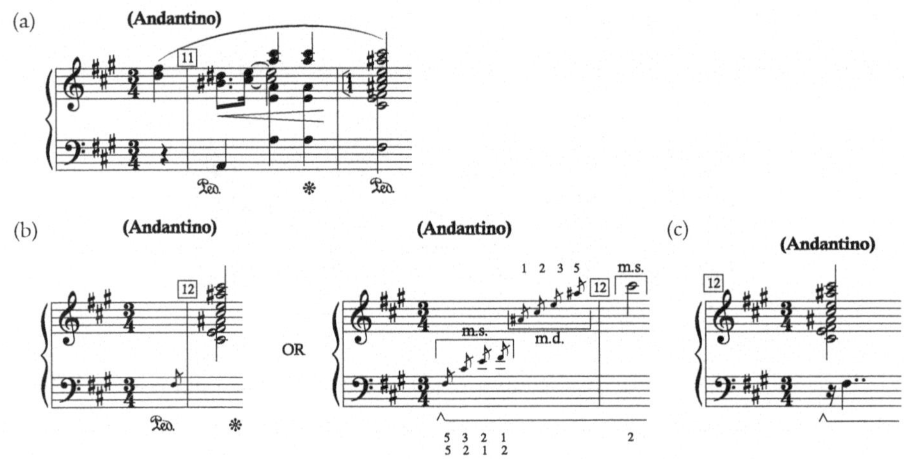

EXAMPLE 6.44. *MacDowell, "Sung Outside the Prince's Door," from* Forgotten Fairytales, *op. 4 no. 1, mm. 9–12*
a. Printed version
b. Chord division before the beat
c. Chord division on the beat

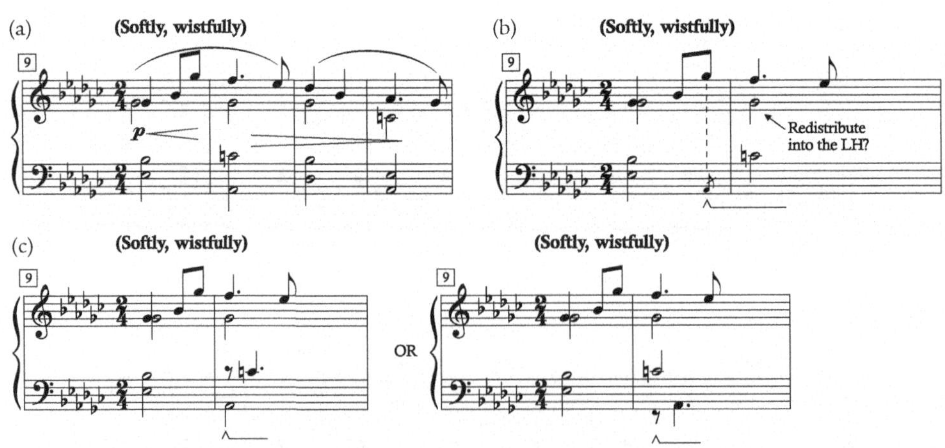

Where there are quick notes just preceding a rolled chord, it is best to execute the roll on the beat. In Example 6.45, rolling both the right-hand chord in measure 8 and the left-hand chord in measure 9 on the beat simplifies the pedaling.

EXAMPLE 6.45. *Brahms, Intermezzo in E♭ Major, op. 117 no. 1, mm. 7–10*

Difficult decisions must be made when many tenths are encountered in an exposed lyrical context. At the starred chords in measures 173 and 177 of Example 6.46, the right-hand tenths should probably be rolled on the beat as Chopin would have preferred.[11] However, on the third beat of measure 180, rolling the right-hand chord on the beat would significantly delay the arrival of the melody note. A pianist might instead choose to play the lower notes (G3 and C4) of the right hand slightly before the beat, using a quarter- or half-pedal change to clarify the harmony prior to the third beat. A second solution might be to play the second left-hand chord in that measure as an anticipation to the third beat along with the G3 of the right hand, changing the pedal at the same time. Sometimes, practical considerations trump all other concerns: in measures 171 and 179, the left-hand rolls could be placed on the beat because of the difficulty of cleanly pedaling the harmonic changes. In the case of the left-hand roll at measure 171, after playing the F and A, the pianist could silently retake the F with the fifth finger before measure

EXAMPLE 6.46. *Chopin, Sonata in B♭ Minor, op. 35, i, mm. 169–183*

[11] Ekier and Kaminsky specifically state that in the exposition, the F♮ of the right hand should sound simultaneously with the D♭ octave in the left hand at m. 41 and all analogous places, implying that right-hand chords should be rolled on the beat.

172 so that the pedal could be changed on the chromatic quarter notes in measure 172. In any case, the tempo and the elasticity of the rubato allow for a flexible and somewhat improvisatory treatment.

Simultaneous Divisions or Rolls

To preserve rhythmic or textural aspects of a piece or to achieve a more compact sound, chords in two hands can be rolled simultaneously. In these situations, small-handedness can influence decisions about the fingering, pedaling, and sequence of the notes. For instance, if a pianist's hands are too small to reach the chords in measure 24 of Example 6.47a, the chords could be rolled simultaneously before the beat in contrary motion as shown in Example 6.47b so as not to interrupt the steady tolling of the low "bells" or the arrival of the melody note. The pedal must be held through all three measures to prevent aural gaps.

EXAMPLE 6.47. *Mompou,* Fountain and Bell, *mm. 22–24*
a. Printed version
b. Simultaneous contrary motion roll

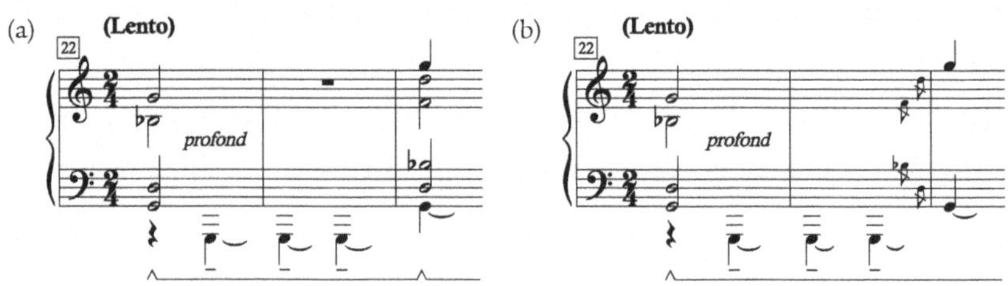

If a hand is too small to reach and hold notes of rolled chords, pianists must often adjust the fingering and pedaling. The simultaneous rolls in the first measure of Example 6.48 should be played before the beat, and the damper pedal can be deployed up to an eighth note early to prevent undesirable melodic gaps. To forestall excessive side-to-side swiveling at the wrist, alter the fingerings as indicated.

EXAMPLE 6.48. *Rebikov, "Une histoire triste qui finit bien," from* Tableaux pour enfants, *op. 37, mm. 21–24*

In Example 6.49, repetitions of the fifth finger may facilitate the execution of the extended simultaneous rolls.

EXAMPLE 6.49. *Chopin, Etude in E♭ Major, op. 10 no. 11, mm. 1–3*

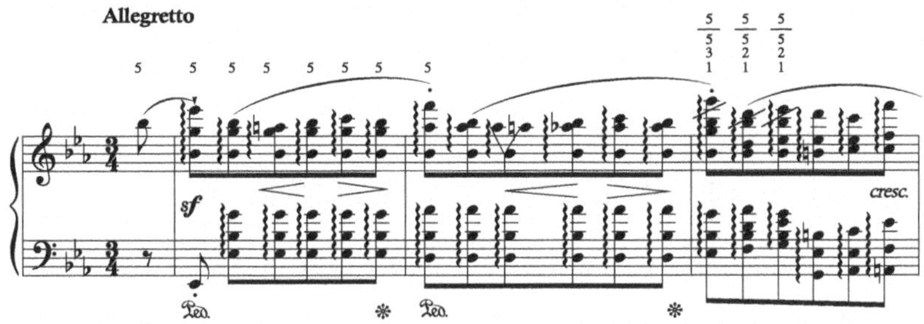

As discussed in Chapter 5, adjacent notes can often be played with the thumb to alleviate uncomfortable stretches. In Example 6.50, it may be necessary for someone with small hands to play the circled notes in measure 79 simultaneously with the thumb instead of rolling them one after the other.

EXAMPLE 6.50. *Chopin, Concerto no. 2 in F Minor, i, mm. 79–80*

RELEASING NOTES

If muscles are repeatedly and forcefully contracted to play large chords and are not rested sufficiently, fatigue ensues.[12] Muscles must be provided with moments of respite and should alternate between periods of work and rest. By "release," the authors mean the cessation or reduction of muscular effort to depress the keys. Releases may be accompanied by a lift away from the key surface, but this is not always the case. In fact, the keys may remain slightly depressed simply from the relaxed weight of the finger, hand, and/or forearm. In linear textures such as melodies, arpeggios, and broken chords, releasing notes also frees the arm to align more optimally behind the playing finger.

[12] Fatigue is extensively discussed in Chapter 2. Refer especially to Figures 2.2a and 2.2b.

This discussion is divided into three subsections: releasing notes in chords and multi-layered textures; releasing notes in melodic intervals, arpeggios, and broken chords; and releasing notes in broken octaves. In the excerpts that follow, all staccato marks are those that appear in the published score. The authors have inserted breath marks or circles to bring the reader's attention to potential release points in the music.

Releasing Notes in Chords and Multi-Layered Textures

If there are several consecutive chords to play, it can be helpful to imagine rests between them in the form of slight rebounds or releases. This will allow the hand to periodically contract to a smaller shape. Example 6.51 can help pianists cultivate this sense of muscular release.

EXAMPLE 6.51. *Benda, Sonatina no. 3 in A Minor, i, mm. 1–4*

In fact, having enough stamina to play the fast passages of consecutive chords in Example 6.52 and deliver the dramatic crescendo at the climax of the phrase may well depend on planning small releases after each chord. Additionally, a few well-timed redistributions of notes into the left hand will prevent the right wrist from swiveling to reach distant notes with the thumb.

EXAMPLE 6.52. *Mendelssohn,* Variations sérieuses, *op. 54, Variation 3, mm. 49–52*

When there is no time to relax between subsequent keystrokes, however, several chords can be grouped together. The repeated sixths and thirds shown enclosed in boxes in Example 6.53 are incorporated within a single upward gesture of the right hand, with a large muscular release back to baseline placed at the end.

EXAMPLE 6.53. *Mozart, Viennese Sonatina no. 6 in C Major, iii, mm. 1–4*

Periodic breaths should also routinely be inserted in passages containing dotted rhythms or tied notes (Example 6.54). In addition to allaying fatigue, these rest points allow the pianist to articulate the rhythm with precision, even when the hand spreads to play the thick *fortissimo* chords in measures 5–7.

EXAMPLE 6.54. *Beethoven, Sonata in C Minor, op. 13 ("Pathetique"), i, mm. 1–7*

In passages requiring both power and speed, large muscular releases can be timed to correlate with phrasing. These large releases, marked by vertical lines in Example 6.55, should not disrupt the rhythmic continuity.

EXAMPLE 6.55. *Tcherepnin, Bagatelle, op. 5 no. 10, mm. 94–102*

In multi-layered textures where one hand must play both melody and accompaniment, it is often necessary to release notes in one part early in order to comfortably play the other part. Small-handed players should guard against holding melodic notes to full value if this stretches the hand to its maximum (Example 6.56a). Schubert's G♭ Major Impromptu, op. 90 no. 3 could be practiced as shown in Example 6.56b until the release reflex becomes instinctive.

EXAMPLE 6.56. *Schubert, Impromptu in G♭ Major, op. 90 no. 3, m. 1*
a. Printed version
b. Practice strategy

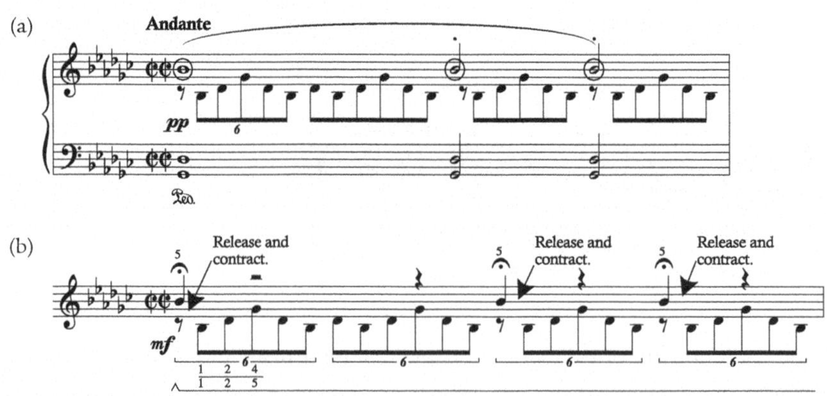

In Example 6.57, selected notes of the melody or the harmony should be released early if stretches between inner fingers become an unnecessary burden.

EXAMPLE 6.57. *Brahms, Rhapsody in B Minor, op. 79 no. 1, mm. 1–8*

In Example 6.58, letting go of the circled notes of the right hand soon after they have been caught in the pedal allows the hand to change to a more comfortable fingering for the tremolo.

EXAMPLE 6.58. *Heller, 25 Etudes for Piano, op. 45 no. 10, mm. 1–3*

Similarly, in the last movements of the Beethoven sonatas, opp. 53, 109, and 111, melodic notes can be released early so that the hand can contract to play the trills (Example 6.59). Respected Beethoven interpreters such as Artur Schnabel and Rudolf

EXAMPLE 6.59. *Beethoven, Sonata in E Major, op. 109, iii, Variation VI, mm. 165–166*

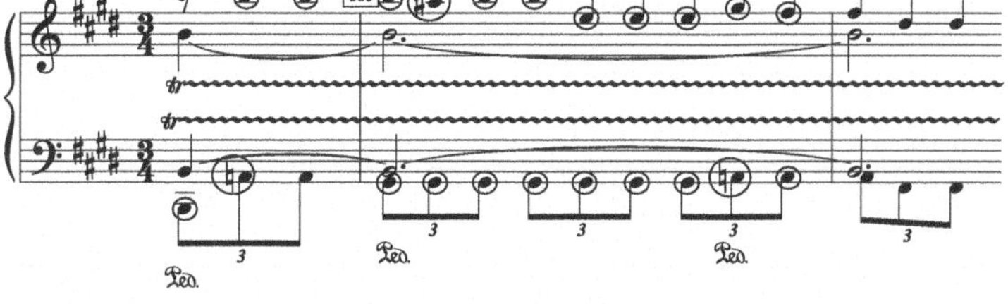

Serkin have stated that the trills should be maintained continuously, but it may be necessary for some pianists to interrupt them momentarily if the hand is too small to reach the notes of the melody and the trill at the same time.

Holding long notes for full value in one part may in fact limit one's ability to shape a beautiful line in another. In Schumann's "Träumerei," only the notes marked with asterisks need to be held for full value because the remaining ones can be prolonged in the pedal (Example 6.60). In measure 6, the right-hand thumb can be used to play both the G and A and then gradually shifted over so that the G is released. The bracketed notes may be redistributed.

EXAMPLE 6.60. *Schumann, "Träumerei," from* Kinderszenen, *mm. 1–7*

If there is no alternative to spreading the hand, then the player should seek to release entire chords or selected notes of chords as soon as possible. Releasing becomes more important as the number of voices in the texture increases. Example 6.61 shows a four-part texture with a long cantabile melody in the right hand and three voices in the left. Holding out the quarter notes of the tenor countermelody and the dotted quarters of the bass line for full value while simultaneously playing the accompanimental triplets in the middle may exert a significant strain on the small hand. Contrary to the advice

EXAMPLE 6.61. *Chopin, Nocturne in F Major, op. 15 no. 1, mm. 1–4*

of some Chopin editors, as long as the voicing of the melody is carefully delineated, it may be preferable to release the left-hand thumb notes as soon as they are secured in the pedal.[13]

In fact, failing to release chords in strenuous passages such as the one shown in Example 6.62 exposes the small-handed pianist to a common risk factor for injury: playing loudly for long periods of time with the hand spread wide. Lavish applications of pedal can take over the work of connecting the slurred notes while augmenting the sound level to its maximum. Instead of attacking or hitting the keys from the air with the dead weight of the arm, the chords can be played using arm impulses starting at the key surfaces.

EXAMPLE 6.62. *Chopin, Ballade no. 1 in G Minor, mm. 209–212, right hand*

Similarly, in the coda of Debussy's *L'isle joyeuse,* releasing all chords immediately is crucial to sustaining a high volume of sound to the very end of the piece (Example 6.63). Letting go of lower chord tones in the left hand will enable the hand to quickly contract into a more compact position behind the notes of the melody, opening up possibilities for alternative fingerings or even note redistributions.

EXAMPLE 6.63. *Debussy,* L'isle joyeuse, *mm. 220–225*

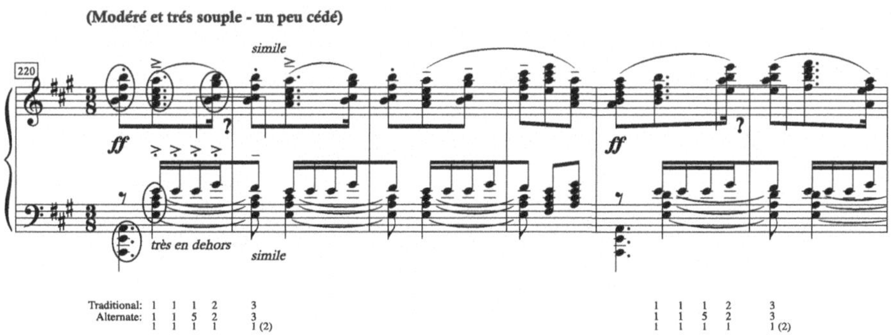

[13] Theodore Kullak states: "Be careful not to play  as if  stood in its place." Theodore Kullak, ed., *Chopin Nocturnes* (Berlin: Schlesinger, 1881; reissued New York: Schirmer, 1881). An alternative solution might involve redistributing the notes of the tenor countermelody in the first two measures into the right-hand thumb.

Releasing Notes in Melodic Intervals, Arpeggios, and Broken Chords

Large melodic intervals, extended arpeggios, and broken chords can tax the hand if they contain uncomfortable stretches between inner fingers. Even major seconds can be problematic if played by adjacent fingers in combinations of black and white notes. Small-handed pianists often underestimate the size of intervals that look small on the printed page.

As an alternative to stretching, a finger can be allowed to release a note after it is played, freeing the forearm and arm to move behind the next playing finger. Example 6.64 shows a simple pedagogical work useful for practicing these adjustments. Because the tempo is moderate, there is adequate time to align the arm optimally behind each note as it is played.

EXAMPLE 6.64. *Kuhlau, Prelude in G Major, mm. 1–3*

At faster tempos, small degrees of wrist deviation can be used in tandem with note releases to improve comfort. In the right-hand part of Example 6.65, the wrist can start out in a position of ulnar deviation, straighten out to a more neutral position, and finally move to a position of radial deviation.[14] Plotting the passage in blocks as shown can clarify these positions. The transition from group to group should be kept as smooth as possible.

EXAMPLE 6.65. *Haydn, Sonata in F Major, L. 3/H. XVI: 9, Scherzo, mm. 1–4*

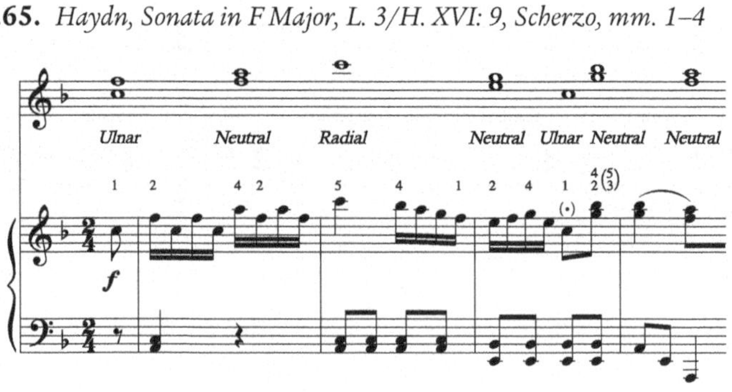

If notes are released and connected with the pedal instead of the fingers, there is time to align fingers more optimally even behind widespread notes. Releasing also may make

[14] Ulnar deviation is a lateral deviation of the wrist such that the fingers point toward the fifth finger side of the wrist; radial deviation is a lateral deviation of the wrist such that the fingers point toward the thumb side of the wrist.

it possible to use more traditional fingerings such as the ones shown in Example 6.66 for the left hand.

EXAMPLE 6.66. *Brahms, Intermezzo in A Minor, op. 76 no. 7, mm. 9–12*

To resolve performance issues associated with rapid arpeggiated figures that lie outside the easy reach of the hand, problem stretches and possible release points need to be systematically identified. The hand should not be frozen into a rigid shape; it must retain the flexibility to alternately open and close with each register shift. In Example 6.67, the melody and/or any of the accompanying notes can be released early because they are sustained in the pedal. Chopin's Etude, op. 25 no. 1 ("Harp"), which Heller's etude recalls,

EXAMPLE 6.67. *Heller, 25 Etudes for Piano, op. 45 no. 9 ("Celestial Voices"), mm. 1–16, right hand*

would benefit from the same treatment. In Examples 6.68 and 6.69, the thumb notes and notes of large intervals can be released so that the hand can close up in the direction of travel and the arm can take over the work of moving laterally across the keyboard. Vertical spanning and passive spread between the largest intervals may also prove helpful in all three examples. Areas that often are problematic for pianists with small hands, especially those who have trouble reaching larger intervals between 1-2, 2-4, 3-5, and 1-4, are marked with brackets.

EXAMPLE 6.68. *Mendelssohn, Praeludium in E Minor, mm. 9–10*

EXAMPLE 6.69. *Chopin, Etude in F Major, op. 10 no. 8, mm. 47–50*

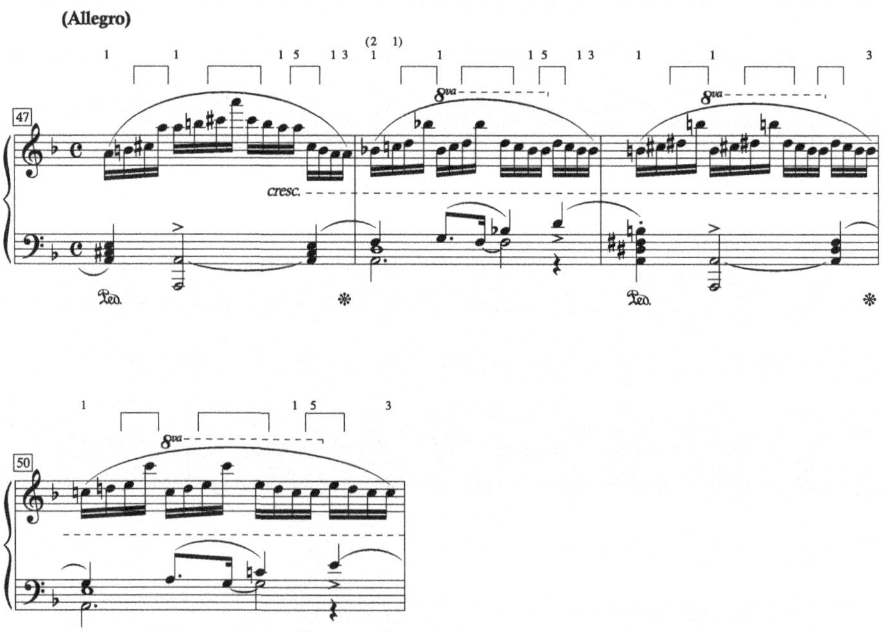

There are many instances in the advanced literature where a small-handed pianist faced with difficult choices might decide that a traditional fingering is best. The topmost notes of the circled arpeggios in Example 6.70 could certainly be redistributed into the right hand, but the result is rhythmically clumsy. Therefore, it may be preferable to play them as written, releasing the thumb notes immediately so that the arm can move freely across the keyboard.

EXAMPLE 6.70. *Chopin, Etude in C♯ Minor, op. 10 no. 4, mm. 35–37*

Releasing Notes in Broken Octaves

For the small-handed pianist, broken octaves pose a special technical problem. It is easy to forget that when broken octaves descend a scale in the right hand starting with the thumb (or ascend a scale in the left hand starting with the thumb) the distance between consecutive octaves is a ninth.

To execute broken octaves, the forearm rotates between the lower and upper notes. It is imperative to release notes on both sides of the rotation to permit freedom of movement. However, this motion is rarely symmetrically balanced. Notes that are on the beat are usually played with an active rotation; intervening offbeat notes are played more softly with a weaker reactive stroke. The pianist should keep the hand flexible and avoid reaching for notes, as counterintuitive as that might seem. Instead, the energy of the rotation carries the forearm through distances that lie outside the reach of the hand.

To prepare for a passage containing hidden ninths and tenths like the one in Example 6.71a, it might be helpful to practice actively rotating to the principal notes while

EXAMPLE 6.71. *Mozart, Concerto in C Major, K.467, i, mm. 238–240*
a. Printed version
b. Practice strategy

(a)

(b)

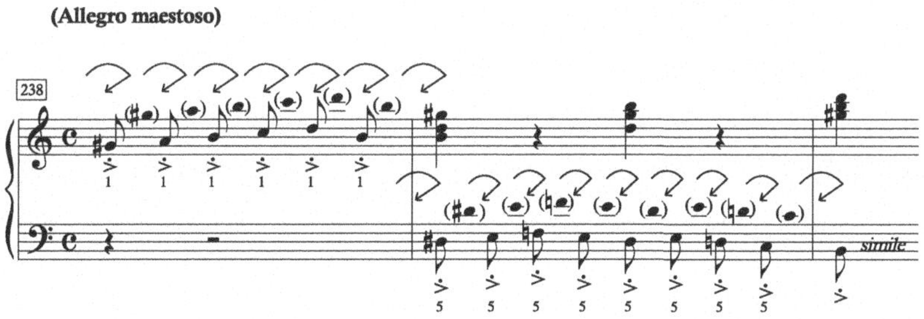

silently fingering the alternate notes at a deliberate tempo, as shown in Example 6.71b. As the tempo increases, the pianist must accelerate the speed of the rotations, reduce the amplitude of the rotations until they are as flat as possible, and keep the fingers close to the keys.

Lengthy passages of broken octaves may have to be mentally and physically subdivided into more manageable groups. In Example 6.72, every half- or whole-note beat can be vigorously accented with a concentrated rotation; the energy of this stroke is then carried through seven or more notes. Even though the dynamic is marked *fortissimo,* the second and fourth sixteenth notes of every beat will be almost inaudible.

EXAMPLE 6.72. *Beethoven, Sonata in C Major, op. 2 no. 3, i, mm. 85–86*

USING ROTATION FOR POWER

As explained in Chapter 3, forearm rotation delivers power to the keyboard by maximizing the speed of key attack. To play loud chords using rotation,

- Position the upper arm in its neutral position for piano playing.
- Lift the forearm, allowing the wrist to flex naturally in response so that the fingers can remain close to the key surfaces.
- Rotate the forearm quickly into the keys to play the desired chord.

It is usually more comfortable to rotate in the direction of the thumb (pronation). The wrist should not collapse at the moment of impact. The pianist should release the active muscular work of the forearm and arm as soon as (or sometimes fractionally before) contact is made with the keybed so that the forearm rebounds upward. During the rebound phase of the movement, the fingers may lift off the keyboard entirely or may retain some contact.

This motion should be practiced with smaller chords, such as three-note triads before working up to four-note chords. For thick chords where maximum volume is needed, each chord is played with a vigorous pronation (Example 6.73). The hand returns to its compact position as often and as soon as possible, releasing keys quickly after they have sounded.

EXAMPLE 6.73. *Rachmaninoff, Prelude in C♯ Minor, op. 3 No 2, mm. 46–48*

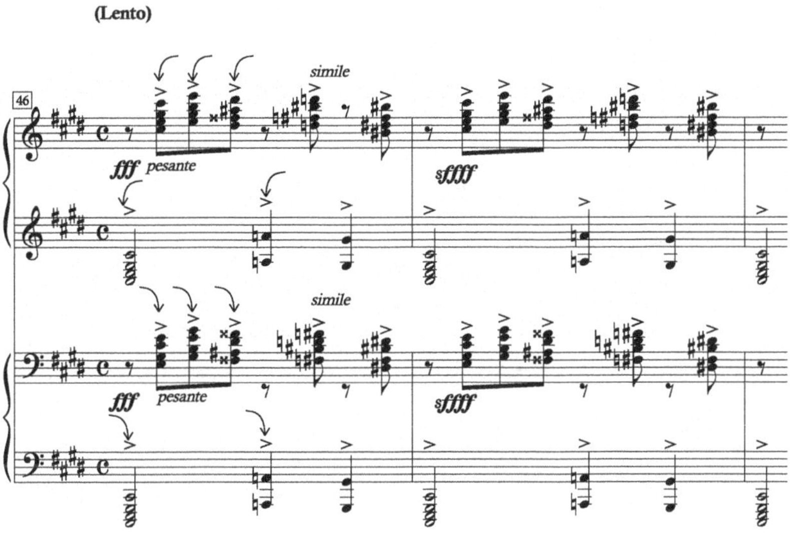

Rotational movements are also broadly useful for projecting fifth-finger or thumb melodies. For instance, forearm rotation gives power to the fifth fingers of both hands to play the staccato notes in the traveling broken octaves in the first movement of Beethoven's "Pathetique" sonata (Example 6.74). Instead of stretching the hand to reach for the sevenths, octaves, and ninths marked on the example, the outward momentum of the rotation itself throws the relaxed hand over the required distance.

EXAMPLE 6.74. *Beethoven, Sonata in C Minor, op. 13 ("Pathetique"), i, mm. 93–99*

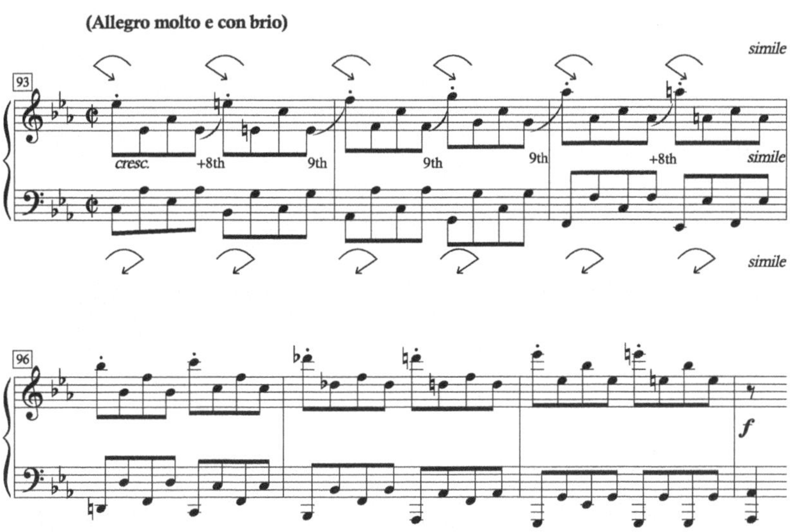

Just as rotation can provide extra energy for voicing the fifth finger side of the hand through supination, it can also be implemented on the thumb side of the hand through pronation. In Example 6.75, a strong rotation to the thumb notes will help to project the melody over the accompaniment. A strict physical legato is unnecessary because the pedal is used throughout; notes can thus be released at will to ease stretches between inner fingers. The pianist should refrain from instinctively reaching for distant notes, such as the B♮ in measure 2, instead relying on the rotating forearm to cover the gap.

EXAMPLE 6.75. *Heller, 25 Etudes for Piano, op. 45 no. 9 ("Celestial Voices"), mm. 1–4*

TAKING ADVANTAGE OF SKELETAL SUPPORT

Small-handed pianists may have hands, fingers, and arms that are shorter and/or lighter than those of larger-handed players. If so, they should attempt to channel a greater proportion of the force delivered to the key through the bones of the fingers and hands, minimizing the necessary degree of muscular contraction.

Loud single notes and small chords are prime opportunities for implementation of this strategy. To project the circled notes in the bass and soprano in Examples 6.76 and 6.77, players should close up the hand, brace the fourth and fifth fingers together to form a solid unit, and keep the hand arches and wrist firm at the moment of key impact.

EXAMPLE 6.76. *Schumann, Waltz, op. 124 no. 4, mm. 1–8*

EXAMPLE 6.77. *Chopin, Ballade no. 1 in G Minor, op. 23, mm. 208–212*

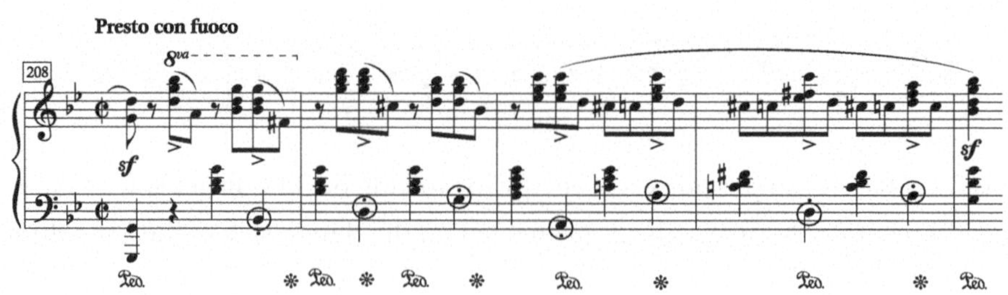

Even in chords that lie comfortably within the span of the hand, the top note can be voiced more easily by channeling the weight of the arm through the bones of the fingers (Example 6.78). The fingers should be curved and aligned such that the joints between the phalanges are completely stable at the moment the sound is activated.

EXAMPLE 6.78. *Schumann, "Northern Song," from* Album for the Young, *op. 68 no. 41, mm. 1–4*

PLAYING STACCATO CHORDS AND OCTAVES

Small-handedness can significantly impair a pianist's ability to play widespread staccato chords and octaves. Because the fingers are cantilevered out from the hand to an extreme degree, they operate at a mechanical disadvantage. Less force is transmitted to the keys. Factors such as the weight of the arms, the length of the arms, the width of the shoulders, the thickness of the fingers and fingertips, and the flexibility of the hand also come into play.

Furthermore, pianists with smaller hands are sometimes unable to play octaves except by using unusual hand positions. Players may work at the very edges of the white keys or lock into an unnaturally high wrist position. Ortmann specifically identified both as makeshift strategies that are impractical for playing octaves at high speeds.[15] The first gives rise to large in-out motions as the player moves to and from white- and black-note octaves. The second forces the middle fingers to hyperextend to avoid striking notes within the octave. Small-handed pianists should proceed with extreme caution if the repertoire they are playing contains large numbers of rapid octaves or chords that are beyond or just within the reach of the hand.

Repeated Chords and Octaves

There are three levers of the arm—the hand, the forearm, or the whole arm—that can be used to play repeated chords or octaves. The speed and dynamic level of the passage determines the lever use that is most appropriate for the task. As a rule, smaller levers generate more speed, whereas larger levers deliver power.[16]

[15] Ortmann, *The Physiological Mechanics of Piano Technique,* 313.

[16] Principles of leverage are extensively discussed in Chapter 2.

When playing repeated chords, pianists can choose from four different coordinations: hand staccato, forearm staccato, whole arm, or arm impulse. Hand staccato, sometimes referred to as wrist staccato, is useful for playing fast repeated chords or octaves. This motion uses flexion and extension at the wrist joint and is produced by the muscles of the forearm. Forearm staccato is produced by combining the forearm with the hand and fingers into a single lever by stabilizing the wrist, using the elbow as a fulcrum. It is activated by the upper arm muscles and is useful when more force is needed than can be produced with the hand. The whole arm—formed by combining the fingers, hand, forearm, and upper arm—provides the most power but greatly limits speed. Thus, the whole arm is only practical for producing chords or octaves when there is adequate recovery time in between. Finally, arm impulses, explained in Chapter 3, coordinate activity between all the levers of the arm, with the joints remaining flexible. This touch can provide endurance and power for repeated chords played at faster tempos. These coordinations are explored in Activity 6.1.

ACTIVITY 6.1
PLAYING STACCATO CHORDS AND OCTAVES

To achieve efficiency and comfort in playing repeated chords and octaves, it is important to cultivate a feeling of muscular release. The following preliminary activity aims to cultivate releasing muscular effort all the way back down to the baseline level that preceded the motion. Although there may not be an opportunity in all musical situations for pianists to release back to baseline after every chord or octave, it is important to build an awareness of how such a complete release feels in the body.

Start the first activity at a moderate tempo and soft dynamic, hands separately, and work gradually to increase speed and volume as well as interval size. Transpose to different keys. First play the activity using hand staccatos. Progress to forearm staccatos, whole arm staccatos, and then to arm impulses. Notice the upper tempo limit for each action, as well as the different qualities of sound and sensation. At faster tempos, it may be necessary to alternate between different levers. For example, arm impulses could be used once every four strokes, with hand staccatos used for the other three. Additionally, the repeated chords might also be accompanied by a slow raising and lowering of the forearm and/or upper arm to distribute the workload to different muscle groups and alleviate fatigue.

To play staccato chords that move up or down the keyboard, the arm must lead the direction of travel. Practice the next activity using hand staccatos, forearm staccatos, whole arm staccatos, and arm impulses, noticing the upper tempo limit for each action. Gradually increase speed, volume, and interval size.

Also practice using the levers in combination—a larger lever followed by a series of smaller levers. For instance, play one forearm staccato followed by three hand staccatos, one arm staccato followed by three forearm staccatos, or one arm impulse followed by three hand staccatos.

Small-handed pianists must sometimes use a longer lever than a larger-handed pianist to provide the force or endurance necessary to play a strenuous passage. This is because larger muscles are recruited to the task as smaller muscles become fatigued. However, pianists may subconsciously and unnecessarily default to using long-lever attacks exclusively rather than a combination of long and short levers, unduly limiting the speed of execution and overtaxing the anatomy with disastrous physical and musical results.

In Example 6.79, hand staccatos are appropriate for playing the offbeat right-hand octaves, which are rapid and light.

EXAMPLE 6.79. *Albeniz, "Prelude," from* Songs of Spain, *op. 232 no. 1, mm. 25–28*

Example 6.80, however, shows a passage of thicker chords that might require supplementation from larger levers—forearm staccatos, arm impulses, or a combination of the two. Other fast passages of alternating staccato chords—the end of the exposition

EXAMPLE 6.80. *Villa-Lobos, "O Policinelo," from* Prole do Bebê, *Book. 1, mm. 1–2*

in the first movement of Beethoven's sonata, op. 10 no. 2, or the opening of Liszt's Transcendental Etude in F Minor, for instance—can be treated in like manner. The two touches sound very different. Forearm staccatos have an energetic, percussive quality because they attack the keys from above. Arm impulses are sonorous and less percussive because strokes are activated on or close to the key surfaces.

Lengthy passages of repeated chords are often less tiring when played as a series of arm impulses as long as the tempo permits. Example 6.81 shows the first four measures of Bartok's Bagatelle, op. 6 no. 5, which opens with 26 measures of staccato chords in the right hand that are followed by 20 measures of repeated chords in the left hand.

EXAMPLE 6.81. *Bartok, Bagatelle, op. 6 no. 5, mm. 1–4*

Traveling Staccato Chords and Octaves

If staccato chords travel along the keyboard instead of staying in one place, the forearm and/or upper arm must provide the lateral movement. Use light rebounding hand staccatos accompanied by lateral shifts of the forearm to play traveling sixths and octaves at slow to moderate tempos (Examples 6.82 and 6.83). It is important not to fall into the habit of holding unused fingers excessively high over the keys, but rather to raise these fingers only enough to avoid hitting unwanted notes. Sometimes, they can even be allowed to brush the keys as long as the joints are kept loose.

EXAMPLE 6.82. *Kuhlau, Sonatina in G Major, op. 55 no. 2, ii, mm. 18–19*

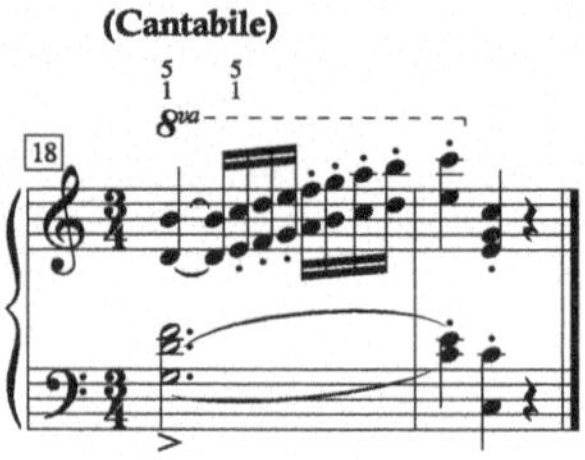

EXAMPLE 6.83. *Burgmüller, Etude in E Major, op. 105 no. 9, mm. 1–3*

Hand staccato alone, however, may not generate sufficient momentum, volume, or endurance for passages of thicker chords at faster tempos. Example 6.84 might necessitate a forearm staccato with a subsequent series of lighter hand staccatos.

EXAMPLE 6.84. *Beethoven, Sonata in C Major, op. 2 no. 3, iv, mm. 1–4*

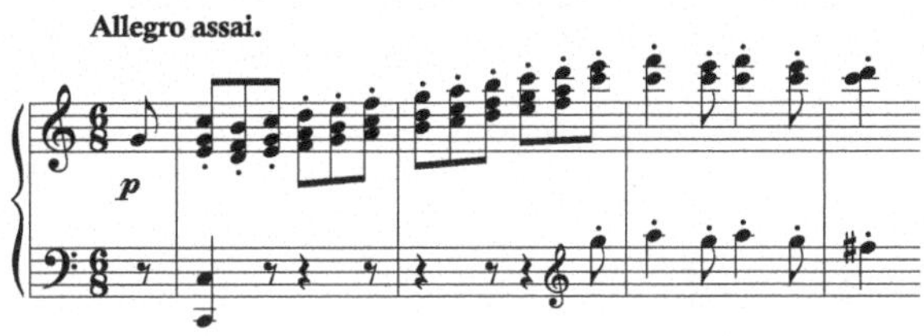

The left-hand octave passage in Example 6.85 is awkwardly placed directly in front of the body and alternates between black and white octaves. Experimental combinations of hand staccatos, forearm staccatos, and arm impulses should be tested until the desired musical result is achieved. For example, a pianist could group the octaves into threes or sixes, using arm impulses or forearm staccatos for the primary beats of each group and hand staccatos for the inner beats. Holding the elbow closer to instead of away from the body may improve stability, and minimizing the amplitude of the vertical movements of the hand and arm will facilitate speed.

EXAMPLE 6.85. *Heller, 25 Etudes for Piano, op. 45 no. 25 ("Epilogue"), mm. 9–12*

Legato Octaves

Legato octaves are a misnomer, especially for the small-handed player. Legato grows out of sensitive dynamic shaping rather than physical connection. Long pedals and broad shaping gestures of the forearm and upper arm will help to erase small gaps in the line. The default octave fingering for most players with small hands is 1-5; 1-4 is occasionally used for black-note octaves, but 1-3 is generally out of reach. In Example 6.86, for instance, 1-5 could be used to play all the chords and octaves of the right-hand melody.[17] The black-key octave (C♯) in measure 28 could possibly be fingered with 1-4. Small moments of respite can be gained by releasing the lower octave note and accompanimental chord tones as soon as possible

[17] See Example 7.38 for a discussion of gestural shaping in connection with the same passage.

EXAMPLE 6.86. *Beethoven, Sonata in F Major, op. 10 no. 2, i, mm. 19–30*

Despite the legato phrase marking, the quick sixteenth-note octaves in Example 6.87 require the use of hand staccato; 1-5 or a combination of 1-4 and 1-5 is appropriate, since most small-handed pianists would find the 1-3 suggested in many editions impossible to execute. Accompanying oval gestures and timely applications of the pedal can help to shape the phrase and ward off fatigue.

EXAMPLE 6.87. *Granados, "May Song," from* Cuentos de la juventud, *op. 1 no. 3, mm. 15–17*

A similar approach is effective in Example 6.88. To project the tone with focus, the shoulder joint, upper arm, and forearm must provide a strong and stable base for overcoming the inertia and resistance of the keys. The size of forward and backward movements between white and black keys should be kept to a minimum, as should the amplitude of the up-and-down motions. Pianists should remember to extend or hold up the middle fingers only to the point necessary to avoid hitting unwanted notes. At this tempo, cleanly pedaling each octave is impossible. Instead, use half pedaling or flutter pedaling to produce the illusion of an unbroken line.

EXAMPLE 6.88. *Schumann,* Papillons, *no. 1, mm. 7–14*

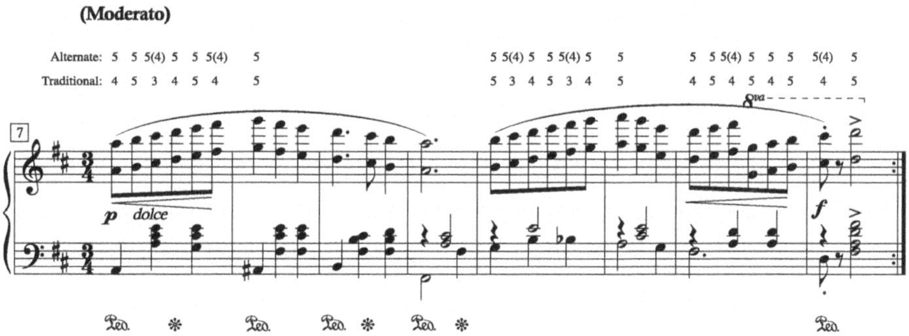

GROUPING NOTES INTO LARGER GESTURES

Larger motions of the forearm and upper arm can be used to shape repetitive series of smaller motions. For instance, dropping and then gradually lifting the arm through a series of parallel chords distributes the work of playing them among different muscle groups. This strategy could be applied to the parallel sixths in Kuhlau's Sonatina in G Major (Example 6.82), the series of first-inversion triads in the opening of the last movement of Beethoven's sonata, op. 2 no. 3 (Example 6.84), or the rising octaves in Schumann's *Papillons* (Example 6.88).

The small-handed player can use these larger gestures to physically map and locate groups of notes along the keyboard. In the last movement of Beethoven's "Moonlight" Sonata, the notes played by the right-hand thumb spell out an ascending C♯-minor chord. These notes can be organized or embedded into a single underarc shape of the arm that peaks with the two *sforzando* chords in the second measure (Example 6.89). Kinesthetic

EXAMPLE 6.89. *Organizing notes within a large arc in Beethoven, Sonata in C♯ Minor, op. 27, no. 2 ("Moonlight"), iii, mm. 1–3*

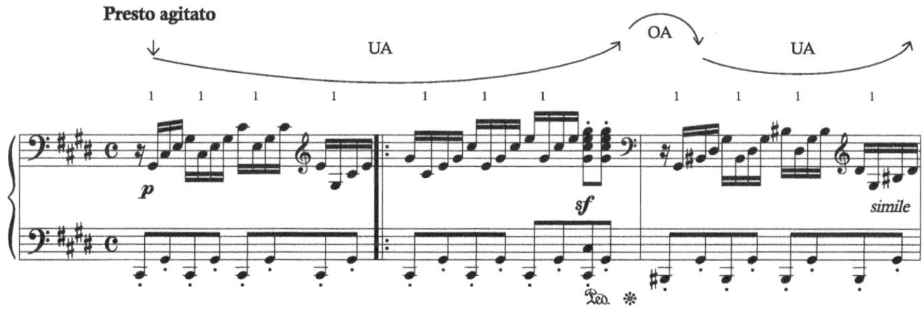

memory of where the forearm and arm are positioned along that curve improves the accuracy and solidity of the passage.

The same method can be used to organize broken octaves. In long passages, the rapid forearm rotations used to play the broken octaves can be incorporated within an underarc or down/up gesture of the arm. In measure 238 of Example 6.90, however, the ascending pattern of octaves is broken by a drop of a tenth on the second half of the last beat. The player should focus on the larger arm gesture, incorporating the broken octaves on D and B into the rising or lifting phase of the arc. The motion, which will crest at the chord at the beginning of measure 239, will absorb or lessen the disruptive effect of the sudden change in the direction of the line. In measure 239 of the same example, the left hand begins a series that ascends through D♯, E, and F before beginning a long descent. Hidden ninths that pianists might overlook are marked in the score. Again, the change of direction can be smoothed over by using the arm to shape a small overarc transitioning into a lengthy underarc.

EXAMPLE 6.90. *Mozart, Concerto in C Major, K.467, i, mm. 238–240*

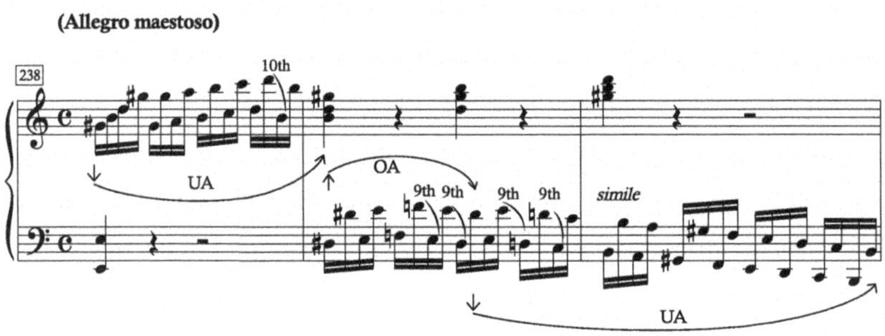

Small-handed pianists may forget that the arm can release and lift weight independently of the action of the fingers. In loud passages where fatigue is an issue, notes can be organized into impulse groupings where the arm drops and lifts in regular cycles. Often, the arm will be thrown at a note or chord on a major beat and then smaller levers such as the hand or fingers will activate in the rebound to fill the spaces, much the way a stone thrown at the correct speed and angle will ricochet across water. In a piece such as the Chopin A Minor Etude, op. 10 no. 2, the arm is thrown sharply downward on the chords followed by an immediate but small upward rebound. The bottom two notes are immediately released to permit the hand to contract, and then the third, fourth, and fifth fingers are set in motion to fill in the connecting notes. The fingerings in Example 6.91 are those of the composer.

EXAMPLE 6.91. *Chopin, Etude in A Minor, op. 10 no. 2, mm. 1–2*

Whenever possible, small-handed pianists should imagine a choreography that integrates smaller oval shapes, fan shapes, or simple rotations into curvilinear gestures of the arm and even the torso.

OMITTING AND REVISING NOTES

If a score is perceived as an unbreakable covenant, omitting and/or changing the notes can seem presumptuous at best or unthinkable at worst. However, many composers acknowledge that the score is only an approximate rendering of a musical idea and that they need the help of performers to translate it into sound.[18] There may be solid arguments for omitting or revising notes in certain circumstances:

- A problematic passage cannot be satisfactorily performed by using other strategies.
- The change would produce an arguably superior musical result than if the passage were performed as written.
- Playing the notes as written is difficult at the stipulated tempo, leading to awkward breaks in the temporal flow of the piece.
- Playing the notes as written would require hours of fruitless practice that could lead to injury.

---

[18] In the nineteenth century in particular, changing attitudes of composers toward their own compositions can often be traced by examining editions of the same work issued at different times or by studying recordings, editions, or memoirs prepared by their students. Performers of the time were expected to assert their individuality as creative artists by arranging and improvising upon the general blueprint of the score. Kenneth Hamilton, *After the Golden Age: Romantic Pianism and Modern Performance* (New York: Oxford University Press, 2008), 179–207.

However, pianists should ensure that omissions and changes are as musically inconspicuous as possible by selecting notes that are doubled elsewhere or those that function in more subservient roles. These changes should be confined to short isolated sections of music. If the piece requires extensive revision and if none of the conditions described above is present, it may be wiser to select something else to perform from the vast trove of piano literature.

Thick chords with multiple doublings are prime candidates for note revision or omission because they challenge the small hand in many ways. Fingers forced to stretch to the limit of their range are not in a position of optimal alignment, cannot be structurally supported, and fatigue easily. For someone with small hands, omitting notes from large chords can therefore often result in a sound that is noticeably louder, not softer, than before. It is worth noting that the four-note octave chord that is most problematic for many small-handed pianists is the first-inversion chord, for example, E–G–C–E reading from bottom to top. Pianists who can play this chord in its root position (C–E–G–-C) may find it impossible to reach the first inversion because of the stretch of a fourth between the second and fourth fingers (G–C) as well as the stretch of a sixth between the second and fifth fingers (G–E), a situation that is exacerbated if there are combinations of black and white notes. Pianists should be alert to sudden unexplained stoppages or tempo changes during a performance that might signal a problem with just one or two chords in a passage.

For a player with a limited hand span, nothing can be more frustrating than to be confronted with a first-inversion chord at the very beginning of a piece. Omitting the bottom A of the right hand in the opening measure of Beethoven's Sonata in F Major, op. 10 no. 2 may allow a small-handed pianist to articulate the rhythm of the opening motive with greater clarity (Example 6.92).

EXAMPLE 6.92. *Beethoven, Sonata in F Major, op. 10 no. 2, i, mm. 1–2*

Example 6.93 shows an excerpt containing two four-note chords, both in the first inversion. These chords are out of step with the smaller shapes found in the rest of the piece. For pianists who find them difficult to play, the bottom note could be omitted.

EXAMPLE 6.93. *Heller, 25 Etudes for Piano, op. 45 no. 2 ("Avalanche"), mm. 61–68*

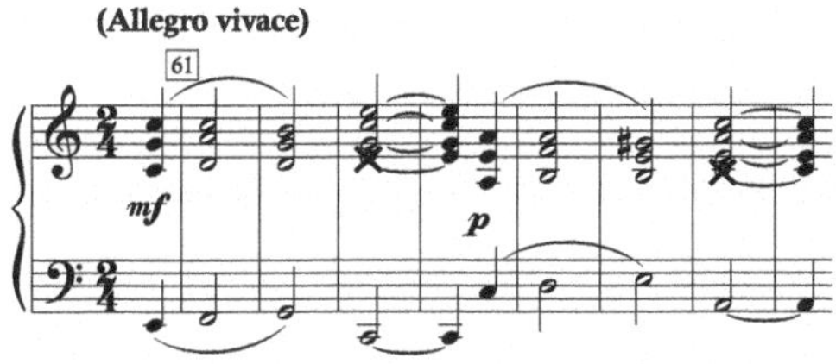

Similarly, pianists who find the chords of Example 6.94 challenging because of the stretch between fingers 2 and 5 can play them as a series of parallel sixths instead, fingered by the thumb and either 4 or 5.

EXAMPLE 6.94. *Joplin,* The Entertainer, *mm. 5–8*

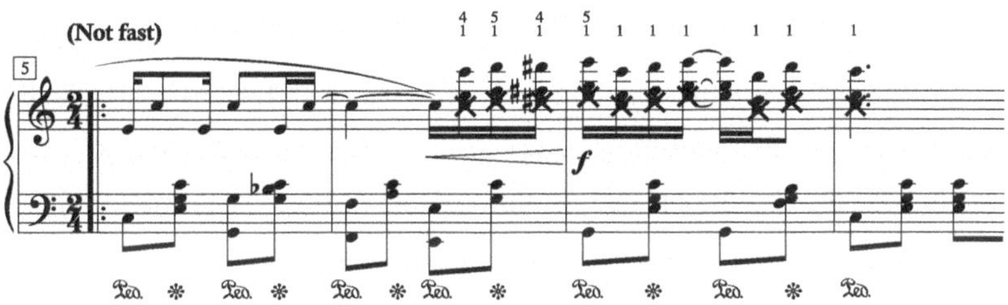

Problems will often arise when the texture suddenly thickens, as when a two-note texture quickly expands to three or four voices. In Example 6.95, the fortissimo unison octaves of the opening leap up to full four-note chords in the right hand. Omitting the lower octave allows the hand to retain a more compact skeletal position and to maintain the volume level.

EXAMPLE 6.95. *Schubert, Valse Noble in A Minor, D. 969, op. 77 no. 9, mm. 1–4*

The music of Chopin can be particularly challenging in this regard. For instance, many pianists stumble on the left-hand chords in measure 17 of Chopin's Prelude in E Minor, op. 28 no. 4 (Example 6.96). The chords are preceded by a large leap that lands directly in front of body. In the first chord, A–C–F♯–A, the top A can be omitted because it is doubled in the bass. In the second chord, G–B–D♯–F♯, the D♯ can be omitted because it is featured prominently in the melody. Voice leading must be carefully considered. For example, on the third beat of measure 17, omitting the top F♯ and playing the D♯ instead would create parallel octaves with the soprano melody. In the final three chords of the same piece, octave doublings at the bottom of the right hand could also be omitted to avoid uncomfortable stretches between inner fingers.

EXAMPLE 6.96. *Chopin, Prelude in E Minor, op. 28 no. 4, mm. 16–19*

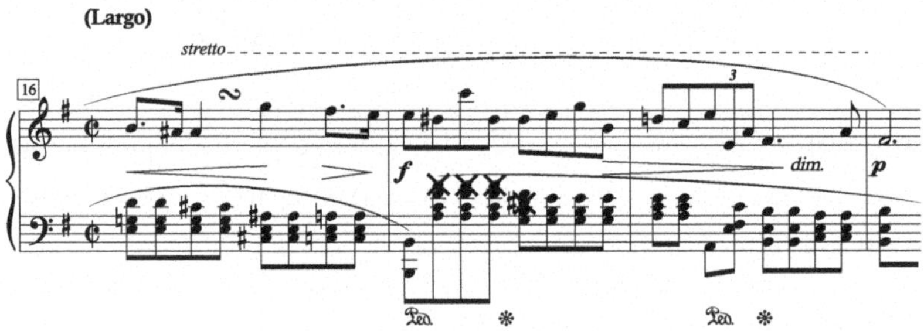

Example 6.97 contains numerous four-note chords in the right hand, many of them in the first-inversion position discussed above, while most of the left-hand chords contain only three notes. The right-hand part moves around two internal pedal points—D♭ and A♭ at the beginning, changing to F and B♭. Since these notes are doubled in the left hand, one could choose to omit one of them in the right hand. Another approach is to eliminate all of the bottom notes in the right hand since they double the melody. This may be the superior choice because the overall stretch of the hand would be reduced to a fifth, sixth, or seventh. In fact, when there are four-note chords in both hands, very often the bottom note of the right and/or the top note of the left can be omitted without significant alteration to the overall sound.

EXAMPLE 6.97. *Sibelius, Romance, op. 24 no. 9, mm. 56–59*

This principle can be broadly applied throughout the Romantic literature. In Rachmaninoff's famous C♯ minor Prelude, the A section returns at the end scored with thick four-note chords in both hands (Example 6.98). Players can easily omit the bottom note of the right-hand chords and the top note of the left-hand chords with negligible musical impact.[19]

[19] An added benefit is that the passage will be transformed into an exact repetition of a redistributed version of the notes, except that the hands will be two octaves apart instead of one. See the discussion of Example 4.1.

Example 6.99 shows another representative case where omission of the bottom octave in the right hand of measures 80–83 would greatly improve ease of execution without any attendant negative consequences; the chord pattern in the right hand of the two preceding measures, measures 78–79, would merely be repeated up an octave in measures 80–81 and up two octaves in measures 82–83.

EXAMPLE 6.98. *Rachmaninoff, Prelude in C♯ Minor, op. 3 no. 2, mm. 45–47*

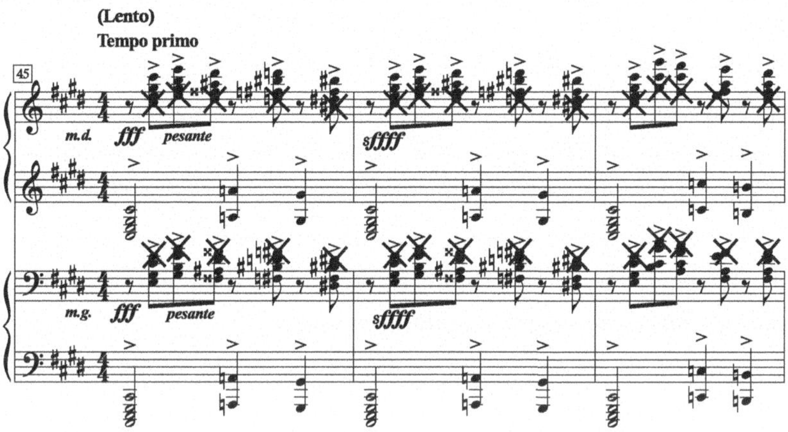

EXAMPLE 6.99. *Liszt, Transcendental Etude no. 10 in F Minor, mm. 78–83*

Because trying to play at full power with the hand outstretched for long periods of time can be extremely fatiguing, note omissions and revisions used in conjunction with redistributions can provide the hand with moments of valuable respite. At times, it can be helpful for pianists to finger a problematic section as if they had a tenth at their disposal and then simply leave out the unreachable notes. In the first measure of Example 6.100, performers could imagine how it might feel to finger the octaves of the second beat with 1-3 moving to 1-5 and then leave out the bottom note of any 1-3 octaves that cannot be moved into the left hand.

EXAMPLE 6.100. *Liszt, Paganini Etude no. 6, Variation 3, mm. 68–71*

If a sonority has been fully established in the pedal, it is then possible to omit notes that simply serve to amplify or prolong that sonority. Rolling the consecutive ninth chords in measures 40–41 of Grieg's "Wedding Day at Troldhaugen" would be awkward because of the quick back-and-forth directional changes that would be required of the smaller hand (Example 6.101). Instead, secure the first rolled chord with the pedal and then leave out the top notes for two chords that follow. This solution maintains the flavor of the harmony while preserving the rhythmic motive.

EXAMPLE 6.101. *Grieg, "Wedding Day at Troldhaugen," from* Lyric Pieces, *op. 65 no. 6, m. 40*

Chords that are manageable when played singly can become extremely problematic if a number of them have to be executed in quick succession, particularly in dotted rhythms. Either there is insufficient time to reposition the fingers or the added amount of joint fixation required for rapid repetitions can strain a hand that is already stretched to full capacity. Omitting the lowest note of the right-hand sixteenth-note chords in measure 52 and again in measure 56 (Example 6.102) promotes rhythmic vitality and can provide time for the thumb to release, briefly relax, and then reengage on the following quarter-note chord.

EXAMPLE 6.102. *Grieg, "Wedding Day at Troldhaugen," from* Lyric Pieces, *op. 65 no. 6, mm. 52 and 56, right hand*

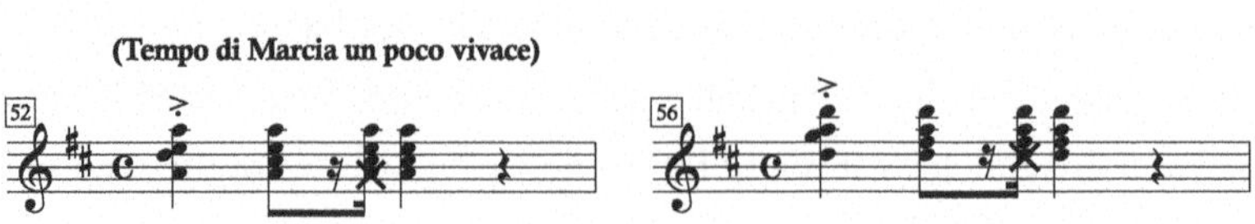

Chopin's polonaises and ballades feature athletic and virtuosic flourishes typical of those genres. Omitting notes may be necessary to enable a small-handed pianist to project the dotted rhythms with vigor and dynamic intensity (Example 6.103).

EXAMPLE 6.103. *Chopin, Polonaise in C♯ Minor, op. 26 no. 1, mm. 3–4*

Sometimes, skipping over notes buried in the middle of a thick chordal passage will pass unnoticed. At other times, note omissions can appreciably change the depth and color of the sound. Pianists will consequently need to weigh the musical costs against the physical benefits, which may in turn put into question the suitability of the repertoire.

Another persistent challenge concerns rapid shifts between large chords. A pianist may practice difficult sections for hours and see little significant improvement. By omitting notes, however, a pianist can sometimes sidestep the problem by eliminating the leaps entirely. In Example 6.104, consider leaving out the lower G in the right hand. Passages of this type can be found in the works of Brahms, Schubert, Rachmaninoff, and many others.

EXAMPLE 6.104. *Tchaikovsky, "February," from* The Seasons, *op. 37a no. 2, mm. 85–88*

Alternately omitting one note at a time from the left- and right-hand bitonal octaves in Tcherepnin's *Bagatelle,* op. 5 no. 1 still preserves the grating dissonance of the passage (Example 6.105).

EXAMPLE 6.105. *Tcherepnin, Bagatelle, op. 5 no. 1, m. 6*

Similarly, in the last two measures of Rachmaninoff's Prelude in B♭ Major, op. 23 no. 2, eliminating one note alternately from the right and left hands allows each hand to cover two sixteenth-note chords in a single placement instead of just one, as shown in Example 6.106.

EXAMPLE 6.106. *Rachmaninoff, Prelude in B♭ Major, op. 23 no. 2, mm. 60–61*

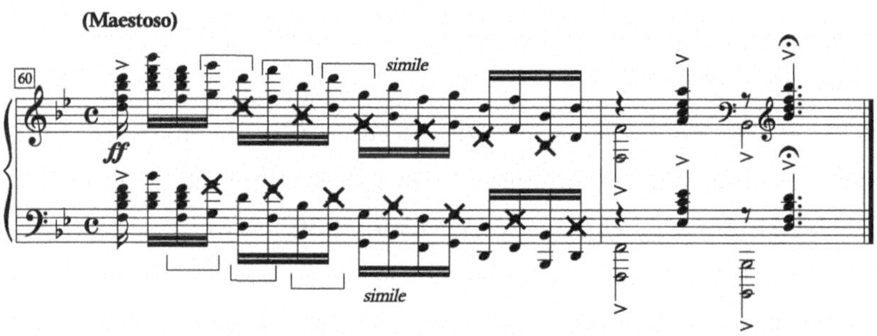

Chains of consecutive large intervals in accompanimental passages sometimes leave few options for the small-handed pianist. The most expeditious way to play the following passage from Chopin's F-Major Nocturne might be to simply leave out the top notes of the rolled tenths in order to preserve the serenity of the mood and to allow ample time for the bass to be caught securely in the pedal (Example 6.107). The omitted note will be played just two sixteenth notes later.

EXAMPLE 6.107. *Chopin, Nocturne in F Major, op. 15 no. 1, mm. 45–48*

Where omission is not a good solution, it may be necessary to transpose notes to a different octave or play an alternate chord tone. The sound of much nineteenth- and twentieth-century music is based on a tenth instead of an octave. Often, the only practical solution is to convert these intervals to thirds, sixths, or octaves.

The middle section of Rachmaninoff's famous C♯ minor Prelude, op. 3 no. 2 contains one tenth in a stream of rapid triplets (Example 6.108). Instead of awkwardly interrupting the flow to reach down for the problematic F♯ with the right hand or redistributing the F♯ into the left hand, an A can be substituted for the F♯. Even though this simple change generates parallel octaves between the As and the G♯s in the right-hand part, it preserves the rhythmic flow and tempo of the passage while simplifying its execution, allowing the pianist to concentrate on the dynamic inflection and projection of mood and melody.

EXAMPLE 6.108. *Rachmaninoff, Prelude in C♯ Minor, op. 3 no. 2, mm. 14–16*

Similarly, the tenths in the left-hand passage at the very end of Chopin's "Black Key" etude, which are impossible for a small-handed pianist to play at a rapid speed, are easily converted into octaves (Example 6.109).

EXAMPLE 6.109. *Chopin, Etude in G♭ Major, op. 10 no. 5 ("Black Key"), mm. 80–81, left hand*

The voices in the left-hand part of movement 3 from Schumann's *Davidsbundlertänze* expand briefly via contrary motion to tenths before contracting back to octaves (Example 6.110). Rolling all five chords would sound clumsy and be difficult to play. One can either convert the tenths (enclosed in boxes) to parallel thirds by moving the bass up an octave, or substitute octaves for the tenths by moving the top note of the left-hand chords down a third to double the bass notes.

EXAMPLE 6.110. *Schumann,* Davidsbundlertänze, *mve. 3, mm. 24–32*

A practical solution for playing the left-hand tenths in the quick and dramatic opening of Liszt's *Sonnetto 104 del Petrarca* or the right-hand tenths in the ninth piece of Prokofiev's *Vision fugitives* is to convert them to thirds (Examples 6.111 and 6.112).

EXAMPLE 6.111. *Liszt, Sonnetto 104 del Petrarca, mm. 1–4*

EXAMPLE 6.112. *Prokofiev,* Visions fugitives, *no. 9, mm. 1–2*

Readers might derive comfort from knowing that some composers recognized the challenges their music posed to small-handed pianists and were receptive to tasteful revisions that alleviated those difficulties. Chopin is a case in point. The most recent PWM edition shows alternate versions that Chopin noted in a student copy for the difficult accompaniment texture in the middle section of the Nocturne, op. 15 no. 1 (Example 6.113). In measure 25, Chopin eliminated one note, but in measures 27 and 29, he cleverly redistributed the notes of the tremolo.[20]

EXAMPLE 6.113. *Chopin, Nocturne in F Major, op. 15 no. 1, mm. 25, 27, and 29, right hand, alternate versions*

[20] Ekier and Kaminski, *Performance and Source Commentary to the Nocturnes of Fryderyk Chopin* (Krakow: PWM, 2010), 4.

A similar strategy could be used to rewrite the left-hand part of the *doppio movimento* section of Chopin's Nocturne in C Minor, op. 48 no. 1, especially in measure 50 of Example 6.114.

EXAMPLE 6.114. *Chopin, Nocturne in C Minor, op. 48 no. 1, mm. 49–50*
a. Printed version
b. Revision of left-hand part

(a)

(b)

The decision to revise notes might be also be motivated by concerns about voicing. In the third measure of the third variation from Mendelsohn's *Variations sérieuses,* the expansion of the right-hand part just prior to the downward descent of parallel first-inversion chords makes a difficult passage even more problematic (Example 6.115a). Revising the part as shown in Example 6.115b eases the stretch in the hand and makes it possible to voice the soprano more evenly.[21]

EXAMPLE 6.115. *Mendelssohn,* Variations sérieuses, *op. 54, Variation 3, mm. 51–52*
a. Printed version
b. Revision of right-hand part

(a)

(b)

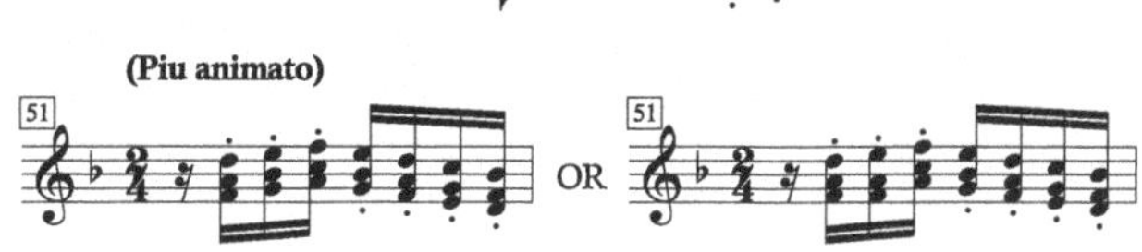

[21] The second option, shown in Example 6.115b, cuts down on the number of shifts between black and white notes with the second finger. This solution was proposed by Catherine Kautsky in a personal communication with the authors, April 2016.

As much as one would prefer to play the score as written, small revisions might be necessary for musical reasons. At climactic moments, a pianist might not be physically able to strike the keys with sufficient force if the hand is in an extended position. The *fortissisimo* chord tremolo at the end of Debussy's *L'isle joyeuse* could be revised by transposing notes into different octaves, slightly rearranging the notes, or as previously suggested in Chapter 4, redistributing notes between the left and right hands (Example 6.116).

EXAMPLE 6.116. *Debussy,* L'isle joyeuse, *m. 252–253*
a. Printed version
b. Alternatives

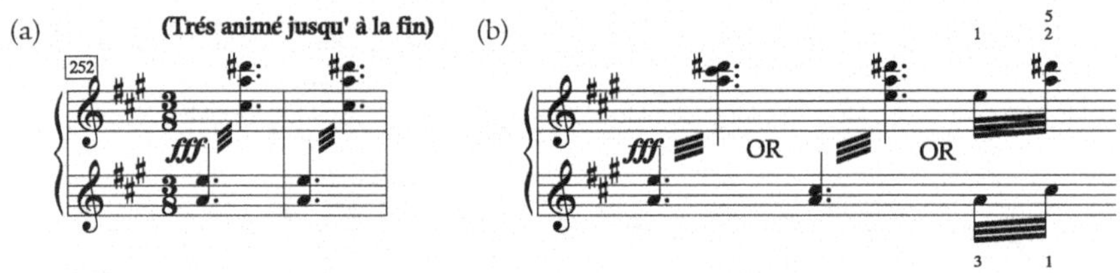

While omitting notes can be a strategy for dealing with large chord shapes, adding a note to a large chord can occasionally prove helpful. Rolling the awkward ninth in Example 6.117 would impede the steady rhythmic flow of the sixteenths. Instead, if the chord is otherwise out of reach, the right-hand thumb can fully extend to encompass both the A♭ and the B♭ next to it. The difference in terms of sound is scarcely perceptible.

EXAMPLE 6.117. *Beethoven, Sonata in C Minor, op. 13 ("Pathetique"), ii, mm. 11–12*

Omitting or revising notes can be a creative art. Bringing a score to life sometimes means briefly relinquishing one's allegiance to a note-for-note rendition in favor of a more flexible approach that still stays true to the spirit of the music. Many composers would agree that the creative process is an enterprise that is shared with the performer. The speed, dynamic level, and/or rapid directional changes of a passage can create impossible situations for small-handed pianists, situations that are non-existent for those with larger hands and that no amount of practice will resolve. Small revisions to the score may be required to maintain the tempo, flow, dynamic level, voicing, or character of a particular passage, not to mention the physiological and psychological poise of the performer.

CONCLUSION

The examples in this chapter speak to the numerous challenges facing small-handed pianists in terms of reach and power as well as the multi-faceted nature of those challenges. Complex problems do not admit to easy solutions. Rather, they are best approached from multiple perspectives. Therefore, the search for practical solutions that balance concerns about physical efficiency with concerns about musical values should be an ongoing pursuit for all pianists.

APPLICATION PROBLEMS

The following excerpts contain problems pertaining to reach and power. Find possible solutions by exploring the strategies presented in this chapter or by devising your own. Be sure to follow the basic principles of effective technique presented in Chapters 2 and 3.

1. Beethoven, Arrangement of Country Dance, WoO 114 no. 10, mm. 1–2: Which would be more effective for covering the large melodic interval in the right hand in measure 1—an underarc or a fan shape? Note how the choice of fingering can influence your decision.
2. Bach, Minuet in D Minor, BWV Anh. 132, mm. 9–12: For the leaps covering a tenth in the right hand, which do you prefer and why: (a) keeping the wrist in a position of ulnar deviation (turning the hand toward the fifth finger) in the circled passages and using the arm to move between notes; (b) using a complete oval gesture; or (c) using a fan-shaped motion and deviating the wrist from side to side (to the fifth finger for the A and to the thumb for the F)?
3. Debussy, "Pagodes," mm. 19–21 and "Soiree dans Grenade," mm. 98–105 from *Estampes*: What releasing strategies would facilitate the performance of these passages?
4. Tchaikovsky, "February," from *The Seasons*, op. 37a no. 2, mm. 9–12: Will rotation enable you to play this passage with accuracy and power? In which direction do the rotations move?
5. Rachmaninoff, Prelude in G Minor, op. 23 no. 5, mm. 70–73: Which notes could be justifiably omitted from this passage and why?

SEVEN

Musical Solutions for Technical Problems

THE STRATEGIES DISCUSSED IN PREVIOUS CHAPTERS CAN ALL BE described as physical interventions for counteracting the negative effects of small-handedness. The paramount goal of musical study, however, is to become more than a mere player of the instrument. Music is a language, and those with small hands struggle every day with central aspects of thinking and speaking in sound—how to forge a legato line from widely separated notes, how to deliver a powerful yet sonorous tone, how to voice textures spanning wide reaches of the keyboard. Chapter 7 seeks answers to these difficulties by looking to musical rather than technical solutions. At the heart of this discussion lies a paradox—that the underlying mechanics of the piano often conflict with the very effects pianists hope to achieve. Legato is sometimes best brought about by releasing notes, not holding on to them. Louder sounds result from increasing the acceleration of the hammer, not from dropping into the keys with as much weight as possible. Small-handed pianists must thus be keenly aware of what they want to express and consider the means by which they can best communicate their thoughts, emotions, and ideas in sound.

This chapter explores the following topics:

- Legato
- Fortissimo playing
- Pedaling
- Voicing
- Rhythmic inflection, tempo modification, and rubato
- Gestural shaping

LEGATO

Many pianists labor under the erroneous assumption that legato necessarily depends on a close physical link between adjacent notes, that is, keeping a note fully depressed until the next one is played. Depending on the context, physically connecting notes may be impractical or even impossible for a small-handed player, as in the case of octave melodies, tenor melodies played between the thumbs of both hands, or widespread left-hand accompaniments. While finger legato may be entirely feasible in music of the Baroque and Classical periods, it begins to be problematic and even counterproductive in music of the nineteenth century. Holding notes can place undue strain on muscles, tendons, and joints. It can lock the hand into place, forcing the pianist to awkwardly lurch from one note to the next. The pianist's ability to shape the line may be impaired if the arm cannot move into proper alignment smoothly and quickly.

Another common misconception concerns the idea of weight transfer to produce legato. At the end of the nineteenth and the beginning of the twentieth centuries, when the pendulum swung away from finger-centered pedagogies to those focusing on arm weight and relaxation, the notion of using weight transfer—rolling the weight of the arm from one note to the next to enforce a connection between them—became popular. Ortmann pointed out, however, that weight transfer is possible only at relatively slow tempos, and it gradually approaches zero as tempo and dynamic levels increase. His experiments further demonstrated that weight transfer is impossible when passing the thumb under for scales or arpeggios.[1]

Therefore, all pianists, and small-handed pianists in particular, must be careful to distinguish between physical and musical legato. In this light, it may be useful to think of legato not as a physical connection but rather as "an aural effect . . . achieved by thinking of connecting tones rather than fingers."[2] The piano can produce sounds of different lengths, from the shortest of *staccatissimos* to the longest of *tenutos* and everything in between. However, depending upon how closely the notes succeed each other, the involvement of the pedal, the articulation, and the dynamic shaping, even notes that are individually separate can sound as if they belong together. A skillful player can release keys between each and every pitch and still convey a sense of legato.

Legato in Widespread Cantabile Melodies and Arpeggiated Accompaniments

The skills that go into producing a cantabile melody take years to master. To sing over the accompaniment, the tone must be of at least moderate intensity and free of percussive

[1] Otto Ortmann, *The Physiological Mechanics of Piano Technique* (1929; reprint New York: E. P. Dutton, 1962), 134–148; Arnold Schultz, *The Riddle of the Pianist's Finger and Its Relationship to a Touch-Scheme* (New York: Carl Fischer, 1936; reissued 1949), 55–61.

[2] Audrey Schneider, "Dorothy Taubman: There Is an Answer," *Clavier* (September 1983), 19–21.

noises caused by hitting the fingers against the key surfaces. The individual notes must be perfectly matched to each other, because anything unexpectedly short, loud, or soft will disrupt the musical line. The connection between pitches, however, is largely accomplished by the pedal, often aided by larger shaping motions of the forearm and/or arm. It is the pedal that camouflages gaps and "glues" the notes together.

In fact, physical legato will clearly be impossible or impractical if the notes are too far apart for a pianist to reach. The melody shown in Example 7.1, for instance, begins with an expressive upward leap of a tenth. It would be counterproductive to hold the first note of the leap, B♭, until the last possible moment in order to make the gap when moving to the D as short as possible. Instead, the pedal should be applied and B♭ released in a leisurely fashion, allowing plenty of time for the arm to carry the fifth finger to the D♮. A shallow underarc serves to bridge the distance between the two notes.[3] In addition, rather than reaching for the D♮ with the fifth finger outstretched, the pianist can close up the right hand and brace the fourth and fifth fingers closely together so that the landing will be secure.

EXAMPLE 7.1. *Chopin, Nocturne in E♭ Major, op. 9 no. 2, m. 8*

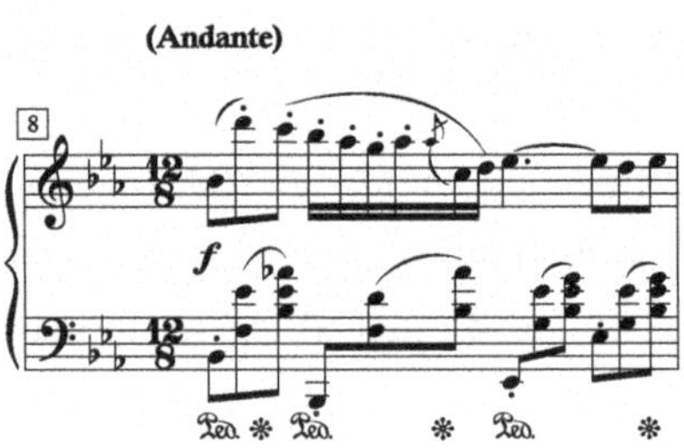

Even when melody notes are within reach, it may still be advisable to release them to improve alignment. Again, the work of connecting notes can be relegated to the pedal. In Example 7.2, to project the tenor melody, the opening F♯ may be released early so the arm can align behind the E♭.[4]

EXAMPLE 7.2. *Liszt, Piano Transcription of* Die Loreley, *S. 532, mm. 1–4*

[3] See Chapters 3 and 6 for detailed explanations and applications of underarcs and overarcs.

[4] Alternatively, a pianist could use passive spread to span between the notes. See Chapters 3 and 6 for detailed explanations and applications of this concept.

Ortmann discusses a cadenza that occurs later in the same piece, pointing out that at tempo, finger legato must be sacrificed to avoid disruptive lateral deviations at the wrist. The shifts (indicated by asterisks in Example 7.3) should be made by rapidly moving the whole arm as a unit without swiveling or twisting the wrist or rotating the forearm. The speed of the passage and the pedal both work to disguise any tonal gaps.[5]

EXAMPLE 7.3. *Liszt, Piano Transcription of* Die Loreley, *S. 532, mm. 21–22*

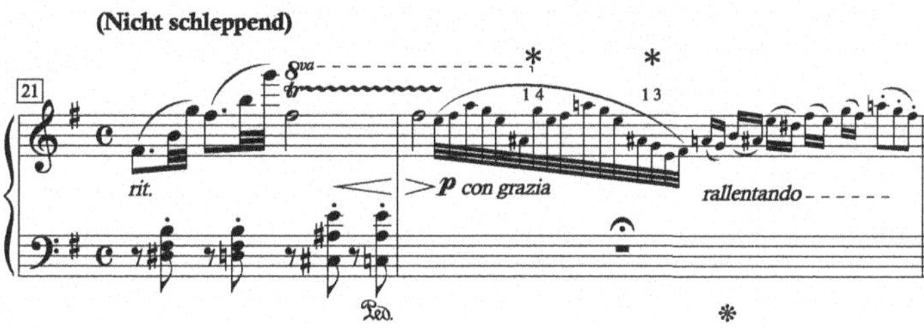

Many Romantic cantabile melodies should actually be played staccato by those with smaller hands. Releasing the melody notes of Schubert's Impromptu in G♭ Major (cited earlier in Example 6.56) and Brahms's Rhapsody in G Minor (Example 7.4) allows the hand to return to a compact position so that the triplet accompaniments can be executed with better rhythmic control. The pedal can be liberally applied, timing the changes to the bass. Under- and overarc gestures can give shape to the phrases.

EXAMPLE 7.4. *Brahms, Rhapsody in G Minor, op. 79 no. 2, mm. 1–4*

Schumann's "Glückes genug," shown in Example 7.5, could be played almost entirely non-legato to give the arm time to align behind the rapidly changing hand

[5] Ortmann, *Physiological Mechanics of Piano Technique*, 282–283.

positions.[6] Instead of trying to hold the eighth notes for full value, pianists should focus on bringing out the melody and shaping the melodic line. Generous use of the damper pedal can supply a wash of color to supplement dynamic shading.

EXAMPLE 7.5. *Schumann, "Glückes genug,"* Kinderszenen, *op. 15, mm. 1–4*

Small-handed pianists should also not hesitate to play alto or tenor melodies in the middle of the texture with multiple detached repetitions of the thumb. In Example 7.6, a pianist may release any of the right-hand notes of the alto line early to relieve uncomfortable stretches. Again, artful use of the pedal and subtle dynamic inflections are critically important to project the line.

EXAMPLE 7.6. *Brahms, Rhapsody in B Minor, op. 79 no. 1, mm. 106–112*

Legato in Octave and Double-Note Passages

To perform passages of legato octaves and double notes, multiple strategies come into play—fingering, early releases, and pedaling. Example 7.7 shows a fingering widely

[6] As stated in Chapter 5, many seconds in the right hand can be played together with the thumb.

recommended by a number of Chopin editors that encourages pianists to connect the upper notes of the right-hand melody using finger substitutions and 1-3 octaves. [7] However, small-handed pianists should play most if not all of the octaves with 1-5. In addition, players can further relieve stresses in the hand by voicing to the top note and releasing one or both notes quite soon after they have sounded. Connecting the octaves with the pedal instead of the fingers makes little to no difference in the aural effect.

EXAMPLE 7.7. *Chopin, Nocturne no. 1 in B♭ Minor, op. 9 no. 1, mm. 19–22*

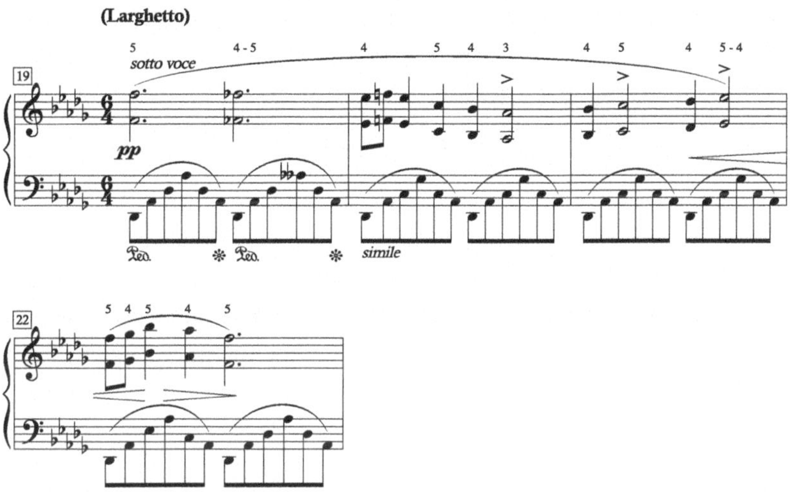

On the other hand, small-handed pianists should be mindful that notes cannot be released indiscriminately. For instance, in measure 10 of Example 7.8, the high E♭ should ideally be held by the fifth finger at least until after the pedal change on the second beat to prevent a significant gap in the melodic line.

EXAMPLE 7.8. *Beethoven, Sonata in C Minor, op. 13 ("Pathetique"), ii, mm. 9–12*

Chains of consecutive intervals can be played as a series of short detached chords sustained by the pedal even when they are located under slur marks. Releasing notes allows a

[7] Hermann Scholtz, ed., *Sämtliche Pianoforte-Werke,* Band 1 (Leipzig: C. F. Peters, n.d. [ca. 1905]). Other editions containing this fingering include those prepared by Theodor Kullak, Karl Klindworth, Carl Mikuli, Rafael Joseffy, and Alfred Cortot.

small hand to return to a more compact position after every chord, and these moments of respite are important in restoring a sense of comfort and control. Care should be taken to eliminate percussive attacks on the keys, however, and the chords should be dynamically matched (Example 7.9).

EXAMPLE 7.9 *Chopin, Nocturne in D♭ Major, op. 27 no. 2, mm. 75–77*

FORTISSIMO PLAYING

Although achieving a beautifully supported tone at soft dynamic levels is as much of an artistic problem as playing loudly, small-handed pianists are more likely to feel that they lack the power or endurance to make it through *fortissimo* passages.[8] Many will assume that they must hit the keys more forcefully to increase the volume of sound. However, pounding is not the best way to achieve a healthy fortissimo since the violent impact of the fingers against the keys degrades tone quality and has a deleterious effect on the body. As discussed in Chapter 2, fingers can deliver more force if they are at an angle approaching 90 degrees in relation to the key, that is, when they are curved. Unfortunately, small-handed pianists often need to play with flatter fingers to reach the notes of large chords and shapes. Those with small hands are also prone to keybedding—pressing downward into the keybed well after the sound has been produced. This is perhaps due to the perceived need to stabilize the fingers to generate a big sound. Keybedding, however, leads to undue tension in the muscles of the hand and forearm that can radiate upward into the shoulders, neck, and upper back.

Alternative approaches to fortissimo playing involve looking at the problem from both mechanical and artistic viewpoints. From the mechanical point of view, consider the following:

- Since volume is a product of both speed and mass, pianists can focus their attention on maximizing key speed—the rate of acceleration of the hammer toward the string—rather than weight. Rotations of the forearm, sometimes working in combination with rotations of the upper arm, are ideally suited to this task because these movements are swift and powerful.

[8] Problems of power and fatigue are discussed at length in Chapters 3 and 6.

- Another method of generating key speed is to suddenly unleash the energy of the entire arm through to the fingers in a whip-like key attack, starting right at the key surface. Afterward, instantly release muscular effort to baseline to avoid keybedding.
- When possible, take advantage of the bones of the hand to transmit force instead of trying to generate power exclusively through muscular contraction. This may mean extending the fingers straight out from the knuckles and playing from a higher wrist position so that the bones are skeletally aligned in as vertical a position as is practical for piano playing.
- If stretch is not a factor, as for example when a single note or a small chord needs to be played loudly, use as compact a hand shape as possible. Two or even three fingers can be used to play a single note, or a finger may be pressed closely against another, acting as a brace for additional support. Small-handed pianists with sufficient hand spans may choose to play octaves using the thumb plus the fourth and fifth fingers together at dramatic climaxes when maximum sound output is desired.
- The strong fingers of the hand should be favored, especially the thumb. However, any finger can be fortified by aligning it with the arm or by bunching it together with another finger or fingers (1-3, 2-3, or 1-2-3 for example).

From the musical point of view, remember these points:

- Volume is often built from the bass. Avoid overemphasis of the melody to the exclusion of the supporting bass parts.
- The damper pedal increases the richness and overall volume of tone. It can be liberally applied when stylistically appropriate, particularly to prolong fundamental bass notes underlying a passage.
- Pianists need to take advantage of resting places in the music (ritards, rests, fermatas, caesuras, and cadences) in order to release muscles and prevent the onset of fatigue.
- Dynamic markings do not represent absolute values; rather, dynamic levels should be calibrated so they fall within the capacity of the individual player. Small-handed pianists often overexert themselves in pieces that are skewed toward the upper end of the dynamic spectrum. Long passages of sustained fortissimo have a high physical cost and tend to be musically static as well. Subtle dynamic fluctuations inserted within lengthy fortissimo passages can allow for muscle recovery. When feasible, it is usually more effective to drop the sound back and gradually rebuild the fortissimo. One pianist to emulate in this regard is Liszt. According to Alan Walker, Liszt

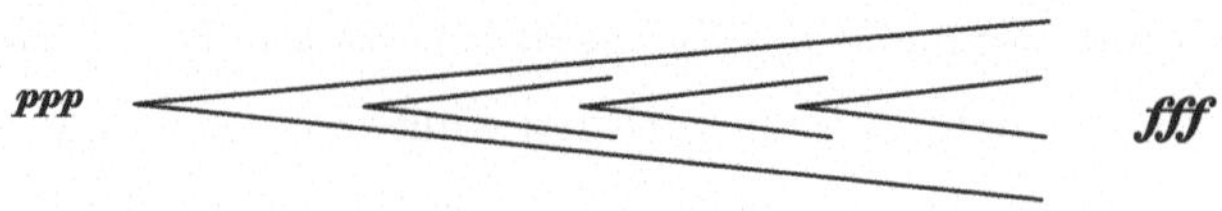

FIGURE 7.1. *Incremental crescendo. Credit: Matthew Bailey, graphic artist. © University of Nebraska-Lincoln. Used by permission. All rights reserved.*

felt that "even a simple crescendo or diminuendo could be transformed . . . into a vehicle of great expressive power" and consequently advised his students to practice building crescendos in incremental stages as shown in Figure 7.1.[9]

All of these strategies come into play when performing climactic passages such as the one shown in Example 7.10:

- For the staccato notes of measure 44, the wrists of both hands are set relatively high and the wrist and fingertips are stabilized. The forearm, hand, and fingers function as a single lever with the elbow as a fulcrum to deliver force to the key. Sound is built from the bass using long pedals to structure the crescendo.

EXAMPLE 7.10. *Grieg, "Wedding Day at Troldhaugen," from* Lyric Pieces, *op. 65 no. 6, mm. 44–50*

[9] Alan Walker, *Franz Liszt: The Virtuoso Years, 1811–1847*, Vol. 1 (New York: Alfred A. Knopf, 1983), 311.

- In measures 45-48, the second and third fingers are clustered together for extra support to play the first beats. Each measure can be shaped as a "Liszt" crescendo by scaling the dynamic level back on the second beat and building to the downbeat of the following measure. Starting at the surface of the keys, the pianist can use a whip-like arm attack to increase key speed for the loud chords and can insert rests immediately after each chord for muscle recovery. Crashing down upon the keys from a great height should be avoided. Long pedals will sustain the volume of sound.
- In measures 49-50, oval motions can be employed to shape the phrase. Feeling the underlying half-note or whole-note macrobeat as the basic unit of pulse will lead to more fluidity and ease, with smaller levers of the forearms and fingers taking over in the smaller subdivisions.

PEDALING

In his seminal book on pedaling, Joseph Banowetz observed that there were many factors to consider when making pedaling choices, among them melody, harmony, articulation, phrasing, tempo, dynamics, color, and desired overall musical effect.[10] Added to this list should be the potential of the pedals to help small-handed pianists overcome a wide array of handicaps. Two of the three standard pedals of the modern grand piano—the damper and sostenuto—are relevant to this discussion: the damper pedal can disguise potential gaps caused by unusual fingerings or connect notes that have to be released because they are out of the reach of the hand, and both pedals can be used to increase the volume of dramatic passages or sustain long bass notes.

Damper Pedal

The function of the damper pedal in relationship to legato was discussed at the beginning of this chapter and throughout Chapter 6. These sections showed how short applications of pedal can be used to connect notes or chords that cannot be reached with the fingers; specific issues with regard to the timing of the pedal were noted, particularly in relationship to rolled chords.

An equally important attribute of the damper pedal is its ability to amass sound. Early in its history, it was used to produce music of an ethereal quality attained by mixing conflicting harmonies together in the same pedal. The original pedalings of famous pianist-composers from the period can often be startling in terms of their length when applied to the modern piano. Famous examples include the first movement of Haydn's Sonata in

[10] Joseph Banowetz, *The Pianist's Guide to Pedaling* (Bloomington: Indiana University Press, 1985), 11. Banowetz's system of pedal markings will be used in this chapter.

C Major, H. 50/L. 60 (the "open pedal" sections in the development and recapitulation); the third movement of Beethoven's Sonata in C Major ("Waldstein"), op. 53 (the opening rondo theme); the first movement of Beethoven's Sonata in D Minor, op. 31 no. 2 (the recitative sections); and Chopin's Prelude in A Major, op. 28 no. 7.

However, Romantic and Impressionist composers such as Chopin, Liszt, Debussy, and Ravel widely exploited the pedal's ability to augment sound for the purpose of making dramatic musical gestures. Banowetz states that in the past this pedal was in fact frequently referred to as the "loud" or "forte" pedal even though the increased richness of tone that results when the dampers are lifted from the strings may be a more important benefit than any actual increase in volume.[11] Examples 7.11 and 7.12 show long pedals of one to six measures stipulated by the composer for prolonging the fundamental bass in order to maximize the output of sound.[12]

EXAMPLE 7.11. *Chopin, Ballade no. 3, op. 47, mm. 213–216*

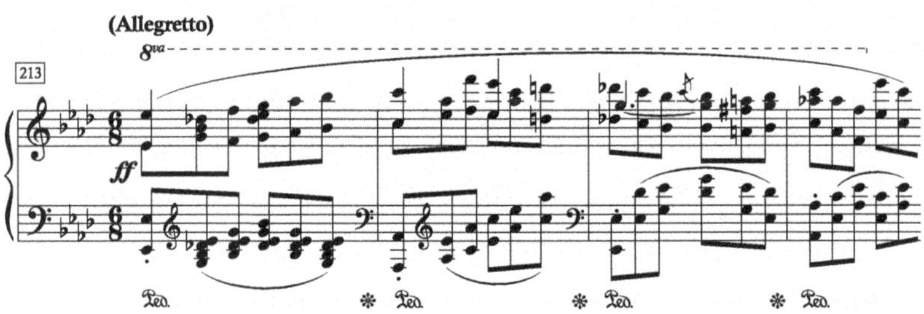

EXAMPLE 7.12. *Liszt, "Funérailles," from* Harmonies poétiques et religieuses, *mm. 1–6*

[11] *Ibid.*

[12] Dean Elder has written about the prolonged pedalings that Walter Gieseking used in a wide variety of Impressionist works, including Debussy's *Préludes, Estampes, L'isle joyeuse,* and *Pour le Piano* suite and Ravel's *Miroirs.* See Dean Elder, "Gieseking's Pedaling in Debussy and Ravel," in Banowetz, *Guide to Pedaling,* 230–287.

While pedaling must be adjusted according to the tone decay of individual pianos, the lesson that can be extracted is that small-handed pianists who are too conservative in their use of the pedal cannot gain full advantage from its use where power is at a premium. What Examples 7.10, 7.11, and 7.12 have in common are long forte or fortissimo bass notes that are sustained through non-harmonic passing notes and chords. In music of this type, contrary to conventional teaching that mandates a different pedal for every chord change, pedaling is sometimes independent of the harmony.[13] Small-handed pianists searching for ways to generate more sound should consider changing the pedal according to the bass, not the harmony. If too much dissonance accumulates, then partial pedal changes and/or flutter pedaling are often preferable to full pedal changes in order to retain a higher volume level.[14]

Pianists with small hands should also use the pedal to bolster the effects of long crescendos. In Example 7.13, it is possible to play the first two and a half measures in a single pedal in order to prolong the fundamental bass, E♭. A shallow half-pedal may be applied in the middle of measure 31 to allow for a slight reduction in sound level in preparation for the lengthy upward crescendo. As the scale in the right hand ascends, the pedal can physically mirror the progression of the crescendo and be depressed gradually to its maximum. If desired, a sudden release of the pedal at the very end of measure 32 will allow for a slight breath before the beginning of the next phrase, which starts on the downbeat of measure 33.

EXAMPLE 7.13. *Griffes, Scherzo, op. 6 no. 3, mm. 29–33*

As part of a general strategy for conserving energy, small-handed pianists should turn more frequently to the damper pedal to augment the sound of *sforzandos* and other bold accents. The famous pianist Alicia de Larrocha was quoted as saying that instead of deploying accented pedals right with the notes, she preferred to engage the pedal "before the note is struck in order to obtain maximum sonority."[15] This habit could be developed

[13] Banowetz, *Guide to Pedaling,* 194.

[14] For a detailed discussion of what happens during the downward journey of the damper pedal as well as partial and flutter pedaling, see Banowetz, *Guide to Pedaling,* 14–15 and 78–87.

[15] Mark Hansen, "The Catalan School of Pedaling," in Banowetz, *Guide to Pedaling,* 227.

from an early stage of piano study in pieces such as Schumann's "Hunting Song," "Wild Rider," or "Knecht Ruprecht" from the *Album for the Young*.

An additional function of the damper pedal is to alleviate problems of stretch. Many ostensibly simple pieces can be literally out of reach for small-handed pianists because they contain so many tenths. In Example 7.14 by MacDowell, half pedals can prolong the bass and provide enough clarity for the harmonic and melodic changes.[16]

EXAMPLE 7.14. *Half pedaling in MacDowell, "Sung Outside the Prince's Door," from* Forgotten Fairytales, *op. 4 no. 1, mm. 1–4*

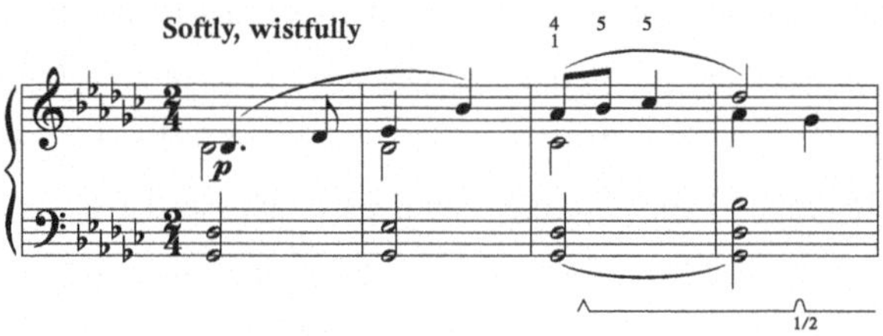

Example 7.15 shows a problematic situation where both the melody and accompaniment are in one hand. Many pianists will find it difficult to reach and hold the half notes in the first two measures of the right hand (C–E♭) for full value while at the same time voicing and phrasing the notes of the melody. A full pedal applied over the first two beats would overly prolong the high C and obscure the clarity of the two repeated Fs. However, a skillfully applied half-pedal change on the second beat would enable the right hand to release both half notes and use a more accommodating 5-2-1 fingering to articulate the opening motive.[17]

EXAMPLE 7.15. *Beethoven, Sonata in E♭ Major, op. 31, no. 3, i, mm. 1–4*

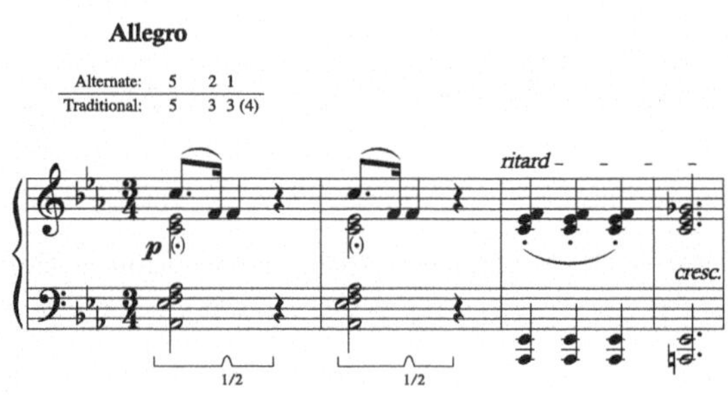

Pianists of all hand sizes struggle with the question of what to do with long sustained bass notes in Impressionist repertoire. Some pianists turn to the sostenuto pedal even though neither Debussy nor Ravel had one. For instance, Walter Gieseking, a widely

[16] Another solution would be to secure the low G♭ in the sostenuto pedal before the piece begins and hold it for four measures, a technique that is discussed in more detail later in this section.

[17] Alternatively, the pianist could release just the E♭ in the middle and use 5-3-2.

admired interpreter of French piano music, reportedly used the sostenuto to hold the low A in the "Prélude" from the *Pour le piano* suite, which freed his left hand to play the double-stemmed quarter notes. When no sostenuto was available, Gieseking held the low A in his left hand and played the rest of the notes with the right hand (Example 7.16).[18] For small-handed pianists, the entire passage is problematic for several reasons. The low A cannot be held for any length of time before the melodic notes in the tenor test or exceed the reach of the hand; the use of the sostenuto is risky to deploy because of the speed of the passage; and the redistribution of the sixteenths into the right hand becomes unsupportable at measure 14 with the introduction of double notes and tenths. Instead, pianists can substitute a simpler solution: securing the opening octave at measure 6 in the left hand with a full damper pedal, immediately releasing the fingers, and then making shallow quarter-pedal changes on the melody notes to clarify the texture. This will allow the bass to sustain throughout and envelopes the passage in an atmospheric haze entirely appropriate to the style. Partial pedal changes and/or flutter pedal are often simpler and more reliable than using the sostenuto.

EXAMPLE 7.16. *Debussy, "Prélude," from* Pour le piano, *mm. 6–7*

Sostenuto Pedal

As previously stated, the sostenuto pedal can be useful for prolonging long notes but might be cumbersome to deploy. Nevertheless, small-handed pianists may wish to use it so that stretches in the hand can be released. On occasion, the sostenuto can be activated significantly ahead of time. For the passage in Example 7.16, the low A could be secured six measures earlier, before the piece even begins, and held for 24 measures.[19]

Banowetz further suggests that the sostenuto might be useful for prolonging organ pedal points in Bach that would be impossible even for large-handed pianists. For the

[18] Dean Elder, "Gieseking's Pedaling in Debussy and Ravel," in Banowetz, *Guide to Pedaling*, 273–274.

[19] For the return of the A section later in the piece at measure 97, securing the sostenuto pedal early is not an option. To use the sostenuto, one would have to quickly release the A played in the left-hand thumb, immediately secure the sostenuto pedal, and then proceed with the rest of the passage using the damper pedal as usual. The tempo would slow down slightly, as these complex actions take a certain minimum amount of time.

end of the Fugue in A Minor from the first book of the *Well-Tempered Clavier,* he suggests securing the low A with the sostenuto in the rest right after the fermata in measure 80 and holding it for 8 measures to the end of the piece (Example 7.17).[20]

EXAMPLE 7.17. *J.S. Bach, Fugue in A Minor,* Das Wohltemperierte Klavier, *Book 1, BWV 865, mm. 80–87*

For small-handed pianists, the most relevant use of the sostenuto is its ability to prolong notes or chords so that the hand is freed to play other notes that are difficult or impossible to reach. Banowetz cites a classic case of a reach of a tenth that would be expedited by use of the sostenuto (Example 7.18). The low G would be secured in the sostenuto before the piece begins and released at the end of measure 2.[21]

EXAMPLE 7.18. *Beethoven, Rondo in G Major, op. 51 no. 2, mm. 1–3*

---

[20] Banowetz, *Guide to Pedaling,* 129.

[21] *Ibid.,* 104.

In Example 7.19, the A on the first beat of measure 1 in the left hand is often redistributed into the right hand by pianists who cannot reach the tenth between that note and the preceding grace note. However, by silently securing the low F with the sostenuto pedal before the piece begins, the left hand is free to release the grace note so it can move to play both the E♭ and A. Voicing is also facilitated since the right hand no longer has to project a melody that is located between two softer sustained chord tones. At the end of measure 2, the long held F in the right hand could be released a little early and the two B♭s—the one held by the right-hand thumb and the other in the left hand on the last eighth note of the measure—could be caught by the sostenuto pedal. This gives both hands complete freedom to play the notes of measure 3 in a way that will facilitate the projection of the melody and the subordination of the accompanimental parts.

EXAMPLE 7.19. *Schumann,* Kreisleriana, *vi, mm. 1–3*

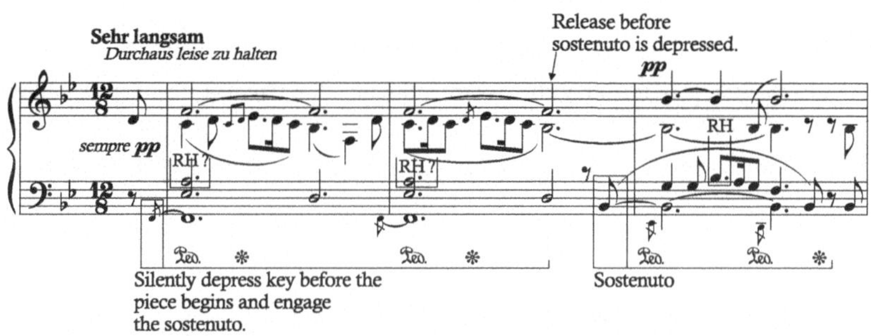

The sostenuto can prolong certain sonorities so that other notes can be played staccato, a situation that occurs in many works by Bartok. A small-handed pianist might choose to use the sostenuto in Example 7.20 specifically to relieve the left hand of the strain of holding the diminished octave when the minor seconds enter in measure 27.

EXAMPLE 7.20. *Bartok, "Free Variations," from* Mikrokosmos, *Book 6 no. 140, mm. 24–34*

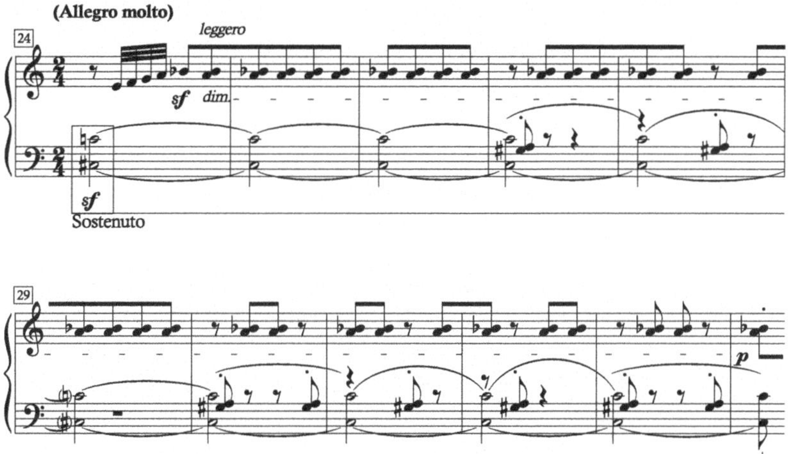

In Example 7.21, after releasing the open fifth chord of the left hand in measure 55, the pianist can take a bit of time to secure the high C in the right hand with the sostenuto pedal. This frees the right hand to play the repeated Gs.

EXAMPLE 7.21. *Bartok, "Six Dances in Bulgarian Rhythm," from* Mikrokosmos, *Book 6 no. 149, mm. 54–58*

In Example 7.22, using the sostenuto to secure the held notes in the right- and left-hand thumbs allows a pianist with small hands to attack the notes of the melody with confidence and power, particularly the accented low G in the left hand in measure 33.

EXAMPLE 7.22. *Bartok, "Six Dances in Bulgarian Rhythm," from* Mikrokosmos, *Book 6 no. 150, mm. 31–35*

The sostenuto pedal can also allow for note redistributions that would be advantageous for small-handed pianists. Securing the upper C♯ of the left-hand octave at the beginning of measure 17 allows the bottom notes of the right hand to be distributed into the left, relieving the stretches between inner fingers of the right hand caused by the parallel dominant seventh chords (Example 7.23).

EXAMPLE 7.23. *Debussy, "La soirée dans Grenade," from* Estampes, *mm. 17–21*

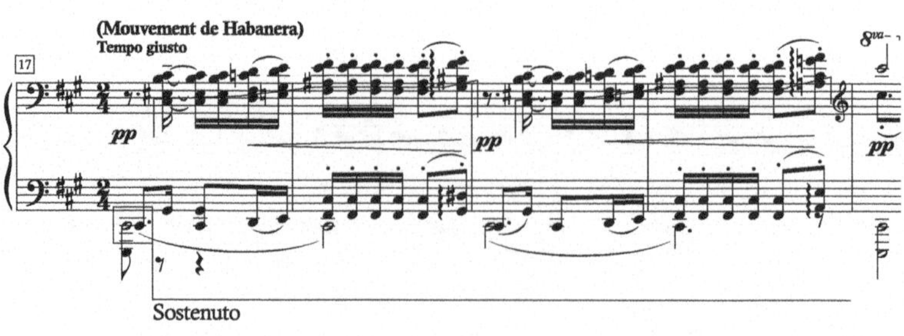

VOICING

Voicing—dynamically favoring one note over another—usually produces a musical result superior to playing simultaneously sounded notes at equal volume levels. Voicing can provide color and clarity to thick chordal passages, help distinguish a melody from its accompaniment, or allow the ear to follow several competing contrapuntal lines of texture. Depending upon the style and context, the dynamic differentiation between the parts may be relatively large or small. Because playing softer is generally less taxing, voicing provides both physical and musical advantages for small-handed pianists. When playing large chords, for example, rather than applying equal force to all the notes, those with small hands can selectively voice only the most important notes. These are, in approximate descending priority order, the melody and the harmonic bass, notes that establish the modality of the chord or its distinctive quality (the major or minor third, sevenths or other extended intervals, dissonant notes), and notes that are not doubled.

In order to voice a note, you must increase the speed of that particular keystroke. This can be done in a variety of ways. For example,

- The knuckle of the playing finger can be stabilized to a greater degree than others.
- The note can be played by a finger in a more upright skeletal position.
- It can be played with two or more fingers or by a finger that is braced by an adjoining finger.
- It can be played with the forearm and upper arm aligned more directly behind it.
- It can be played with a finger that is naturally stronger. The thumb is a prime candidate.
- It can be played fractionally earlier or later than the other notes.
- Forearm rotation can be directed to the favored note. Though any finger can be supplemented by rotation, the thumb or fifth finger are ideally suited to it due to their position at the ends of the hand.

In melody/accompaniment textures, voicing to favor the melody and harmonic bass note while subordinating all the others involves subtle dynamic distinctions. In Example 7.24, despite the overall ***p*** dynamic, both the melody and the bass would be projected in a full cantabile tone, perhaps ***mp*** or ***mf***, while the large chords on the second and third eighth notes of each dotted quarter beat could be played in the ***p*** or ***pp*** range. Ortmann established that widely outstretched fingers lose their ability to produce fine dynamic distinctions, so it is musically as well as physiologically important to use releasing or fingering strategies to maintain as compact a hand position as possible.[22]

---

[22] Ortmann, *Physiological Mechanics of Piano Technique*, 320.

EXAMPLE 7.24. *Chopin, Nocturne in E♭ Major, op. 9 no. 2, mm. 1–2*

Similarly, in Example 7.25, even though the overall dynamic level is specified as ***pp***, the melody could be voiced at ***mp***, the low G♭ of the bass at ***p***, the upper D♭ of the left hand at ***pp*** since it does not contribute to the color of the chord, and the internal triplets of the right-hand accompaniment at ***p*** or ***pp***. Thus, the melody will not be swamped by competition with the other notes.

EXAMPLE 7.25. *Schubert, Impromptu in G♭ Major, op. 90 no. 3, mm. 1–2*

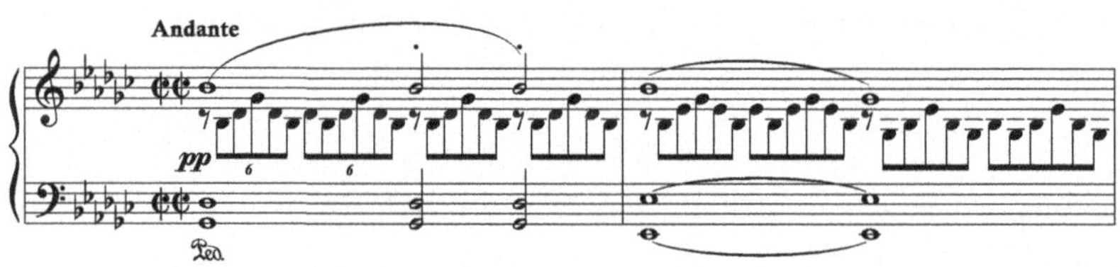

For the small-handed, it is crucial to voice double-note chords and octaves, a point noted earlier in this chapter. In the previously cited Example 7.4, the left-hand octaves can be voiced to favor either the top or bottom note instead of playing both at equal volume. The melody notes in the right hand would be projected at high intensity, but the second and third notes of the right-hand triplets would only be played at ***p*** or ***pp***, even though the overall dynamic is indicated as ***f***. In Example 7.9, voicing the top note of the parallel sixths at ***p*** or ***mp*** but the lower note at ***pp*** eases the physical strain of playing this lengthy series of double sixths. In Example 7.10, it is not necessary to play all the notes in the thick chords of measures 45-48 at full volume, especially since the depression of the damper pedal allows the strings to vibrate in sympathy with the harmony.

All these approaches work even better if the hand is in a more compact, less spread out position. Therefore, the small-handed pianist should implement strategies discussed in previous chapters—redistribution, refingering, early releases, note omissions or revisions, and like—to keep the hand as close to neutral as possible.

RHYTHMIC INFLECTION, TEMPO MODIFICATION, AND RUBATO

Rhythmic inflections in music, like the cadences of language, convey rich emotional and structural messages. Artistic playing, even in pieces that are primarily rhythmic, is virtually never strictly metronomic. Rather, it contains microscopic breaks or tempo fluctuations that reflect phrasings, note groupings, or muscular responses. In the right context, deliberately broadening the tempo over a wide leap or using rubato to bend time in an arching melody can add to the expressive impact of a performance. Rhythmic inflection is a musically valid way of resolving technical obstacles.

A small-handed pianist may deliberately decide to take more time to move through musical figures that stretch the hand. For example, if a chord is just within reach, it may take longer to position the fingers to reach the correct notes. If the chord is harmonically or melodically distinctive, the resulting agogic accent may be a desirable outcome. If not, then other courses of action such as note omissions or note redistributions can be explored. In other instances, a pianist might take extra time to accommodate a fingering with extra shifts or to align the arm behind loud notes. Inserting breaths or resting points within strenuous passages is another situation that may require the pianist to temporarily broaden the tempo.

Rhythmic inflections can also provide the time needed to move through leaps between melodic intervals or chords. For instance, the leaps at the beginnings of Examples 7.1 and 7.2 could be highlighted by an expressive rubato. Time taken in one location can usually be made up elsewhere.

Where the tempo must remain steady because taking time would sound ungainly or out of place, however, microbeats can be rhythmically compressed within larger macrobeats. In Example 7.26a, pianists can cluster the sixteenths in handfuls instead of spacing them evenly over the quarter-note beat. The passage could be reimagined as a succession of quickly rolled chords directed or accelerated toward the beat, as shown in Example 7.26b.[23] As a consequence, small gaps will open up for shifting and resting the hand.

EXAMPLE 7.26. *Beethoven, Sonata in C♯ Minor, op. 27 no. 2 ("Moonlight"), iii, mm. 1–2*
a. Printed score with added acceleration and breath marks
b. Conceptualizing the right–hand part as a series of rolled chords

(a)

(b)

[23] Catherine Kautsky suggested reimagining the even sixteenth notes as triplets directed toward the primary beats. Unpublished comments to the authors, April 2016.

Example 7.27 shows how reconceptualizing and compressing the notes in Chopin's F Major Etude, op. 10 no. 8, opens up spaces so the arm can shift from register to register, at the same time endowing the figure with a fleetness that might otherwise be absent. Note the slight change in the location of the breath marks when the arpeggio changes direction in the second measure. Time is provided exactly where it is needed; however, one must take care to avoid accenting the thumb. Note that at Chopin's stipulated tempo of 88 to the half note, the mathematical difference between notes that are a sixteenth or a thirty-second in duration is fractional.

EXAMPLE 7.27. *Chopin, Etude in F Major, op. 10 no. 8, mm. 1–2*
a. Printed score with added breath marks
b. Conceptualizing the right–hand part as a series of rolled chords
c. Approximate rhythm of the right–hand part

(a)

(b)

(c)

In Example 7.28, in lieu of elaborate fingering strategies, small-handed pianists could release at the points indicated and simply compress the inner arpeggiated notes.[24]

EXAMPLE 7.28. *Chopin, Etude in A♭ Major, op. 25 no. 1 ("Aeolian Harp"), mm. 26–27*
a. Printed score with added breath marks
b. Approximate rhythm of the right–hand part

(a)

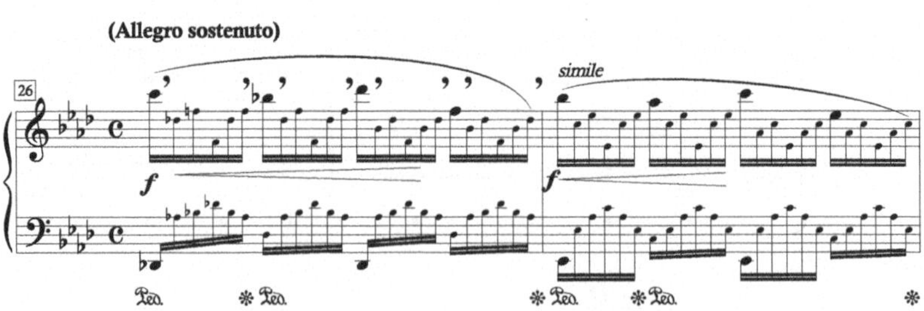

(b)

[24] A possible fingering approach to this passage is shown in Example 5.34.

A slight amount of rushing toward the end of the bracketed groupings in Example 7.29 yields enough time for the left arm to shift over the right. Fan shapes can be used in the left hand to facilitate the quick changes of direction.

EXAMPLE 7.29. *Scarlatti, Sonata in C Minor, K. 116, mm. 108–115*

Gaps between series of broken chords can be just as problematic as arpeggios. The compressions noted in Example 7.30 will give a pianist time to prepare for the fortissimo rotations on the primary beats.

EXAMPLE 7.30. *Beethoven, Sonata in C Major, op. 2 no. 3, i, mm. 97–98*
a. Printed score with added breath marks
b. Approximate rhythm of the right-hand part

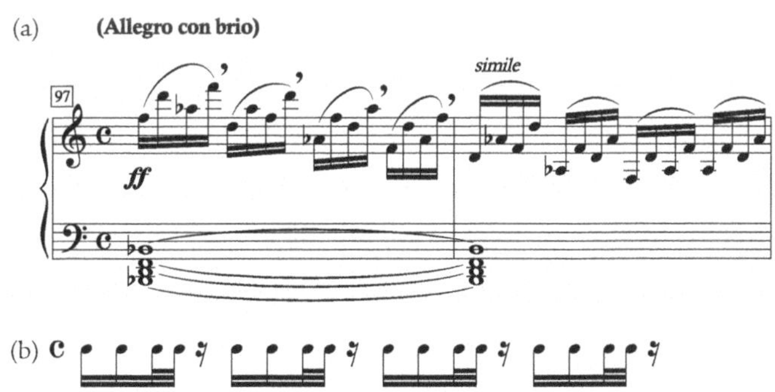

Similarly, compressing the right-hand chords slightly together in the coda of Chopin's G Minor Ballade (Example 7.31) heightens the anxious and impetuous effect of the music and opens up small moments where the muscles can rest.

EXAMPLE 7.31. *Chopin, Ballade no. 1 in G Minor, op. 23, mm. 208–212*
a. Printed score
b. Approximate rhythm of the right-hand part

(a)

(b)

Rhythmic compression can resolve conflicts between musical and technical goals, as for example when the need to maintain a steady beat clashes with the need for extra time. This happens frequently at cadences, when a thick final chord may be preceded by fast notes. The space allows extra time to position the fingers on the proper keys and to focus on voicing and tone quality. Examples 7.32 and 7.33 show two examples from Beethoven with a rough approximation of the suggested rhythmic execution.

EXAMPLE 7.32. *Beethoven, Sonata in C Minor, op. 10 no. 1, i, mm. 21–22*
a. Printed score
b. Approximate rhythm of the right-hand part

(a)

(b)

EXAMPLE 7.33. *Beethoven, Sonata in D Minor, op. 31 no. 2 ("Tempest"), i, mm. 119–121*
a. Printed score
b. Approximate rhythm of the left–hand part

In fact, rhythmic compression may be the only alternative when the pianist is playing with an ensemble and must keep a steady beat. In Example 7.34, playing the bracketed notes more like sixteenth notes than eighth notes creates time gaps within which to move between pitches that are up to a twelfth apart.

EXAMPLE 7.34. *Beethoven, Piano Trio in D Major, op. 70, no. 1 ("Ghost"), iii, piano part, mm. 35–49*

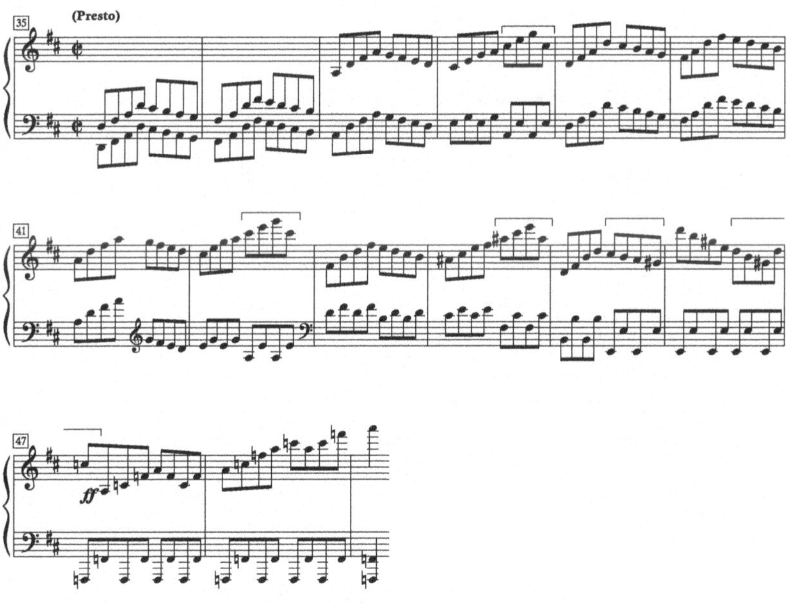

To maintain speed and tempo in Example 7.35, the player can sharply rotate the arm to the accented thumb note and let the momentum of the stroke carry the hand through the remaining three sixteenths, compressing the last two to gain time for the thumb to prepare for the next accent. The entire shoulder joint must remain free along with the collarbone. Any backward/forward, down/up, and side-to-side movements need to be minimized.

EXAMPLE 7.35. *Chopin, Etude in C♯ Minor, op. 10 no. 4, mm. 3–4*
a. Printed score marked with breathing spaces
b. Approximate rhythm of the right–hand part

(a)

(b)

Instead of compressing notes so more time can be taken to execute a leap, a pianist may prefer to start a phrase slightly under tempo and then quickly build up momentum to accelerate right through the leap. In this case, the first note is used as a springboard to propel the arm to the second pitch (compare Example 7.36 with the solution shown in Example 7.29).

EXAMPLE 7.36. *Scarlatti, Sonata in C Minor, K. 116, mm. 108–115*

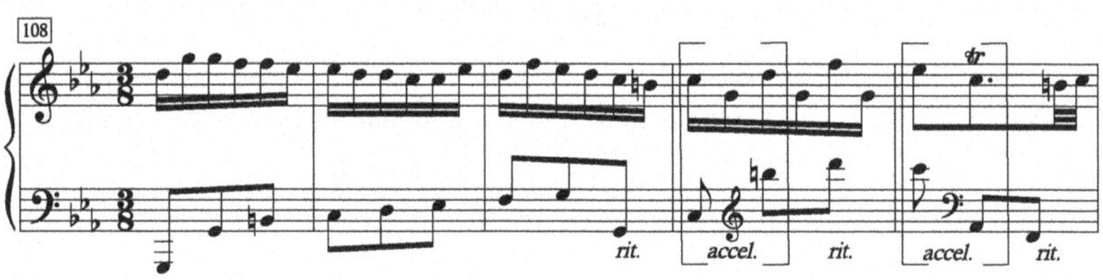

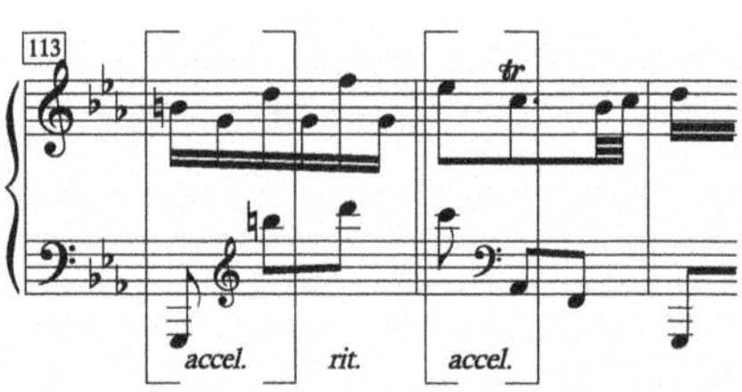

GESTURAL SHAPING

Just as the motions of smaller levers such as the fingers and hand can be subsumed into the movements of progressively larger levers, individual sound events can be molded into broader patterns through the use of larger gestures. In fact, the solution to recalcitrant technical problems is often to think in terms of longer phrases or lines.

Since larger levers move more slowly than smaller ones, the shaping motions they circumscribe in space, such as broad over- and underarcs or drop/lift motions, also tend to be larger and take more time to unfold.[25] For example, the octaves in Example 7.37 could be executed as quick hand staccatos integrated within a single underarc gesture of the arm. This strategy has both technical and musical advantages. It is tiring for pianists with small hands to play with the hand at maximum extension or to repeat the same movement many times. The arc slightly alters the position of the hand and arm for every chord, thus distributing the work. Since the arc provides an aural connection between the octaves, a fingering can be chosen that will accommodate the hand comfortably and reduce backward/forward movements between black and white keys. The end of the phrase tapers naturally as the weight of the arm is gradually lifted away from the keyboard.

EXAMPLE 7.37. *Schumann,* Papillons, *no. 1, mm. 1–8*

Shallow arc-shaped motions of the forearm guided by the upper arm and enabled by a flexible wrist can help to connect and inflect octave melodies, which most small-handed pianists would have to play with multiple repetitions of the fifth finger and thumb. This solution prevents the clunky vertical sound that results from separate downward drops on each chord. Example 7.38 shows how overlapping over- and underarcs can smooth

EXAMPLE 7.38. *Beethoven, Sonata in F Major, op. 10 no. 2, i, mm. 19–22*
a. Printed score
b. View of arm gesture from the frontal plane

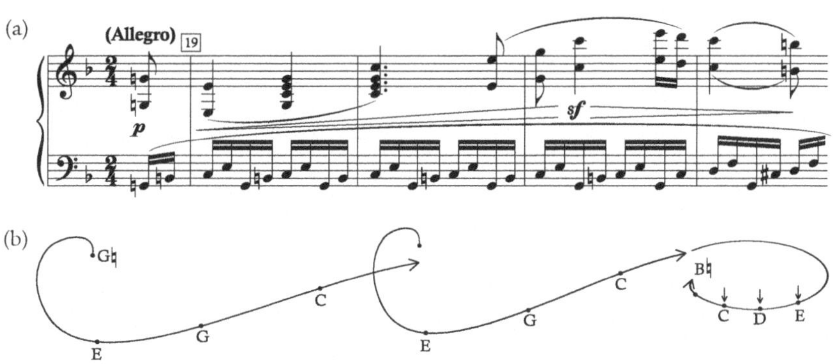

[25] Since it is impossible to accurately represent the complexities of movement at the piano in two dimensions, all illustrations in this section are only approximate representations. Most of the examples chosen involve

over the contour of the octave melody. Three quick hand staccatos for the octaves at the end of measure 21 leading into the first beat of measure 22 (indicated by vertical arrows in Example 7.38b) are integrated within the phrasing provided by the larger arm gesture.

Similarly, the melody notes of Example 7.39, which will need to be released early by the fingers, could be encompassed within overlapping overarc and underarc arm motions. In both Examples 7.38 and 7.39, the physical gesture marries the visual to the aural by underscoring the phrasing of the melody.

EXAMPLE 7.39. *Beethoven, Sonata in C Minor, op. 13 ("Pathetique"), ii, mm. 9–12*
a. Printed score
b. View of arm gesture from the frontal plane

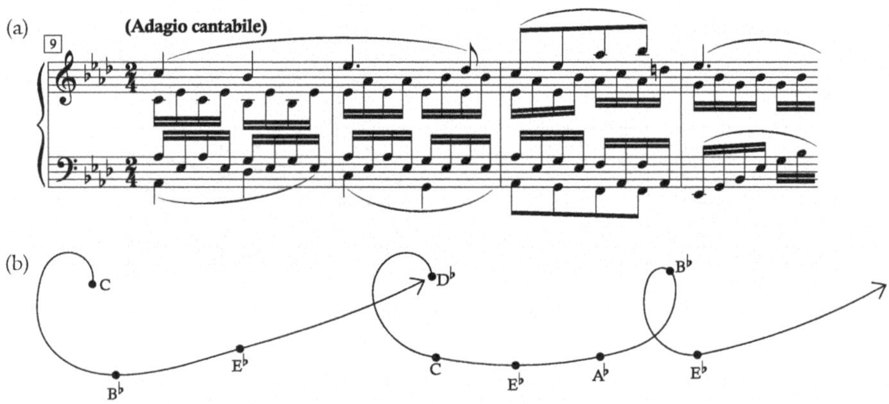

Gestural shaping can greatly facilitate the execution of lengthy and rapid scale or arpeggio passages, a matter of concern for all pianists. The five-octave scale at the end of Chopin's fourth Scherzo is a case in point (Example 7.40). One could first drop the weight of the arm forcefully on the E-major chord, release the chord immediately, and use the momentum of the stroke to ricochet the right hand down to the F♯. Then, the

EXAMPLE 7.40. *Chopin, Scherzo no. 4 in E Major, op. 54, mm. 961–967*

ascending passages so as to minimize the confusion of trying to illustrate a gesture for a descending passage: descending passages follow the layout of the keyboard by beginning on the right and proceeding to the left, whereas the score, which signifies the progression of time, reads from left to right.

fingers would take over for the playing of the scale, incorporating the notes into a sweeping underarc gesture of the arm, supported by a sympathetic movement of the torso.

Several forearm fans or rotational motions can also be subsumed into oversized ovals. Example 7.41b shows a representation of the larger gesture one might use to execute Example 7.41a.

EXAMPLE 7.41. *Debussy, "Reflets dans l'eau," from* Images, *Book 1, m. 23*
a. Printed score
b. View of arm gesture from the frontal plane

(a)

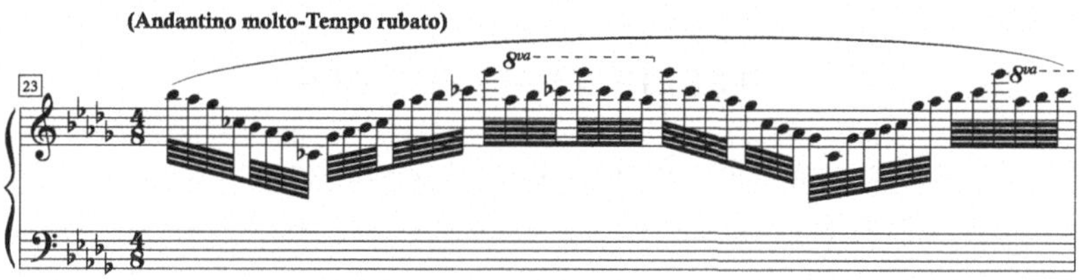

(b)

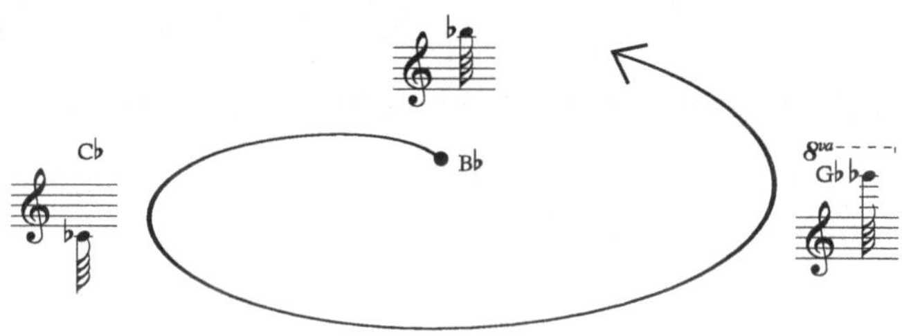

Drop/lift motions can also be used to physically organize difficult passages. Since the arm is heavy and thus moves relatively slowly, drops must be spaced comfortably far apart. For instance, instead of dropping equal weight on every eighth note of the melody in Example 7.42, each drop/lift cycle can be spread over two, three, or even four beats. Fingers, hand, and forearm can all work somewhat independently of the upper arm to realize the musical objectives of the performer.

EXAMPLE 7.42. *Grieg, "Wedding Day at Troldhaugen," from* Lyric Pieces, *op. 65 no. 6, mm. 49–50*

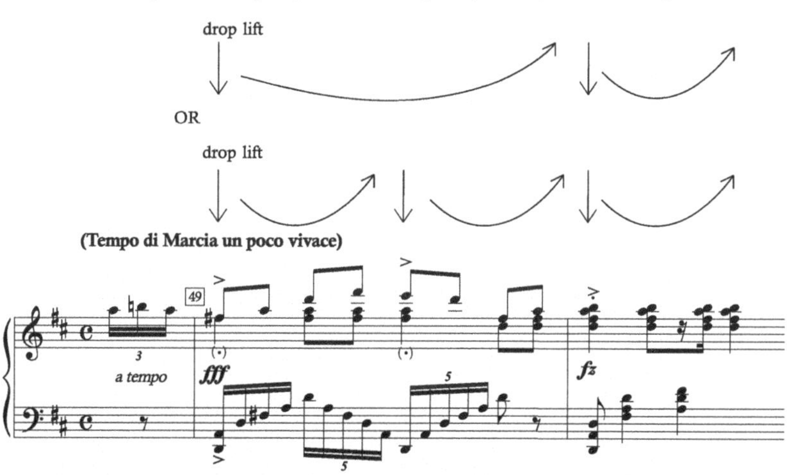

CONCLUSION

In a sense, the strategies of all the previous chapters converge into this one in terms of articulating an overarching goal—that of helping pianists to attain greater artistic freedom at the instrument. Despite the considerable challenges of small-handedness, the solution to technical obstacles is found not only in physical accommodations but also in learning how to bend and mold the elements of rhythm and tone—in learning how to make the notes on a page come alive. This is the work of a lifetime.

REFLECTION POINTS

1. Which of the strategies discussed in this chapter do you find particularly compelling? Why?
2. Do any of these musical strategies challenge what you have been taught in the past? In what way?
3. Can these strategies be applied to repertoire you are currently playing?

EIGHT

Integration and Exploration

THE PATH TO BECOMING A FINE MUSICIAN IS RARELY STRAIGHTFORWARD. It is more often a series of negotiations, filled with twists and turns, surprising tangents, abrupt dead ends, and stubborn roadblocks. All pianists must overcome barriers of one kind or another to succeed, but for those with small hands, the hurdles are particularly high. Many give up or are lost along the way. Small-handed pianists must accordingly know precisely what they can and cannot do with their hands. They must develop the ability to identify obstructions and anticipate possible threats. They must build upon their strengths and work around their weaknesses. By mastering a wide array of adaptive skills, they can escape the maze of small-handedness.

Small-handedness as an issue of pianistic concern has its roots in the nineteenth century, but it continues to be misunderstood centuries later. As explained in Chapter 1, its scope is multi-dimensional, encompassing aspects of ethnicity, gender, and age. The extent of ignorance about the dilemma of pianists who find that their hands are too small for the conventional keyboard, who lose their tactile connection with the instrument, and who expose themselves to increased risks of injury is still extraordinarily high. This book aims to provide some guidance, using principles gathered from the biological and physical sciences to devise healthful technical and musical solutions. The application of strategies such as redistribution, refingering, note releases, gestural shaping, omissions, and the like is demonstrated in the central chapters of this book. However, real-world musical contexts are fluid and dynamic, and these alternatives must be fully integrated to be truly effective. Once problems are diagnosed, multiple strategies may be required to address the difficulties inevitably encountered when playing repertoire of greater length and artistic complexity.

DIAGNOSIS AND INTEGRATION

Diagnosing issues related to small-handedness begins with a heightened awareness of kinesthetic cues. The tactile feedback that pianists receive from their own bodies should not be relegated to the background; rather, this feedback should play a critical role in structuring all musical decisions. Foremost among the warning signs that should never be ignored are fatigue and pain. Pianists should also be alert to situations where the hand feels stretched, where power is lacking, where inaccuracies pile up, or where tempo slackens. Indeed, small-handed pianists may be their own best teachers where problems of reach are concerned. In certain situations, they may need to overrule advice that they know to be physically inadvisable or ineffectual, even if it comes from a respected teacher or peer.

Tables 8.1 to 8.7 focus on broad categories where obstacles encountered by pianists with small hands can be found—chords, arpeggios, leaps, octaves, broken intervals,

TABLE 8.1 Chords

Problem	Adaptive Strategies
Fatigue	• Avoid keybedding (3). • Insert muscular releases between as many chords as possible (2, 6). • Return the hand to neutral between chords instead of remaining maximally extended for long periods of time (2, 3, 6). • For series of chords, reduce backward/forward movements and the vertical amplitude of the strokes (2, 3, 6). • Play less important notes more softly (7). • Release inner chord tones as soon as possible (6). • Group several chords within a single arm gesture (6). • Start under tempo and gradually accelerate into tempo (7). • Manage crescendos wisely (7). • Omit or revise carefully selected notes (6). • As a last resort, select different repertoire (3, 6).
Power	• Redistribute notes between the hands to spread the workload and/or eliminate inner stretches (4). • Use fingerings that allow the hand to remain more compact (5). • Direct force through the bones of the hand (2, 3, 6). • Use a larger lever (2, 3, 6). • Maximize key speed (2, 3, 6, 7). • Use impulses of the arm beginning right at the key surface (2, 3, 6, 7). • Build volume from the bass (7). • Increase volume by using longer applications of the damper and/or sostenuto pedals (7). • Build crescendos in incremental stages (7). • Selectively omit or rewrite notes according to the musical context (6).

TABLE 8.1 *(Continued)*

Problem	Adaptive Strategies
Reach	• Redistribute notes between the hands (4). • Use passive spread (3, 6). • Use vertical spanning (3, 6). • Roll notes or break chords into smaller groups of notes (6). • Selectively omit or rewrite notes (6). • Use a piano equipped with a reduced-size (ergonomically scaled) keyboard (1). • As a last resort, choose different repertoire (3, 6).
Speed	• Use smaller levers (2, 3, 6). • Avoid keybedding (3). • For a series of chords, reduce backward/forward movements and the vertical amplitude of the strokes (2, 3, 6). • Use a mixture of light rebounding hand strokes incorporated into larger arc-shaped arm gestures (3, 6, 7). • Release chords quickly and connect them with the pedal if necessary (6, 7).

TABLE 8.2 Arpeggiated passages or rolled chords

Problem	Adaptive Strategies
Fatigue	• Avoid keybedding (3). • Allow the hand to close up behind the arpeggio or rolled chord (3, 6). • Refinger the notes to favor stronger fingers (5). • Insert frequent resting points between note groups or phrases (6). • For long passages, reduce backward/forward movements and stay close to the keys (2, 3, 6). • Integrate smaller movements into larger gestures of the arm (3, 6). • Omit or revise notes according to the musical context (6).
Power	• Redistribute notes into smaller groups divided between the hands (4). • Refinger the notes to favor stronger fingers (5). • Use positional fingerings to achieve maximum power (5). • Release notes as soon as possible and move through distances with the arm (6, 7). • Use oval- or fan-shaped motions (3, 6). • Play the notes in arm impulse groups, rhythmically compressing them as needed to allow more time for shifts (3, 6, 7). • For long passages, reduce backward/forward movements and stay close to the keys (2, 3, 6). • Eliminate swiveling or twisting motions (3, 7). • Connect groups with longer applications of the pedal (7).

(*Continued*)

TABLE 8.2 *(Continued)*

Problem	Adaptive Strategies
Reach	• Redistribute notes between the hands (4). • Refinger arpeggios or chords, breaking them into smaller shapes and adding extra shifts (5). • Repeat the same finger to recover space in the hand (5). • Move longer fingers over shorter ones (5). • Substitute the thumb for inner fingers (5). • Use arcs and oval motions to span long distances (3, 6, 7). • Use passive spread to reach stretches between inner fingers (3, 6). • Use vertical spanning (3, 6). • Release notes and move through distances with the arm (6).
Speed	• Consider all possible fingerings, including ones with the thumb on black notes or with finger repetitions (5, 6). • Avoid actively reaching for distant notes with the fingers, and instead move through horizontal distances with the forearm, upper arm, and/or torso (6). • Use rhythmic compression in the notes just preceding a leap to gain more time for the shift (7). • Use positional fingerings to achieve maximum speed (5).

TABLE 8.3 Large leaps, hand shifts, and changes of direction

Problem	Adaptive Strategies
Fatigue	• Redistribute notes to eliminate leaps and share the workload between the hands (4). • Release notes as soon as possible, and blend them together with the pedal if necessary (6, 7). • Reduce the vertical amplitude of the strokes and stay close to the keys (2, 3, 6). • Omit or revise notes (6). • Select different repertoire (3, 6).
Power	• Use rotation to increase key speed (3, 6). • Use arcs or ovals to enhance skeletal alignment (3, 6). • Add hand staccato, forearm staccato, or arm impulses (3, 6).
Speed	• Redistribute notes between the hands to eliminate leaps (4). • Release notes played by the fingers and move through horizontal distances with the arm (6). • Use oval- or fan-shaped motions (3, 6). • Eliminate swiveling or twisting motions (3, 7). • Reduce backward/forward movements as well as the vertical amplitude of the strokes (2, 3, 6). • Omit or revise notes (6).

TABLE 8.3 *(Continued)*

Problem	Adaptive Strategies
Accuracy	• Reduce the size of backward/forward movements as well as the vertical amplitude of the strokes (2, 3, 6). • Use oval-shaped motions to move between distant notes (3, 6). • Use fan-shaped motions of the forearm for rapid, frequent changes of direction (3, 6). • Avoid actively reaching for notes with the fingers and instead find them with the arm aligned behind the playing finger (2, 3, 6). • Incorporate the notes into a larger overall gesture (3, 7).

TABLE 8.4 Octaves and other harmonic double notes

Problem	Adaptive Strategies
Fatigue	• Redistribute notes to divide the workload between the hands (4). • Insert muscular releases between chords as frequently as possible (2, 6). • Release one or both notes of the interval as soon as possible (6). • Group several strokes within a single arm gesture (3, 6). • Selectively omit or revise notes (6). • As a last resort, select different repertoire (3, 6).
Power	• Redistribute notes between the hands to allow for compact hand positions (2, 3, 4). • Refinger to favor stronger fingers and/or ultra-compact hand positions (5). • Use 1-5 as a default hand position for larger double intervals (5). • Selectively omit or revise notes (6). • As a last resort, select different repertoire (3, 6).
Speed	• Redistribute notes between the hands (4). • Release notes as quickly as possible and connect them with the pedal (6, 7). • Use smaller levers for quicker strokes (2, 6). • Group several strokes within a single arm gesture (3, 6). • Selectively omit or revise notes (6). • As a last resort, select different repertoire (3, 6).
Accuracy	• Redistribute notes between the hands to eliminate jumps or uncomfortable stretches (4). • Release one or both notes of the interval as quickly as possible (6). • Selectively omit or revise notes (6). • As a last resort, select different repertoire (3, 6).

TABLE 8.5 Broken intervals

Problem	Adaptive Strategies
Fatigue	• Release notes quickly (6). • Avoid reaching for distant notes, and instead use fan-shaped motions to carry the hand through large horizontal distances (6). • For a long series of broken intervals, subdivide into more manageable groups, using a larger arm gesture at the beginning followed by smaller rebounds (6). • Use the forearm to rotate between notes (2, 3, 6). • Organize forearm rotations into a larger arc-shape or down/up motion of the arm (6).
Power	• Use arm impulses on selected strong beats (3, 6). • For additional power, use active forearm rotations to strong fingers, especially the thumb (2, 6).
Speed	• Actively rotate in one direction only (6). • Release notes quickly (6). • Avoid reaching for distant notes, and instead use fan-shaped motions to carry the hand through large horizontal distances (6).

TABLE 8.6 Legato

Problem	Adaptive Strategies
Reach	• Redistribute notes between the hands to eliminate large skips or leaps (4). • Release notes and connect with the pedal (6, 7). • In appropriate circumstances, use finger substitution (5). • In appropriate circumstances, retake a released note silently (5).

TABLE 8.7 Voicing

Problem	Adaptive Strategies
Power	• Refinger so that the hand is in a more compact shape and so the fingers can retain the ability to produce fine dynamic gradations (3, 5). • Use 5 or 1 repeatedly to project inner or outer melodies, and connect the notes with the pedal (5, 6). • Align the arm to favor the finger playing the melody, and direct the force of the stroke against the bones (2, 3, 5, 6). • Use rotation to increase key speed for voicing (7). • Cluster several fingers together to play high-intensity single notes, such as structural bass notes (5, 6).

legato, and voicing. The tables list solutions drawn from throughout the book, providing a variety of approaches to each problem. The numbers noted within parentheses identify chapters containing information relevant to a specific strategy. The effectiveness of these tools in a particular context should be evaluated on the basis of sound, comfort, reliability, and ease of execution. Not every approach will work equally well for every pianist. Some problems are fixed rather easily while others may require extended experimentation. If a problem is incompletely resolved, players should ask whether the performance is at least better than it was before or whether further adaptations are needed. If a problem is intractable enough to resist every possible solution, however, the pianist should probably ask whether the piece should be set aside in favor of something more appropriate.

EXPLORATION

Integrating adaptive strategies into a player's technical vocabulary requires an intimate familiarity developed over time. As fluency in these skills develops, however, small-handed pianists will find themselves becoming increasingly adept at individualizing strategies to specifically address their own needs, choosing between feasible options, and using this knowledge to pursue alternative lines of thinking that allow them to fully realize the music they hear and feel within. Limited only by their imaginations, pianists may explore many different paths as they work through difficult technical and interpretive quandaries. Examples 8.1 to 8.6 reconstruct hypothetical journeys pianists might take to arrive at important interpretative decisions and to achieve a synthesis between technique and musical expression.[1]

[1] A list of abbreviations used in these examples can be found at the beginning of the book. The interpretive comments in Example 8.1 refer to teaching suggestions given by Clara Schumann to her daughter Eugenie Schumann. Eugenie Schumann, *Memoirs of Eugenie Schumann*, trans. Marie Busch (London: William Heinemann, 1927), 98–101. The 1-5, 3-2-1 fingering for measure 41 of Example 8.3 was suggested by Catherine Kautsky in a personal communication with the authors, April 2016.

EXAMPLE 8.1. *Intermediate: Schumann, "Hunting Song," from* Album for the Young, *op. 68 no. 7, mm. 1–12, 17–18, and 21–28*

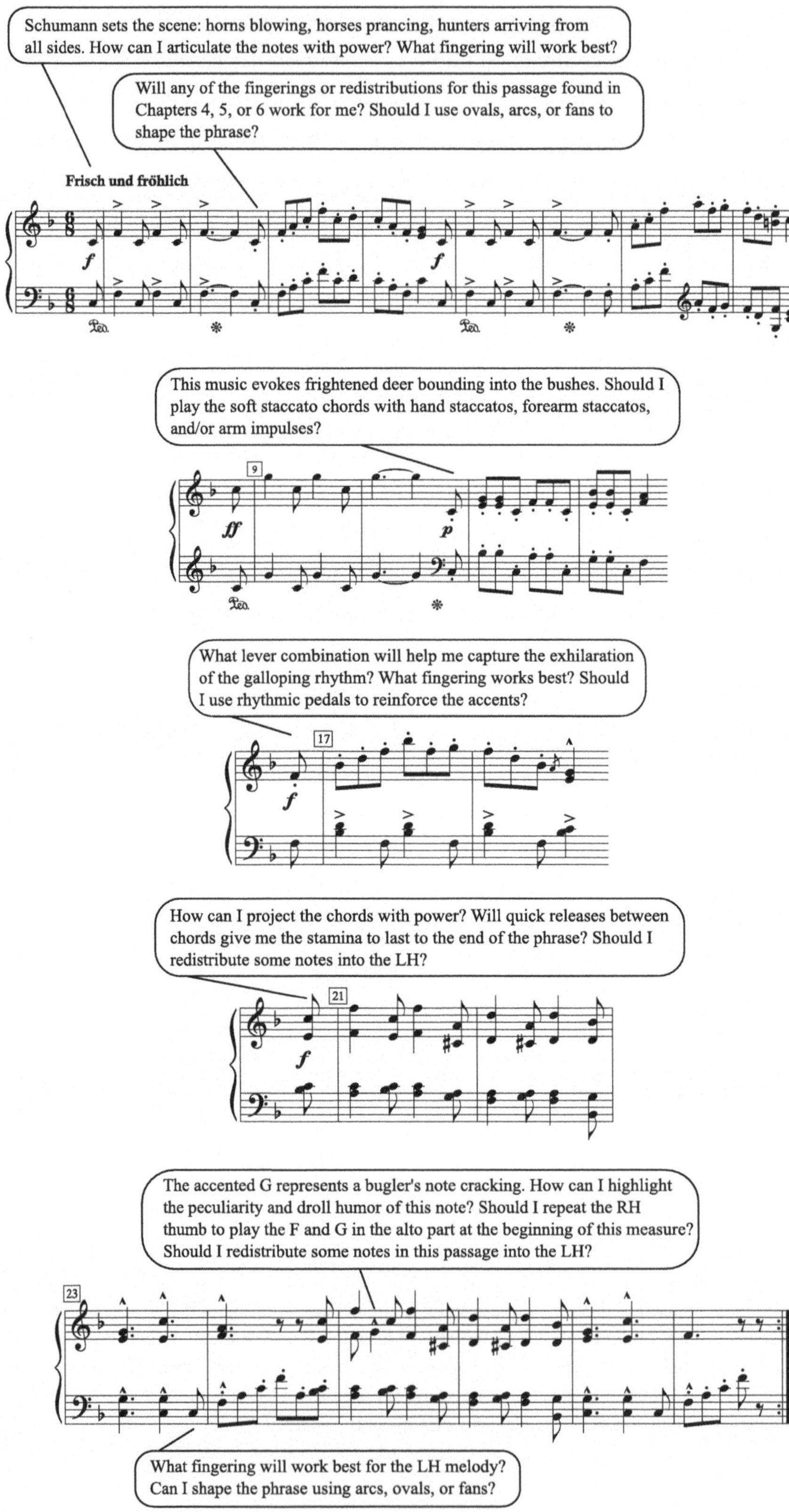

EXAMPLE 8.2. *Late Intermediate: Schumann, "Träumerei," from* Kinderszenen, *op. 15 no. 7, mm. 1–4, 6–7, 10–12, and 21–22*

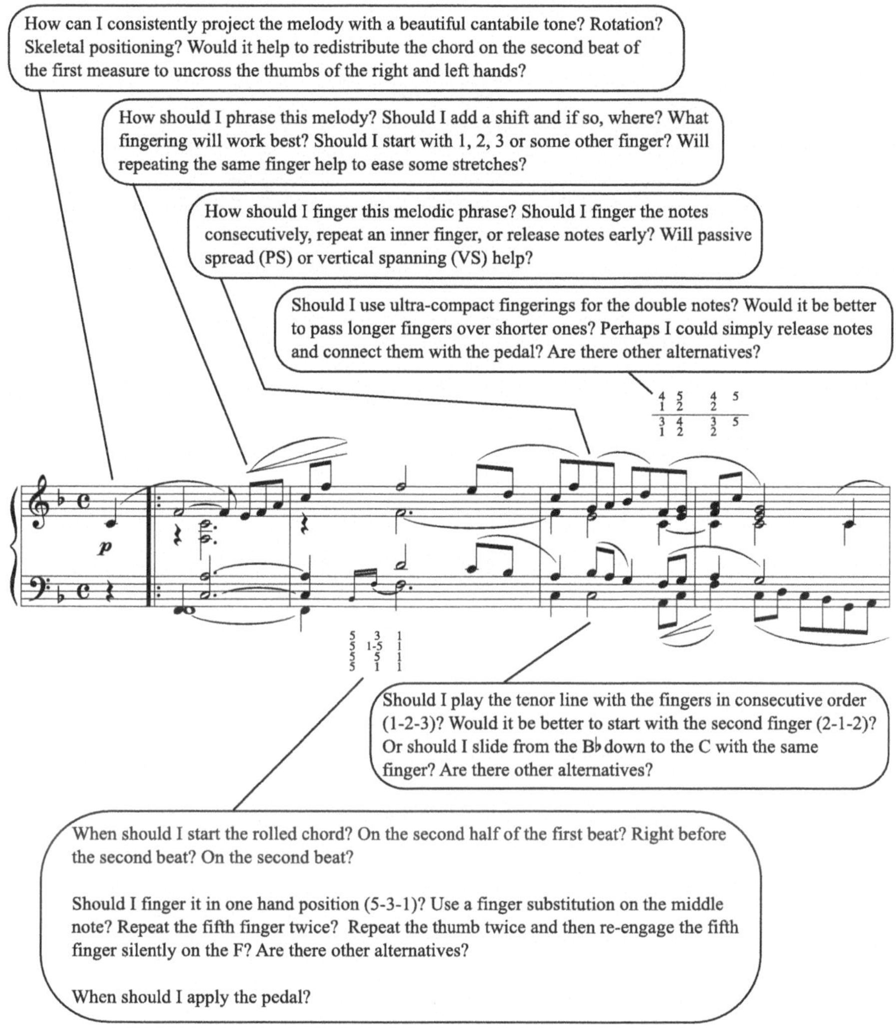

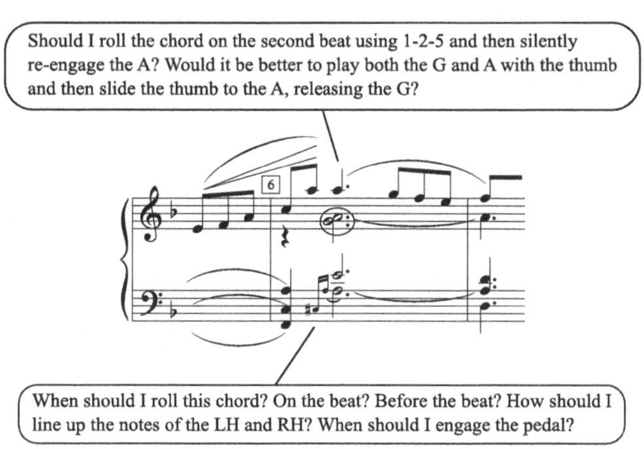

(Continued)

EXAMPLE 8.2. (*Continued*)

EXAMPLE 8.3. *Late Intermediate: Manuel de Falla,* Ritual Fire Dance (for solo piano), *mm. 40–43, 75–80, 115–118, 256–260, and 267–273*

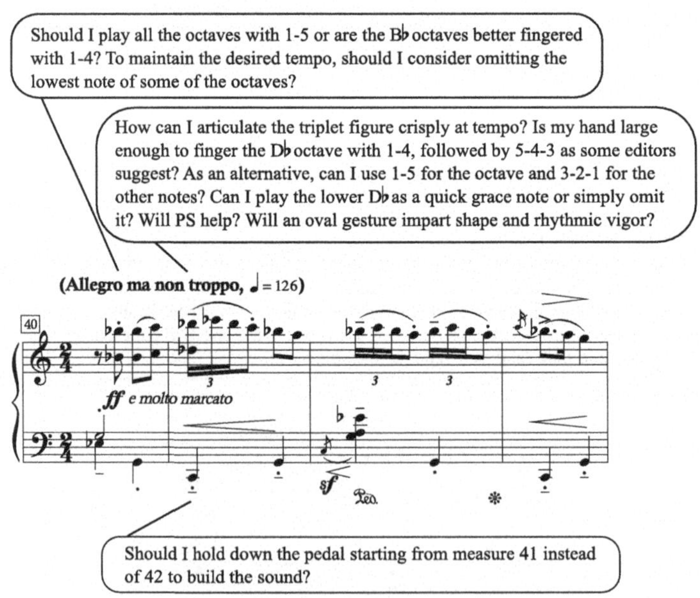

EXAMPLE 8.3. (*Continued*)

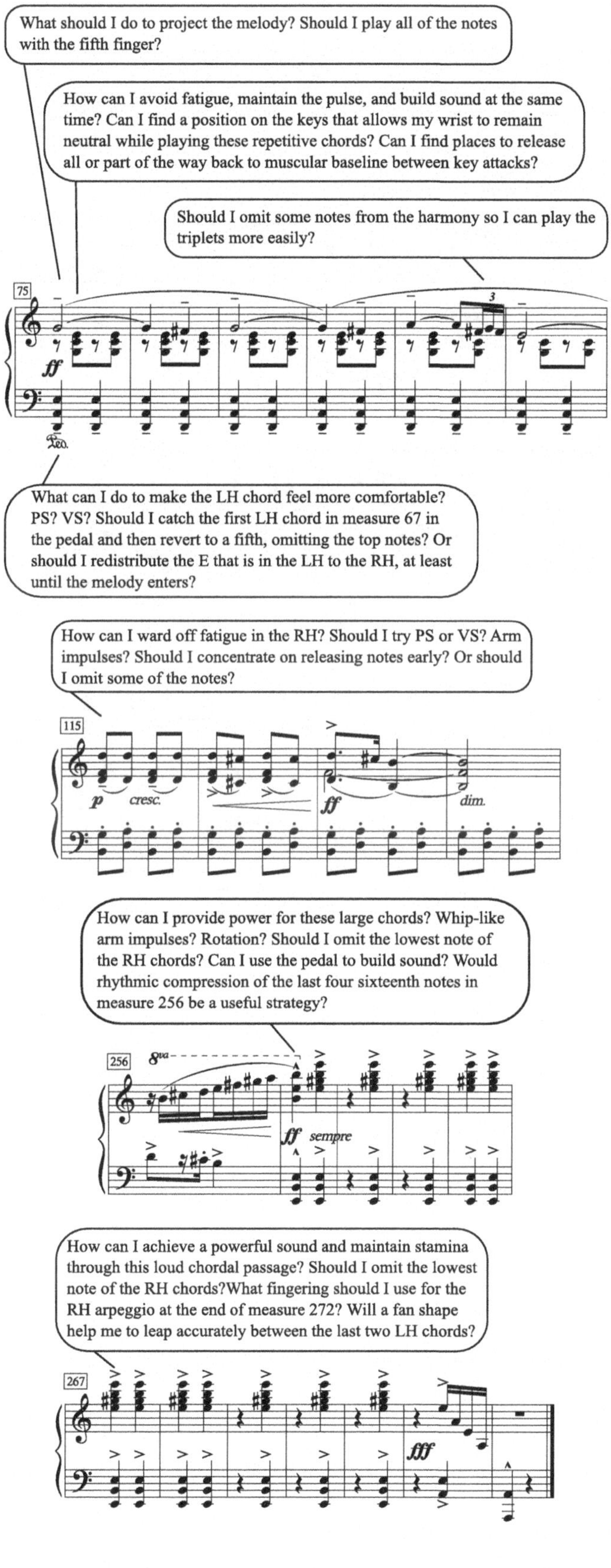

EXAMPLE 8.4. *Early Advanced: Grieg, "Wedding Day at Troldhaugen," from* Lyric Pieces, *op. 65 no. 6, mm. 49–56*

EXAMPLE 8.5. *Advanced: Chopin, Ballade no. 1 in G Minor, Coda, mm. 208–212, 217, 240–242, 255–56, and 258–264*

How can I project maximum sound in the RH and still conserve energy? Rotation? Skeletal positioning? Voicing some notes over others? Arm impulses? Do I need to use VS or PS to reach the notes of the chords? Is there a large gesture I can use to organize the phrase? Arcs? Down/up motions? Would rhythmic compression of the slurred groupings help? Where can I release effort to avoid fatigue?

How can I position my body? Where do my elbows need to be for greatest strength? How are my fingers placed over the notes? Is the wrist angled or not? What joints do I need to stabilize to maintain good posture? Can I minimize the size of my motions?

Can I shape the small components of the phrase dynamically to conserve energy? Can I start a little softer? If so, where should my crescendo begin? Can I start under tempo and accelerate?

Presto con fuoco

How can I leap between the LH chords accurately? Using fans? Ovals? Rotations? Can some chords be played more softly? Is there time to use a more compact fingering for the upper chords in the LH?

How long can I hold the pedal? Do I need to insert some half pedals for clarity?

How should I shift from octave to octave? Using rotation? Fans? Is VS useful for maximizing spread? Which notes or chords can be played more softly to conserve energy?

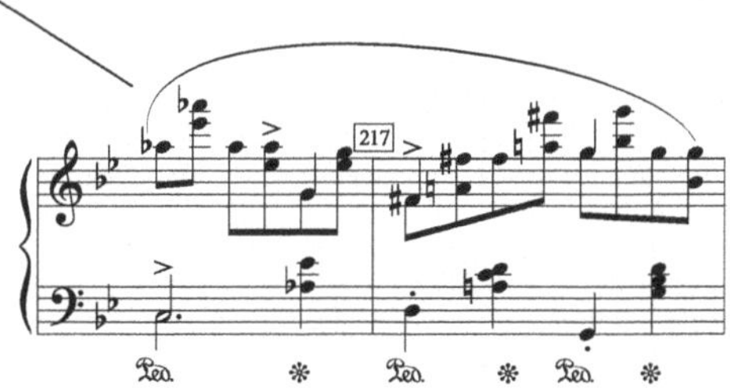

(Continued)

EXAMPLE 8.5. (*Continued*)

Do down/up motions work for these slurred octaves or would hand or forearm staccatos be better?

How can I play this unison scale with speed and power? If I incorporate the notes into a larger arm gesture, what is the shape of this gesture? Are the gestures of the RH and LH identical or are they mirror images of each other? What levers participate? The forearm? The upper arm? The entire torso? What role do the fingers play?

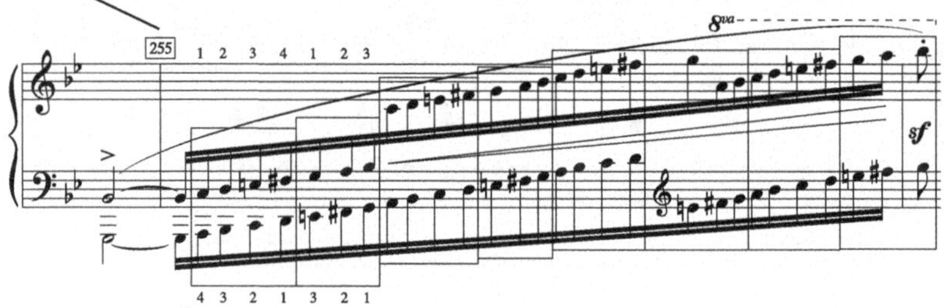

Would the positional fingering noted on the score help me achieve greater speed and power or not?

How can I achieve maximum power? Skeletal positioning? A higher wrist position? Arm impulses? Do I need to change to a smaller lever—hand staccatos or forearm staccatos—when the octaves accelerate? Is there an overall arm gesture I can use to organize the phrase? Does my torso need to change position? Which muscles need to provide stability and which need to remain free?

Should I finger all the octaves with 1-5 or not? How long can I keep the pedal down?

How can I play the chord in measure 263 with maximum power? By using a higher wrist position? Should I cluster the RH fingers together to play the single note in measure 264? Should I cluster 4 and 5 together to play the bottom note of the LH octave in measure 264? Will I achieve maximum sound by releasing the weight of the arm? By using an arm impulse? By using rotation?

EXAMPLE 8.6. *Advanced: Rachmaninoff, Concerto no. 2 in C Minor, i, mm. 63–70*

CLOSING THOUGHTS

Throughout this book, the authors have attempted to show how pianists can navigate the intricate agendas surrounding small-handedness and embrace the physical aspect of their pianism in a reasoned way. Certainly, performers of all hand sizes encounter difficulties at the piano. Pianists with very large hands and thick fingers, for instance, may find it nearly impossible to play between the black keys, negotiate quick scales and intricate figures, fit their hands into narrow spaces, or move their thumbs fluidly between black and white notes—actions that come more naturally to someone with shorter and thinner fingers. Even so, small-handed pianists face pervasive adversities due to the basic mismatch between their hands and the size of the standard keyboard, adversities that have been significantly discounted or underestimated by generations of musicians. The entire pianistic community can join in raising awareness about these concerns, sounding out cautionary tales and celebrating stories of success.

This book represents and builds upon the expertise and collected wisdom of many musicians past and present, but it is not an encyclopedic catalogue. Rather, the authors encourage pianists to confront small-handedness and to explore innovative solutions based on sound technical principles. This book seeks to challenge misperceptions about how the piano should be played—and by whom—in order to model an appreciation for a diversity of musical expression. Above all, it is grounded in the conviction that all musicians deserve a voice, a voice that will be bolder and richer if physical limitations are viewed as opportunities for creative exploration. Robert Schumann is said to have exhorted artists to aspire to send light into the darkness of men's hearts. This lofty goal is certainly within the reach of all pianists, no matter the size of their hands.

Glossary

Abduction moving away from the midline of the body or body segment.

Acceleration in classical physics, the rate of change of velocity per unit of time.

Adduction moving in toward the midline of the body.

Agonist the muscle(s) that primarily causes motion; often works in opposition to an antagonist.

Amplitude the magnitude of vertical distance above or below the level of the piano keys when using gestures at the keyboard.

Anatomic neutral in biomechanics, a position in which a person is standing fully upright, with arms fully extended with palms forward. For the purposes of this book, neutral (not anatomic neutral) is used to define the starting position in which the pianist is seated at the piano and in position for playing, with the hands resting on the keys.

Antagonist a muscle that opposes the agonist in a particular movement. When the agonist contracts, causing the movement to occur, the antagonist either relaxes so as not to impede the agonist or eccentrically contracts to slow down or stop a movement.

Baseline a state of muscular rest.

Biofeedback a process of training performers to bring automatic body functions or movements under voluntary control by using electronic monitoring.

Biomechanics the study of the structure and function of biological systems using classical mechanics, a branch of physics that involves analysis of the actions of forces.

Concentric contraction any contraction where the muscle shortens under load or tension.

Down/Up see Drop/Lift.

Drop/Lift one of the basic vertical gestures used to activate keys. The weight of the forearm is lifted slightly and then dropped into the key. The wrist does not drop below the level of the keybed. The weight of the arm is then lifted and suspended as the subsequent note or notes are played.

Dynamic loading muscle activity characterized by alternate contraction and release back to baseline.

Eccentric contraction any contraction where the muscle lengthens under load or tension.

Electromyography a technique for recording electrical activity of muscles using electrodes.

Ergonomics the study of human interaction with objects and environments.

Extension movement that increases the angle between two adjacent body segments; returns the segment back to neutral following flexion. May also be used to describe a position.

Extensors (hand/finger) muscles or tendons running the length of the back of the forearm that lift the hand or fingers back to neutral position following flexion. They may also be used to lift the hand or fingers beyond neutral position, a movement known as hyperextension.

Fan shape a gesture of the forearm observable in the transverse plane, used to cover distance in which the elbow does not travel through space.

Fixation keeping a joint between two or more levers stable immediately before and at the moment of key impact. Can be followed by immediate release and a natural upward follow-through motion.

Flexion bending that decreases the angle between two adjacent body segments (for example, bending the hand toward the underside of the forearm); the opposite motion of extension. May also be used to describe a position.

Flexors (hand/finger) muscles or tendons running the length of the underside of the forearm that allow for the fingers to be moved inward toward the palm and the hand to be moved toward the underside of the forearm.

Forearm staccato a coordination in which the forearm, hand, and fingers work as one lever, joined by a fixated wrist.

Force push or pull upon an object resulting from that object's interaction with another object.

Frontal plane one of the cardinal planes of motion in which movements made by the body may be observed, visible from behind the pianist.

Fulcrum the point on which a lever pivots.

Hand staccato a touch where the hand is actively flexed and extended by the muscles of the forearm.

Humerus the bone of the upper arm.

Hyperextension extension beyond neutral position or outside the normal range of motion.

Impulse a coordination involving the entire upper limb, where bursts of muscle activity generated by the upper arm are transmitted down the length of the arm with the elbow and wrist remaining flexible and the hand bouncing in response.

Inertia a property of matter by which it continues in its existing state of rest or uniform motion in a straight line unless that state is changed by an external force; the resistance of an object being accelerated when acted on by an external force.

Isometric contraction any contraction of muscles where little or no movement occurs.

Keybed the resting point of the key after it has been fully depressed.

Keybedding continuing to actively push into the piano key after it has reached its full depth of travel.

Key surface the place where the finger touches the key before it has been depressed.

Kinematics the study of classical mechanics that describes the motion of points, objects, and groups of objects without consideration of the causes of motion.

Kinesiology the study of human movement.

Mass a measure of how much matter is in an object; in everyday parlance, mass is often equivocated with weight.

Momentum in classical physics, the quantity of motion of a moving body, measured as a product of its mass and velocity.

Muscular bracing continuous involuntary contraction of muscles in anticipation of (or in response to) a perceived threat or challenge.

Neutralizer one of the roles that muscles may play, in which an unwanted motion is prevented so that the desired motion can occur.

Ovals one of the possible combinations of overarcs and underarcs in which the elbow and/or the wrist may travel through space.

Overarc (OA) a curvilinear gesture of the forearm beginning at the key surface which starts with an upward lift of the forearm that is extended horizontally, ending with a drop back toward the keybed; the opposite of an underarc (UA).

Over-rotation a movement in which the humerus rotates slightly in the shoulder joint to supplement the rotational range of the forearm.

Passive spread allowing the keys to separate the fingers rather than actively abducting the fingers.

Positional fingering fingering in positions or handfuls by repeating the same pattern over and over irrespective of whether the notes are white or black.

Pronation rotation of the forearm toward the thumb side of the hand.

Radial deviation a lateral positioning of the wrist out of neutral, with the fingers pointing toward the thumb side of the wrist.

Redistribution moving pitches notated for one hand into the other hand.

Release cessation of activity in a particular muscle. May also be referred to as muscular release.

Rotation turning motion of the forearm involving the crossing or uncrossing of the two bones of the forearm, the radius and the ulna, along the long axis. This motion would be used in turning a screwdriver. Rotation may be preceded by a short preparatory stroke in the opposite direction.

Sagittal plane one of the cardinal planes of motion in which movements made by the body may be observed, visible from either side of the pianist.

Skeletal positioning aligning various levers and stabilizing the intervening joints so that these levers are braced against each other, allowing force to be more directly transmitted to the keyboard.

Somatic training (or somatics) any approach that focuses the participant's attention on the internal physical perception of movement.

Stabilizer one of the roles that muscles may play in which the muscle contracts to hold a body part immobile while another body part is moving.

Static loading physical exertion in which the same posture or position is held throughout.

Supination rotation of the forearm toward the fifth finger side of the hand.

Swiveling a maladaptive tendency involving repetitive lateral deviations at the wrist within a short period of time.

Tension the development of contraction in a muscle or muscles necessary to perform work.

Transverse plane one of the cardinal planes of motion in which movements made by the body may be observed, visible from above looking downward.

Twisting a maladaptive tendency in which the pianist attempts to connect distant pitches by lifting the upper arm away from the body and rotating the forearm.

Ulnar deviation a lateral positioning of the wrist out of neutral, with the fingers angled toward the fifth finger side of the wrist. When the hand is positioned with thumb on middle C, it is in a position of ulnar deviation.

Ultra-compact fingering a fingering that allows the hand to be used in a smaller-than-neutral position.

Underarc (UA) a curvilinear gesture of the forearm which starts with a downward drop toward the keybed that is extended horizontally, ending with a lift; the opposite of an overarc (OA).

Vector a quantity having direction as well as magnitude.

Vertical spanning a relationship between two fingers which involves flexing one finger while extending the other instead of abducting the fingers apart horizontally.

Whole arm the lever formed by the upper arm, the forearm, the hands, and the fingers, with all the joints in between fixated immediately prior to and at the moment of key attack.

Bibliography

Alexander, F. M. *The Use of the Self*. London: Methuen, 1932.

Arx, Victoria A. von. *Piano Lessons with Claudio Arrau: A Guide to His Philosophy and Techniques*. New York: Oxford University Press, 2014.

Banowetz, Joseph. *The Pianist's Guide to Pedaling*. Bloomington: Indiana University Press, 1985.

Bartolome, Sarah J. "Naturally Emerging Self-Regulated Practice Behaviors among Highly Successful Beginning Recorder Students." *Research Studies in Music Education* 31, no. 1 (2009): 37–51.

Bernstein, Seymour. *With Your Own Two Hands*. New York: Schirmer, 1981.

Boyle, Rhonda. "The Experience of Playing Reduced-Size Piano Keyboards: A Survey of Pianists." *Music Teachers National Association e-Journal* 3, no. 4 (April 2012): 2–20.

Boyle, Rhonda, and Robin Boyle. "Hand Size and the Piano Keyboard: Literature Review and a Survey of the Technical and Musical Benefits for Pianists Using Reduced-Size Keyboards in North America." *Proceedings of the 9th Australasian Piano Pedagogy Conference*, Sydney, Australia, July 2009. Accessed August 5, 2015. www.appca.com.au.

Boyle, Rhonda, Robin Boyle, and Erica Booker. "Pianist Hand Spans: Gender and Ethnic Differences and Implications for Piano Playing." Presented to the Australasian Piano Pedagogy Conference, Melbourne, July 2015. Accessed August 1, 2015. www.appca.com.au.

Burger, Ernst. *Robert Schumann: A Chronicle of His Life in Pictures and Documents*. In *Robert Schumann: New Edition of the Complete Works*, Series VIII: Supplements, Vol. 1 (Main: Schott, 1998), 1831.

Busoni, Ferrucio. *Klavierübung in zehn Büchern*, Vol. 1. Leipzig: Breitkopf und Härtel, 1925.

Campbell, James Methuen. *Chopin Playing: From the Composer to the Present Day*. New York: Taplinger, 1981.

Chung, In-Seol, Jaiyoung Ryu, Nobuki Ohnishi, Bruce Rowen, and Jeff Headrich. "Wrist Motion Analysis in Pianists." *Medical Problems of Performing Artists* 7 (March 1992): 1–5.

Clark, Frances. *Questions and Answers*. Northfield, IL: Instrumentalist, 1992.

Conable, Barbara, and Benjamin Conable. *What Every Musician Needs to Know about the Body*. Portland, OR: Andover Press, 2000.

Cortot, Alfred. *Rational Principles of Pianoforte Technique*. Translated by R. Le Roy-Metaxas. Paris: Editions Salabert, 1930.

Cortot, Alfred, ed. *Chopin: 12 Études, Op. 10*. Paris: Éditions Maurice Senart, 1915.

Deahl, Lora, and Brenda Wristen. "Strategies for Small-Handed Pianists." *American Music Teacher* 52, no. 6 (June/July, 2003): 21–25.

Deahl, Lora, and Brenda Wristen. "What Are Some of the Problems Faced by Small-Handed Pianists?" *Keyboard Companion* 13, no. 3 (Autumn 2002): 16, 18.

De Smet, L., H. Ghyselen, and R. Lysens. "Incidence of Overuse Syndromes of the Upper Limb in Young Pianists and Its Correlation with Hand Size, Hypermobility, and Playing Habits." *Annals of Hand and Upper Limb Surgery* 17, no. 4 (1998): 309–313.

Dugan, Sheila A., and Walter R. Frontera. "Muscle Fatigue and Muscle Injury." *Physical Medicine and Rehabilitation Clinics of North America Journal* 11, no. 2 (May 2000): 385–403.

Eger, Patricia. "A Study of the Problems Encountered by the Pianist with Small Hands and a Compendium of Practical Solutions." DA dissertation, Ball State University, 1982.

Eigeldinger, Jean-Jacques. *Chopin: Pianist and Teacher as seen by His Pupils.* London: Cambridge University Press, 1986.

Ekier, Jan, and Pawel Kaminski, eds. *Performance and Source Commentary to the Nocturnes of Fryderyk Chopin.* Krakow: PWM, 2012.

Ekier, Jan, and Pawel Kaminski, eds. *Performance Commentary and Source Commentary to the Sonatas of Fryderyk Chopin.* Krakow: PWM, 2010.

Farias, J., F. J. Ordonez, M. Rosety-Rodriguez, C. Carrasco, Al Ribelles, M. Rosety, J. M. Rosety, and M. Gomez del Valle. "Anthropometrical Analysis of the Hand as a Repetitive Strain Injury (RSI) Predictive Method among Pianists." *Italian Journal of Anatomy and Embryology* 107, no. 4 (2002): 225–231.

Fink, Seymour. *Mastering Piano Technique: A Guide for Students, Teachers, and Performers.* Portland, OR: Amadeus Press, 2003.

Fraser, Alan. *The Craft of Piano Playing: A New Approach to Piano Technique.* Lanham, MD: Scarecrow Press, 2011.

Fraser, Alan. *Honing the Pianistic Self-Image.* Montreal: Maple Grove Productions, 2010.

Fry, H. J. H. "Overuse Syndrome in Musicians—100 Years Ago: An Historical Review." *Medical Journal of Australia* 145 (December 1986): 620–625.

Fry, H. J. H. "Prevalence of Overuse (Injury) Syndrome in Australian Music Schools." *British Journal of Industrial Medicine* 44 (1987): 37.

Furuya, S., H. Nakahara, T. Aoki, and H. Kinoshita. "Prevalence and Causal Factors of Playing-Related Musculoskeletal Disorders of the Upper Extremity and Trunk among Japanese Pianists and Piano Students." *Medical Problems of Performing Artists* 21, no. 3 (2006): 112–118.

Gallagher, Sean, and John R. Heberger. "Examining the Interaction of Force and Repetition on Musculoskeletal Disorder Risk: A Systematic Literature Review." *Human Factors: The Journal of the Human Factors and Ergonomics Society* 55, no. 1 (2013): 108–124.

Gát, József. *The Technique of Piano Playing.* Translated by Istvan Kleszky. London: Collet's, 1974.

Gelberman, R. H., P. T. Hergernroeder, A. R. Hargens, G. N. Lundborg, and W. H. Akeson. "The Carpal Tunnel Syndrome: A Study of Carpal Canal Pressures." *Journal of Bone and Joint Surgery* 63A (1981): 380–383.

Gopnik, Adam. "Feel Me: What the New Science of Touch Says about Ourselves." *New Yorker* (May 16, 2016): 56–67.

Gordon, Stewart. *A History of Piano Literature.* New York: Schirmer, 1996.

Grant, Nancy S. "Arts Medicine: When Practice Turns to Pain." *Keyboard* (February 1988): 90.

Hall, Susan J. *Basic Biomechanics.* 1st ed. St. Louis: Mosby Year Book, 1991.

Hall, Susan J. *Basic Biomechanics.* 3rd ed. Boston: WCG McGraw-Hill, 1999.

Hamilton, Kenneth. *After the Golden Age: Romantic Pianism and Modern Performance.* New York: Oxford University Press, 2008.

Hillberry, Benjamin M. "Dynamic Effects of Work on Musculoskeletal Loading." In *Repetitive Motion Disorders of the Upper Extremity,* 99–110. Rosemont, IL: Academy of Orthopaedic Surgeons, 1995.

Hinson, Maurice. *Guide to the Pianist's Repertoire.* 3rd ed. Bloomington: Indiana University Press, 2000.

Hofmann, Josef. *Piano Playing with Piano Questions Answered.* 1920. Reprint, New York: Dover, 1976.

Kamolsiri, Tretip. "Even a Star Shines in the Darkness: Overcoming Challenges Faced by Pianists with Small Hands." DMA Dissertation, West Virginia University, 2002.

Kochevitsky, George. *The Art of Piano Playing.* Evanston, IL: Summy-Birchard, 1967.

Kropff, Kris, ed. *A Symposium for Pianists and Teachers: Strategies to Develop the Mind and Body for Optimal Performance.* Dayton, OH: Heritage Music Press, 2002.

Kullak, Theodore, ed. *Chopin Nocturnes.* Berlin: Schlesinger, 1881; reissued New York: Schirmer, 1881.

Kwak, Jun Young. "Awakening the Kinesthetic Senses in Piano Performance: New Approaches to Fingering and Movement Based upon Ferrucio Busoni's *Die Klavier Übung.*" DMA Paper, Texas Tech University, 2013.

Leone, Carol. "Goldilocks Had a Choice." *American Music Teacher* 52, no. 6 (June/July 2003): 26–29.

Lhevinne, Josef. *Basic Principles in Pianoforte Playing.* 1924. Reprint, with a foreword by Rosina Lhevinne, New York: Dover, 1972.

Libin, Lawrence. "The Instruments." In *Eighteenth-Century Keyboard Music,* edited by Robert Marshall, 1–32. New York: Schirmer, 1994.

Lister-Sink, Barbara. *Freeing the Caged Bird* (DVD). Lewisville, NC: Wingsound.

Liebermann, Julie Lyonn. *You Are Your Instrument.* New York: Huiksi Music (Hal Leonard), 1997.

Lockwood, Alan H. "Medical Problems of Musicians." *New England Journal of Medicine* 320 (January 1989): 221–227.

Loesser, Arthur. *Men, Women, and Pianos.* New York: Simon and Schuster, 1954.

Lundberg, Ulf, Roland Kadefors, Bo Melin, Gunnar Palmerud, Pater Hassmen, Margareta Engstrom, and Ingela Elfsberg Dohns. "Psychophysiological Stress and EMG Activity of the Trapezius Muscle." *International Journal of Behavioral Medicine* 1, no. 4 (1994): 354–370.

Magrath, Jane. *The Pianist's Guide to Standard Teaching and Performance Literature.* Van Nuys, CA: Alfred, 1995.

Manchester, Ralph A. "Medical Aspects of Music Development." *Psychomusicology* 7, no. 2 (1988): 147–152.

Manshardt, Thomas. *Aspects of Cortot.* Northumberland, Great Britain: Appian, 1994.

Mark, Thomas. *What Every Pianist Needs to Know about the Body.* Chicago: Gia, 2004.

McPherson, Gary E., and Renwick, James M. "A Longitudinal Study of Self-Regulation in Children's Musical Practice." *Music Education Research* 3, no. 2 (2001): 169–186.

Miller, Michelle. *Minds Online.* Cambridge, MA: Harvard University Press, 2014.

Nag, Anjali, P. K. Nag, and Hina Desai. "Hand Anthropometry of Indian Women." *Indian Journal of Medical Research 117* (June 2003): 260–269.

Neuhaus, Heinrich. *The Art of Piano Playing.* Translated by K. A. Leibovitch. New York: Praeger, 1973.

Ortmann, Otto. *The Physiological Mechanics of Piano Technique.* 1929. Reprint, with an introduction by Arnold Schultz, Toronto: E. P. Dutton, 1962.

Palmeiri, Robert, and Margaret W. Palmieri, eds. *Encyclopedia of Keyboard Instruments,* Vol. 1: *The Piano.* New York: Garland, 1994.

Parakilas, James et al. *Piano Roles: Three Hundred Years of Life with the Piano.* New Haven, CT: Yale University Press, 1999.

Peres Da Costa, Neal. *Off the Record: Performing Practices in Romantic Piano Playing.* New York: Oxford University Press, 2012.

Philipp, Isidore. *Complete School of Technic for the Piano.* Bryn Mawr, PA: T. Presser, 1908.

Philipp, Isidore. *Exercises for the Independence of the Fingers.* Part II. New York: G. Schirmer, 1917, 1944.

Pierce, Alexandra. *Deepening Musical Performance through Movement.* Bloomington: Indiana University Press, 2007.

Pulat, B. Mustafa. *Fundamentals of Industrial Ergonomics.* Englewood Cliffs: Prentice Hall, 1992.

Rempel, David. "Musculoskeletal Loading and Carpal Tunnel Pressure." In *Repetitive Motion Disorders of the Upper Extremity,* 123–132. Rosemont: American Academy of Orthopedic Surgeons, 1995.

Ripin, Edwin M. et al. "Pianoforte." *Grove Music Online.* Oxford University Press. Accessed May 27, 2016. http://www.oxfordmusiconline.com/subscriber/article/grove/music/21631.

Rosen, Charles. *Beethoven's Piano Sonatas: A Short Companion.* New Haven: Yale University Press, 2002.

Rosen, Charles. *The Classical Style: Haydn, Mozart, Beethoven.* New York: W. W. Norton, 1972, 1971.

Rosen, Charles. *The Romantic Generation.* Cambridge, MA: Harvard University Press, 1995.

Sakai, N. "Hand Pain Attributed to Overuse among Professional Pianists." *Medical Problems of Performing Artists* 17 (2002): 178–180.

Sakai, N. "Hand Pain Related to Keyboard Techniques in Pianists." *Medical Problems of Performing Artists* 7 (June 1992): 63–65.

Sakai, N. "Keyboard Span in Old Musical Instruments: Concerning Hand Span and Overuse Problems in Pianists." *Medical Problems of Performing Artists* 23 (2008): 169–171.

Sakai, N., M. C. Liu, F. C. Su, A. T. Bishop, and K. N. An. "Hand Span and Digital Motion on the Keyboard: Concerns of Overuse Syndrome in Musicians." *Journal of Hand Surgery* 31A (2006): 831–835.

Samson, Jim. *Chopin.* New York: Schirmer Books, 1996.

Sanders, Mark S., and Ernest J. McCormick. *Human Factors in Engineering and Design.* 7th ed. New York: McGraw-Hill, 1993.

Sandor, Gyorgy. *On Piano Playing.* New York: Schirmer, 1981.

Sataloff, Robert Thayer, Alice G. Brandfonbrener, and Richard J. Lederman, eds. *Performing Arts Medicine.* 2nd ed. San Diego: Singular, 1998.

Schnabel, Artur, ed. *32 Sonatas for the Pianoforte by Ludwig van Beethoven,* Vol. 2. New York: Simon and Schuster, 1935.

Schnabel, Artur. *My Life and Music.* New York: St. Martin's Press, 1961.

Schneider, Audrey. "Dorothy Taubman: There Is an Answer." *Clavier* (September 1983): 19–21.

Scholtz, Hermann, ed. *Frederic Chopin: Sämtliche Pianoforte-Werke,* Band 1. Leipzig: C. F. Peters, n.d. [1880, reissue 1905].

Schultz, Arnold. Introduction to *The Physiological Mechanics of Piano Technique* by Otto Ortmann. 1929. Reprint, Toronto: E. P. Dutton, 1962.

Schulz, Arnold. *The Riddle of the Pianist's Finger.* New York: Carl Fischer, 1949.

Schumann, Eugenie. *Memoirs of Eugenie Schumann.* Translated by Marie Busch. London: William Heinemann, 1927.

Siek, Stephen. *England's Piano Sage: The Life and Teachings of Tobias Matthay.* Lanhan, MD: Scarecrow Press, 2012.

Snell, Richard S. *Clinical Anatomy for Medical Students.* 5th ed. New York: Little, Brown, 1995.

Steinbuhler, David. "Measuring Hand Span." *Steinbuhler & Company.* Accessed July 2015. www.steinbuhler.com.

Steinbuhler, David. "Our Research." *Steinbuhler & Company.* Accessed July 2015. www.steinbuhler.com.

Swinkin, Jeffrey. "Keyboard Fingering and Interpretation: A Comparison of Historical and Modern Approaches." Claremont Graduate University. Accessed February 10, 2014. http://scholarship.claremont.edu/cgi/viewcontent.cgi?article=1210&context=ppr.

Szabo, Robert M., and Michael Madison. "Carpal Tunnel Syndrome as a Work-Related Disorder." *Repetitive Motion Disorders of the Upper Extremity.* Rosemont: American Academy of Orthopedic Surgeons, 1995.

Taruskin, Richard. *Text and Act.* New York: Oxford University Press, 1995.

The Taubman Techniques. Videotape series produced by Enid Stettner. Lectures presented by Edna Golandsky. Medusa, NY: Taubman Institute of Piano, 1994.

Uszler, Marienne, Stewart Gordon, and Elyse Mach. *The Well-Tempered Keyboard Teacher.* New York: Schirmer, 1991.

Verbalis, John. *Natural Fingering: A Topographical Approach to Pianism.* London: Oxford University Press, 2012.

Wagner, C. H. "The Pianist's Hand: Anthropometry and Biomechanics." *Ergonomics* 31 (1988): 97–131.

Wagner, C. H. "Success and Failure in Musical Performance." In *The Biology of Music Making: Proceedings of the 1984 Denver Conference.* Edited by Franz Roehmann and Frank Wilson. St. Louis: MMB, 1988, 154–179.

Walker, Alan. *Franz Liszt: The Virtuoso Years, 1811–1847.* Vol. 1. New York: Alfred A. Knopf, 1983.

Walker, Alan. "Liszt and the Piano," *Grove Music Online, s.v. "Liszt, Franz."* Accessed May 16, 2014. http://www.oxfordmusiconline.com/subscriber/article/grove/music/48265pg9#S48265.9.

Whiteside, Abby. *Abby Whiteside on Piano Playing.* Portland, OR: Amadeus Press, 1997.

Wirth, Paul. *The Gravi-DVD.* St. Cloud: Central Minnesota Music School, 2007.

Wristen, B. G., M. C. Jung, A. K. G. Wismer, and M. S. Hallbeck. "Assessment of Muscle Activity and Joint Angles in Small-Handed Pianists: A Pilot Study on the 7/8 Keyboard versus the Full-Sized Keyboard." *Medical Problems of Performing Artists* 21 (2006): 3–9.

Wristen, Brenda, and M. Susan Hallbeck. "The 7/8 Piano Keyboard: An Attractive Alternative for Small-Handed Players." *Update: Applications of Research in Music Education* 28 (2009): 9–16.

Wristen, Brenda. "Avoiding Piano-Related Injury: A Proposed Theoretical Procedure for Biomechanical Analysis of Piano Technique." *Medical Problems of Performing Artists* 15, no. 2 (2000): 55–64.

Yoshimura, Eri, Anncristine Fjellman-Wiklund, Pamela Mia Paul, Cyriel Aerts, and Kris Chesky. "Risk Factors for Playing-Related Pain among Piano Teachers." *Medical Problems of Performing Artists* 23, no. 3 (September 2008): 107–113.

Yoshimura, Eri, Pamela Mia Paul, Cyriel Aerts, and Kris Chesky. "Risk Factors for Piano-Related Pain among College Students." *Medical Problems of Performing Artists* 21, no. 3 (September 2006): 118-125.

Activities

Activity 3.1	Cultivating Finger-to-Elbow Alignment	61
Activity 3.2	Using Down-Up (Drop-Lift) Motions to Organize Series of Notes	63
Activity 3.3	Using Down-Up (Drop-Lift) Motions to Organize Series of Double Notes	64
Activity 3.4	Using Overarcs and Underarcs	65
Activity 3.5	Coordinating Oval Gestures	66
Activity 3.6	Using Impulses	70
Activity 3.7	Using Fan-Shaped Forearm Gestures	73
Activity 3.8	Using Vertical Spanning	78
Activity 5.1	Basic Triads and Chord Inversions	111
Activity 5.2	Waltz- or Stride-Bass Textures	112
Activity 5.3	Three-Note Broken Arpeggios	113
Activity 5.4	Fingering Arpeggios in Two-Note Groups	113
Activity 5.5	Variations in Arpeggio Fingerings	114
Activity 5.6	Non-Consecutive Fingerings	122
Activity 5.7	Finger Repetitions and Slides	124
Activity 5.8	Practicing Positional Fingerings	135
Activity 6.1	Playing Staccato Chords and Octaves	182

Examples

Example 1.1	Schumann test chord, "Träumerei," from *Kinderszenen*, op. 15, m. 22	10
Example 1.2	Excerpt of exercise from Isidore Philipp's *School of Technic for the Piano* (1908)	17
Example 3.1	Maladaptive extension of the hand in opposing directions	57
Example 3.2	Maladaptive twisting of the forearm	58
Example 3.3	Maladaptive swiveling visible at the wrist	58
Example 3.4	Using key speed to produce volume	68
Example 3.5	Vertical spanning in ascending pitches	77
Example 3.6	Vertical spanning in descending pitches	77
Example 3.7	Vertical spanning used to play large chords	77
Example 4.1	Rachmaninoff, "Prélude in C♯ Minor," from *Morceaux de fantasie*, op. 3 no. 2, mm. 1–3	84
	a. Printed version	84
	b. Redistribution	84
Example 4.2	Grieg, "Wedding Day at Troldhaugen," from *Lyric Pieces*, op. 65 no. 6, m. 6	85
Example 4.3	Beethoven, Sonata in A♭ Major, op. 110, ii, mm. 1–4	85
Example 4.4	Debussy, "Prélude," from *Pour le Piano*, mm. 40–42	85
Example 4.5	Beethoven, Sonata in C♯ Minor, op. 27 no. 2 ("Moonlight"), i, mm. 8–9	86
Example 4.6	Heller, 25 Etudes for Piano, op. 45 no. 15 ("Warrior's Song"), mm. 1–4	86
Example 4.7	Chopin, Polonaise in C♯ Minor, op. 26 no. 1, mm. 5–12	87
Example 4.8	Chopin, Polonaise in C♯ Minor, op. 26 no. 1, mm. 32–33	88
	a. Printed version	88
	b. Redistribution	88
Example 4.9	Liszt, *Étude en douze exercices*, S. 136 no. 4, mm. 25–27	88
Example 4.10	Rachmaninoff, Prelude in C♯ Minor, op. 3 no. 2, mm. 14–16	89
Example 4.11	Beethoven, Sonata in B♭ Major, op. 22, i, mm. 147–148	89
Example 4.12	Mendelssohn, *Songs without Words*, op. 19 no. 1, mm. 16–19	89
Example 4.13	Brahms, Intermezzo in A Major, op. 118 no. 2, mm. 25–30	90
Example 4.14	Clara Schumann, "Notturno," from *Soireés musicales*, op. 6, mm. 1–6	90
Example 4.15	MacDowell, "To a Wild Rose," from *Woodland Sketches*, op. 51, mm. 17–25	90
Example 4.16	Lowell Liebermann, *Gargoyles*, iii, mm. 50–53	91

Example 4.17	Liszt, *Un sospiro*, mm. 46–51	91
Example 4.18	Brahms, Rhapsody in B Minor, op. 79 no. 1, mm. 81–82	92
Example 4.19	Schumann, *Kreisleriana*, v, mm. 52–55	92
Example 4.20	Chopin, Scherzo no. 1 in B Minor, op. 20, mm. 69–76	93
Example 4.21	Chopin, Scherzo no. 1 in B Minor, op. 20, mm. 85–88	93
Example 4.22	Chopin, Scherzo no. 1 in B Minor, op. 20, mm. 110–113	93
Example 4.23	Chopin, Nocturne in C♯ Minor, op. 27 no. 1, mm. 27–28	94
	a. First Chopin solution	94
	b. Second Chopin solution	94
Example 4.24	Chopin, Scherzo no. 2 in B♭ Minor, op. 31, mm. 118–132	94
Example 4.25	Mozart, Sonata in C Major, K. 279, i, mm. 22–24	95
Example 4.26	Beethoven, Sonata in E♭ Major, op. 31 no. 3, ii, mm. 43–45	95
Example 4.27	Beethoven, Sonata in D Minor, op. 31 no. 2 ("Tempest"), i, mm. 117–120	95
Example 4.28	Beethoven, Sonata in C Major, op. 53 ("Waldstein"), i, mm. 272–274	96
Example 4.29	Beethoven, Sonata in C Minor, op. 111, i, mm. 1–2	96
Example 4.30	Beethoven, Sonata in B♭ Major, op. 106 ("Hammerklavier"), i, mm. 1–4	96
Example 4.31	Beethoven, Concerto no. 5, op. 73 ("Emperor"), i, mm. 438–439	97
Example 4.32	Prokofiev, Sonata no. 2 in D Minor, i, mm. 85–90	97
Example 4.33	Debussy, "Poissons d'or," from *Images*, Book II, mm. 14–15	97
Example 4.34	Debussy, "Reflets dans l'eau," from *Images*, Book I, m. 67	98
Example 4.35	Debussy, "Hommage à Rameau," from *Images*, Book I, mm. 51–53	98
Example 4.36	Schumann, "Hunting Song," from *Album for the Young*, op. 68 no. 7, mm. 1–4	98
Example 4.37	Debussy, "Prélude," from *Pour le Piano*, mm. 43–46	99
	a. Redistribution	99
	b. Parallel voice leading	99
Example 4.38	Liszt, *Un sospiro*, m. 29	99
Example 4.39	Chopin, Nocturne in C♯ Minor, op. 27 no. 1, mm. 83–84	99
Example 4.40	Grieg, "Wedding Day at Troldhaugen," from *Lyric Pieces*, op. 65 no. 6, mm. 45–46	100
	a. Printed version	100
	b. Redistribution	100
Example 4.41	Chopin, Etude in F Major, op. 10 no. 8, mm. 93–95	101
Example 4.42	Muczynski, Prelude, op. 6 no. 1, mm. 61–65	101
Example 4.43	Tchaikovsky, "February," from *The Seasons*, op. 37a, mm. 85–88	101
Example 4.44	Beethoven, Sonata in E♭ Major, op. 27 no. 1, iii, mm. 21–24	102
Example 4.45	Debussy, "Prélude," from *Pour le Piano*, mm.74–76	102
Example 4.46	Debussy, *L'isle joyeuse*, mm. 252–255	102
Example 4.47	J. S. Bach, Sinfonia no. 4 in D Minor, mm. 1–2	102
Example 4.48	Händel, Suite in F Major, HWV 427, iv, m. 23, m. 26, and m. 42	103
Example 4.49	Haydn, Sonata in D Major, L. 50/Hob. XVI: 37, i, mm. 7–8	103
Example 4.50	Debussy, "Danseuses de Delphes," from *Préludes*, Book 1, mm. 18–20	103
Example 4.51	Schumann, *Kreisleriana*, ii, mm. 1–2	103
Example 4.52	Bartok, "Buciumeana," from *Six Romanian Folk Dances*, mm. 17–18	104
Example 4.53	Schumann, *Kreisleriana*, vi, mm. 1–2	104
Example 5.1	Grieg, "Wedding Day at Troldhaugen," from *Lyric Pieces*, op. 65 no. 6, mm. 1–2	111
Example 5.2	Chopin, Waltz in A Minor, op. Posth., mm. 1–4	112
Example 5.3	Clementi, Sonatina in C Major, op. 36 no. 3, i, mm. 10–12	112
Example 5.4	Schumann, "Wild Rider," from *Album for the Young*, mm. 1–4	114
Example 5.5	Mendelssohn, *Rondo Capriccioso*, op. 14, mm. 204–205	114
Example 5.6	Debussy, "Reflets dans l'eau," from *Images*, Book I, mm. 43–44	115
Example 5.7	Chopin, Etude in G♭ Major, op. 10 no. 5 ("Black Key"), mm. 74–76	115
Example 5.8	Chopin, Etude in C Major, op. 10 no. 1, mm. 1–2	116
	a. Chopin's and Cortot's fingerings	116
	b. Alternate refingerings	116

Example 5.9	Beethoven, Sonata in A♭ Major, op. 110, i, mm. 14–17	117
Example 5.10	Bartok, "Absent is My Sweetheart," from *For Children*, Sz. 42 no. 36, mm. 1–4	117
Example 5.11	Mendelssohn, *Rondo Capriccioso*, op. 14, mm. 43–46	118
Example 5.12	Schubert, Waltz in B Minor, op. 18 no. 6, mm. 31–32	118
Example 5.13	Brahms, Waltz in A♭ Major, op. 39 no. 15, mm. 1–4	119
Example 5.14	Chopin, Sonata in B♭ Minor, op. 35, i, mm. 137–140	119
Example 5.15	Chopin, Polonaise in C♯ Minor, op. 26 no. 1, mm. 38–46, right hand	120
Example 5.16	Schumann, "Glückes genug," from *Kinderszenen*, mm. 1–6	121
Example 5.17	Debussy, "Pagodes," from *Estampes*, mm. 1–3	121
Example 5.18	Beethoven, Sonatina in F Major, Anh. 5, i, mm. 19–26	122
Example 5.19	Tchaikovsky, "February," from *The Seasons*, mm. 43–46	123
Example 5.20	Tchaikovsky, "February," from *The Seasons*, mm. 23–25	123
Example 5.21	Haydn, Sonata in D Major, Hob. XVII: D1, iii, mm. 17–18	123
Example 5.22	Debussy, *L'isle joyeuse*, mm. 244–245	124
Example 5.23	Schumann, "Mignon," from *Album for the Young*, mm. 19–22	125
Example 5.24	Grieg, *Norwegian Dance*, op. 35 no. 2 (solo piano), mm. 5–6	125
Example 5.25	Tcherepnin, Bagatelle, op. 5 no. 9, mm. 20–23	125
Example 5.26	Tchaikovsky, "February," from *The Seasons*, mm. 93–96	126
Example 5.27	Chopin, Nocturne in B Major, op. 9 no. 3, mm. 107–108, left hand	126
Example 5.28	Chopin's fingering, Nocturne in E♭ Major, op. 9 no. 2, m. 16	126
Example 5.29	Chopin's fingering, Nocturne in B♭ Minor, op. 9 no. 1, m. 26, left hand	127
Example 5.30	Chopin, Nocturne in G Major, op. 37 no. 2, mm. 1–2, left hand	127
Example 5.31	Chopin's fingering, Nocturne in C♯ Minor, op. 27 no. 1, mm. 13–14, left hand	127
Example 5.32	Grieg, "Wedding Day at Troldhaugen," from *Lyric Pieces*, op. 65 no. 6, m. 25, left hand	127
Example 5.33	Grieg, "Wedding Day at Troldhaugen," from *Lyric Pieces*, op. 65 no. 6, m. 3	128
Example 5.34	Chopin, Etude in A♭ Major, op. 25 no. 1 ("Aeolian Harp"), m. 27, right hand	128
Example 5.35	Mendelssohn, *Rondo Capriccioso*, op. 14, mm. 67–74, right hand	128
Example 5.36	Mendelssohn, *Rondo Capriccioso*, op. 14, mm. 83–86, left hand	129
Example 5.37	Beethoven, Sonata in C Minor, op. 13 ("Pathetique"), ii, mm. 1–8	129
Example 5.38	Brahms, Rhapsody in G Minor, op. 79 no. 2, m. 21, right hand	129
Example 5.39	Chopin, Scherzo no. 1 in B Minor, op. 20, Trio, mm. 305–311	130
Example 5.40	Chopin, Etude in G♭ Major, op. 10 no. 5 ("Black Key"), m. 65, left hand	131
Example 5.41	Chopin, Etude in F Major, op. 10 no. 8, m. 95	131
Example 5.42	Chopin, Etude in A♭ Major, op. 25, no. 1 ("Aeolian Harp"), m. 24	131
Example 5.43	Schubert, Impromptu in G♭ Major, op. 90 no. 3, mm. 82–84	132
Example 5.44	Brahms, Capriccio in G Minor, op. 116 no. 3, m. 9	132
Example 5.45	Liszt, "Harmonies du soir," from the *Transcendental Etudes*, mm. 38–41	133
Example 5.46	Schumann, "Hunting Song," from the *Album for the Young*, op. 68 no. 7, mm. 3–4	133
Example 5.47	J.S. Bach, Prelude in C Minor, from *Das Wohltemperierte Klavier*, Book *I*, BWV 847, mm. 1–2	134
Example 5.48	C.P.E. Bach, Sonata in C Minor, Wq. 48 no. 4, iii, mm. 29–30	134
Example 5.49	Beethoven, Rondo in C Major, op. 51 no. 1, mm. 17–20	134
Example 5.50	Positional scale fingering for the E-major scale in Chopin, Scherzo no. 4 in E major, op. 54, mm. 960–965	137
Example 5.51	Chopin, Etude in F Major, op. 10 no. 8, mm. 93–94	137
Example 5.52	Debussy, "Reflets dans l'eau," from *Images*, Book I, mm. 56–58	138
Example 5.53	Martinu, "Columbine Remembers," from *Puppets*, Book II, no. 3, H. 116, mm. 1–8	138
Example 5.54	Debussy, "Hommage à Rameau," from *Images*, Book I, mm. 72–74	139
Example 5.55	Chopin, Sonata in B♭ Minor, op. 35, i, mm. 169–173	139
Example 6.1	Kabalevsky, *A Slow Waltz*, op. 39 no. 23, mm. 1–8	145
	a. Ovals in the left hand, beginning with an overarc	145

	b. Ovals in the left hand beginning with an underarc	145
	c. Fan shapes in the left hand	145
Example 6.2	J. Philipp Kirnberger, *Les Carillons*, mm. 13–16	146
	a. Ovals in the right hand beginning with an overarc	146
	b. Fan shapes in the right hand	146
Example 6.3	Chopin, Nocturne in C♯ Minor, op. 27 no. 1, m. 1, left hand	146
Example 6.4	Chopin, Nocturne in D♭ Major, op. 27 no. 2, mm. 2–6, right hand	146
Example 6.5	Beethoven, Sonata in C♯ Minor, op. 27 no. 2 ("Moonlight"), i, mm. 6–9	147
Example 6.6	Rebikov, "Une histoire triste qui finit bien," from *Tableaux pour enfants*, op. 37, mm. 21–24	147
Example 6.7	Scarlatti, Sonata in D Minor, K. 1, mm. 23–26	148
Example 6.8	Liszt, *La campanella*, mm. 5–6, right hand	148
Example 6.9	Liszt, *La campanella*, mm. 41–42, right hand	148
Example 6.10	Chopin, Ballade no. 3 in A♭ Major, mm. 165–166, right hand	149
Example 6.11	Chopin, Etude in G♭ Major, op. 10 no. 5 ("Black Key"), mm. 57–58	149
Example 6.12	Chopin, Concerto no. 2 in F Minor, op. 21, i, mm. 101–103	149
Example 6.13	Schumann, Waltz, op. 124 no. 4, mm. 1–8	150
Example 6.14	Chopin, Nocturne in E♭ Major, op. 9 no. 2, m. 1, left hand, and Nocturne in B Major, op. 9, no. 3, m. 1, left hand	150
Example 6.15	Schumann, "Aufschwung," from *Fantasiestücke*, op. 12 no. 2, mm. 1–3	151
Example 6.16	Beethoven, Sonata in C Minor, op. 13 ("Pathetique"), i, mm. 50–59	151
Example 6.17	William Bolcom, *Raggin' Rudi*, mm. 5–8	152
Example 6.18	Brahms, Rhapsody in G Minor, op. 79 no. 2, mm. 9–13	152
Example 6.19	Heller, 25 Etudes for Piano, op. 45 no. 15 ("Warrior's Song"), mm. 1–4	153
Example 6.20	Debussy, "Danseuses de Delphes," from *Préludes*, Book 1, mm. 6–9	153
Example 6.21	Schumann, "Hunting Song," from *Album for the Young*, mm. 1–4	153
Example 6.22	Brahms, Rhapsody in B Minor, op. 79 no. 1, mm. 81–82	154
Example 6.23	Chopin, Prelude in C Minor, op. 28 no. 20, mm. 1–4	154
Example 6.24	Debussy, "Danseuses de Delphes" from *Préludes*, Book I, mm. 1–2	155
Example 6.25	J. S. Bach, Fugue in C Major from *Das Wohltemperierte Klavier*, Book I, BWV 846, mm. 24–27	155
Example 6.26	Schumann, "Happy Farmer," from *Album for the Young*, mm. 1–3	156
Example 6.27	Haydn, Sonata in E Minor, L. 53/H. XVI: 34, i, mm. 30–32	156
Example 6.28	Beethoven, Sonata in C♯ Minor, op. 27 no. 2 ("Moonlight"), iii, mm. 1–9, right hand	157
Example 6.29	Beethoven, Sonata in E♭ Major, op. 31 no. 3, iv, mm. 68–69, left hand	157
Example 6.30	Chopin, Sonata no. 2 in B♭ Minor, op. 35, i, mm. 9–13, left hand	158
Example 6.31	Heller, 25 Etudes for Piano, op. 45 no. 15 ("Warrior's Song"), mm. 1–4	158
Example 6.32	Chopin, Prelude in D♭ Major, op. 28 no. 15, mm. 40–43	158
Example 6.33	Chopin, Nocturne in C♯ Minor, op. 27 no. 1, mm. 3–6, left hand	159
Example 6.34	Mendelssohn, "Venetian Boat Song," from *Songs without Words*, op. 30 no. 6, mm. 1–5	159
Example 6.35	Chopin, Etude in C Major, op. 10 no. 1, mm. 1–2	159
Example 6.36	Ligeti, *Musica ricercata*, vii, mm. 61–62 and 91–92, right hand	160
Example 6.37	Schumann, "Der Dichter spricht," from *Kinderszenen*, mm. 24–25	161
	a. Printed version	161
	b. Dividing the chord before the beat	161
Example 6.38	Schumann, "Träumerei," from *Kinderszenen*, op. 15, mm. 21–22	162
	a. Printed version	162
	b. Dividing the chord before the beat	162
Example 6.39	Brahms, Waltz in E Minor, op. 39 no. 4 (for solo piano), mm. 23–26	162
Example 6.40	Martinu, "Columbine Sings," from *Puppets*, Book II, no. 5, H. 116, mm. 13–16	163

Example 6.41	Ravel, "Alborada del gracioso," from *Miroirs*, mm. 194–195	163
Example 6.42	Chopin, Sonata in B♭ Minor, op. 35, i, mm. 9–16	163
	a. Printed version	163
	b. Rolling the left-hand accompaniment	163
Example 6.43	Chopin, Prelude in A Major, op. 28 no. 7, mm. 11–12	164
	a. Printed version	164
	b. Dividing or rolling the chord before the beat	164
	c. Playing the bass as an afterbeat	164
Example 6.44	MacDowell, "Sung Outside the Prince's Door," from *Forgotten Fairytales*, op. 4, no. 1, mm. 9–12	164
	a. Printed version	164
	b. Chord division before the beat	164
	c. Chord division on the beat	164
Example 6.45	Brahms, Intermezzo in E♭ Major, op. 117 no. 1, mm. 7–10	165
Example 6.46	Chopin, Sonata in B♭ Minor, op. 35, i, mm. 169–183	165
Example 6.47	Mompou, *Fountain and Bell*, mm. 22–24	166
	a. Printed version	166
	b. Simultaneous contrary motion roll	166
Example 6.48	Rebikov, "Une histoire triste qui finit bien," from *Tableaux pour enfants*, op. 37, mm. 21–24	166
Example 6.49	Chopin, Etude in E♭ Major, op. 10 no. 11, mm. 1–3	167
Example 6.50	Chopin, Concerto no. 2 in F Minor, i, mm. 79–80	167
Example 6.51	Benda, Sonatina no. 3 in A Minor, i, mm. 1–4	168
Example 6.52	Mendelssohn, *Variations sérieuses*, op. 54, Variation 3, mm. 49–52	168
Example 6.53	Mozart, Viennese Sonatina no. 6 in C Major, iii, mm. 1–4	169
Example 6.54	Beethoven, Sonata in C Minor, op. 13 ("Pathetique"), i, mm. 1–7	169
Example 6.55	Tcherepnin, Bagatelle, op. 5 no. 10, mm. 94–102	170
Example 6.56	Schubert, Impromptu in G♭ Major, op. 90 no. 3, m. 1	170
	a. Printed version	170
	b. Practice strategy	170
Example 6.57	Brahms, Rhapsody in B Minor, op. 79 no. 1, mm. 1–8	171
Example 6.58	Heller, 25 Etudes for Piano, op. 45 no. 10, mm. 1–3	171
Example 6.59	Beethoven, Sonata in E Major, op. 109, iii, Variation VI, mm. 165–166	171
Example 6.60	Schumann, "Träumerei," from *Kinderszenen*, mm. 1–7	172
Example 6.61	Chopin, Nocturne in F Major, op. 15 no. 1, mm. 1–4	172
Example 6.62	Chopin, Ballade no. 1 in G Minor, mm. 209–212, right hand	173
Example 6.63	Debussy, *L'isle joyeuse*, mm. 220–225	173
Example 6.64	Kuhlau, Prelude in G Major, mm. 1–3	174
Example 6.65	Haydn, Sonata in F Major, L. 3/H. XVI: 9, Scherzo, mm. 1–4	174
Example 6.66	Brahms, Intermezzo in A Minor, op. 76 no. 7, mm. 9–12	175
Example 6.67	Heller, 25 Etudes for Piano, op. 45 no. 9 ("Celestial Voices"), mm. 1–16, right hand	175
Example 6.68	Mendelssohn, Praeludium in E Minor, mm. 9–10	176
Example 6.69	Chopin, Etude in F Major, op. 10 no. 8, mm. 47–50	176
Example 6.70	Chopin, Etude in C♯ Minor, op. 10 no. 4, mm. 35–37	177
Example 6.71	Mozart, Concerto in C Major, K.467, i, mm. 238–240	177
	a. Printed version	177
	b. Practice strategy	177
Example 6.72	Beethoven, Sonata in C Major, op. 2 no. 3, i, mm. 85–86	178
Example 6.73	Rachmaninoff, Prelude in C♯ Minor, op. 3 no. 2, mm. 46–48	179
Example 6.74	Beethoven, Sonata in C Minor, op. 13 (" Pathetique"), i, mm. 93–99	179
Example 6.75	Heller, 25 Etudes for Piano, op. 45 no. 9 ("Celestial Voices"), mm. 1–4	180

Example 6.76	Schumann, Waltz, op. 124 no. 4, mm. 1–8	180
Example 6.77	Chopin, Ballade no. 1 in G Minor, op. 23, mm. 208–212	180
Example 6.78	Schumann, "Northern Song," from *Album for the Young*, op. 68 no. 41, mm. 1–4	181
Example 6.79	Albeniz, "Prelude," from *Songs of Spain*, op. 232 no. 1, mm. 25–28	183
Example 6.80	Villa-Lobos, "O Policinelo," from *Prole do Bebê*, Book 1, mm. 1–2	183
Example 6.81	Bartok, Bagatelle, op. 6 no. 5, mm. 1–4	184
Example 6.82	Kuhlau, Sonatina in G Major, op. 55 no. 2, ii, mm. 18–19	184
Example 6.83	Burgmüller, Etude in E Major, op. 105 no. 9, mm. 1–3	184
Example 6.84	Beethoven, Sonata in C Major, op. 2 no. 3, iv, mm. 1–4	185
Example 6.85	Heller, 25 Etudes for Piano, op. 45 no. 25 ("Epilogue"), mm. 9–12	185
Example 6.86	Beethoven, Sonata in F Major, op. 10 no. 2, i, mm. 19–30	186
Example 6.87	Granados, "May Song," from *Cuentos de la juventud*, op. 1 no. 3, mm. 15–17	186
Example 6.88	Schumann, *Papillons*, no. 1, mm. 7–14	187
Example 6.89	Organizing notes within a large arc in Beethoven, Sonata in C♯ Minor, op. 27 no. 2 ("Moonlight"), iii, mm. 1–3	187
Example 6.90	Mozart, Concerto in C Major, K.467, i, mm. 238–240	188
Example 6.91	Chopin, Etude in A Minor, op. 10 no. 2, mm. 1–2	189
Example 6.92	Beethoven, Sonata in F Major, op. 10 no. 2, i, mm. 1–2	190
Example 6.93	Heller, 25 Etudes for Piano, op. 45 no. 2 ("Avalanche"), mm. 61–68	191
Example 6.94	Joplin, *The Entertainer*, mm. 5–8	191
Example 6.95	Schubert, Valse Noble in A Minor, D. 969, op. 77 no. 9, mm. 1–4	191
Example 6.96	Chopin, Prelude in E Minor, op. 28 no. 4, mm. 16–19	192
Example 6.97	Sibelius, Romance, op. 24 no. 9, mm. 56–59	192
Example 6.98	Rachmaninoff, Prelude in C♯ Minor, op. 3 no. 2, mm. 45–47	193
Example 6.99	Liszt, Transcendental Etude no. 10 in F Minor, mm. 78–83	193
Example 6.100	Liszt, Paganini Etude no. 6, Variation 3, mm. 68–71	194
Example 6.101	Grieg, "Wedding Day at Troldhaugen," from *Lyric Pieces*, op. 65 no. 6, m. 40	194
Example 6.102	Grieg, "Wedding Day at Troldhaugen," from *Lyric Pieces*, op. 65 no. 6, mm. 52 and 56, right hand	194
Example 6.103	Chopin, Polonaise in C♯ Minor, op. 26 no. 1, mm. 3–4	195
Example 6.104	Tchaikovsky, "February," from *The Seasons*, op. 37a no. 2, mm. 85–88	195
Example 6.105	Tcherepnin, Bagatelle, op. 5 no. 1, m. 6	196
Example 6.106	Rachmaninoff, Prelude in B♭ Major, op. 23 no. 2, mm. 60–61	196
Example 6.107	Chopin, Nocturne in F Major, op. 15 no. 1, mm. 45–48	196
Example 6.108	Rachmaninoff, Prelude in C♯ Minor, op. 3 no. 2, mm. 14–16	197
Example 6.109	Chopin, Etude in G♭ Major, op. 10 no. 5 ("Black Key"), mm. 80–81, left hand	197
Example 6.110	Schumann, *Davidsbundlertänze*, mve. 3, mm. 24–32	197
Example 6.111	Liszt, Sonnetto 104 del Petrarca, mm. 1–4	198
Example 6.112	Prokofiev, *Visions fugitives*, no. 9, mm. 1–2	198
Example 6.113	Chopin, Nocturne in F Major, op. 15 no. 1, mm. 25, 27, and 29, right hand, alternate versions	198
Example 6.114	Chopin, Nocturne in C Minor, op. 48 no. 1, mm. 49–50	199
	a. Printed version	199
	b. Revision of left-hand part	199
Example 6.115	Mendelssohn, *Variations sérieuses*, op. 54, Variation 3, mm. 51–52	199
	a. Printed version	199
	b. Revision of right-hand part	199
Example 6.116	Debussy, *L'isle joyeuse*, mm. 252–253	200
	a. Printed version	200
	b. Alternatives	200
Example 6.117	Beethoven, Sonata in C Minor, op. 13 ("Pathetique"), ii, mm. 11–12	200
Example 7.1	Chopin, Nocturne in E♭ Major, op. 9 no. 2, m. 8	204

Example 7.2	Liszt, Piano Transcription of *Die Loreley*, S. 532, mm. 1–4	204
Example 7.3	Liszt, Piano Transcription of *Die Loreley*, S. 532, mm. 21–22	205
Example 7.4	Brahms, Rhapsody in G Minor, op. 79 no. 2, mm. 1–4	205
Example 7.5	Schumann, "Glückes genug," *Kinderszenen*, op. 15, mm. 1–4	206
Example 7.6	Brahms, Rhapsody in B Minor, op. 79 no. 1, mm. 106–112	206
Example 7.7	Chopin, Nocturne no. 1 in B♭ Minor, op. 9 no. 1, mm. 19–22	207
Example 7.8	Beethoven, Sonata in C Minor, op. 13 ("Pathetique"), ii, mm. 9–12	207
Example 7.9	Chopin, Nocturne in D♭ Major, op. 27 no. 2, mm. 75–77	208
Example 7.10	Grieg, "Wedding Day at Troldhaugen," from *Lyric Pieces*, op. 65 no. 6, mm. 44–50	210
Example 7.11	Chopin, Ballade, no. 3, op. 47, mm. 213–216	212
Example 7.12	Liszt, "Funérailles," from *Harmonies poétiques et religieuses*, mm. 1–6	212
Example 7.13	Griffes, Scherzo, op. 6 no. 3, mm. 29–33	213
Example 7.14	Half pedaling in MacDowell, "Sung Outside the Prince's Door," from *Forgotten Fairytales*, op. 4, no. 1, mm. 1–4	214
Example 7.15	Beethoven, Sonata in E♭ Major, op. 31 no. 3, i, mm. 1–4	214
Example 7.16	Debussy, "Prélude," from *Pour le piano*, mm. 6–7	215
Example 7.17	J.S. Bach, Fugue in A Minor, *Das Wohltemperierte Klavier*, Book 1, BWV 865, mm. 80–87	216
Example 7.18	Beethoven, Rondo in G Major, op. 51 no. 2, mm. 1–3	216
Example 7.19	Schumann, *Kreisleriana*, vi, mm. 1–3	217
Example 7.20	artok, "Free Variations," from *Mikrokosmos*, Book 6, no. 140, mm. 24–34	217
Example 7.21	Bartok, "Six Dances in Bulgarian Rhythm," from *Mikrokosmos*, Book 6, no. 149, mm. 54–58	218
Example 7.22	Bartok, "Six Dances in Bulgarian Rhythm," from *Mikrokosmos*, Book 6, no. 150, mm. 31–35	218
Example 7.23	Debussy, "La soirée dans Grenade," from *Estampes*, mm. 17–21	218
Example 7.24	Chopin, Nocturne in E♭ Major, op. 9 no. 2, mm. 1–2	220
Example 7.25	Schubert, Impromptu in G♭ Major, op. 90 no. 3, mm. 1–2	220
Example 7.26	Beethoven, Sonata in C♯ Minor, op. 27 no. 2 ("Moonlight"), iii, mm. 1–2	221
	a. Printed score with added acceleration and breath marks	221
	b. Conceptualizing the right-hand part as a series of rolled chords	221
Example 7.27	Chopin, Etude in F Major, op. 10 no. 8, mm. 1–2	222
	a. Printed score with added breath marks	222
	b. Conceptualizing the right-hand part as a series of rolled chords	222
	c. Approximate rhythm of the right-hand part	222
Example 7.28	Chopin, Etude in A♭ Major, op. 25 no. 1 ("Aeolian Harp"), mm. 26–27	222
	a. Printed score with added breath marks	222
	b. Approximate rhythm of the right-hand part	222
Example 7.29	Scarlatti, Sonata in C Minor, K. 116, mm. 108–115	223
Example 7.30	Beethoven, Sonata in C Major, op. 2 no. 3, i, mm. 97–98	223
	a. Printed score with added breath marks	223
	b. Approximate rhythm of the right-hand part	223
Example 7.31	Chopin, Ballade no. 1 in G Minor, op. 23, mm. 208–212	224
	a. Printed score	224
	b. Approximate rhythm of the right-hand part	224
Example 7.32	Beethoven, Sonata in C Minor, op. 10 no. 1, i, mm. 21–22	224
	a. Printed score	224
	b. Approximate rhythm of the right-hand part	224
Example 7.33	Beethoven, Sonata in D Minor, op. 31 no. 2 ("Tempest"), i, mm. 119–121	225
	a. Printed score	225
	b. Approximate rhythm of the left-hand part	225

Example 7.34	Beethoven, Piano Trio in D Major, op. 70 no. 1 ("Ghost"), iii, piano part, mm. 35–49	225
Example 7.35	Chopin, Etude in C♯ Minor, op. 10 no. 4, mm. 3–4	226
	a. Printed score marked with breathing spaces	226
	b. Approximate rhythm of the right-hand part	226
Example 7.36	Scarlatti, Sonata in C Minor, K. 116, mm. 108–115	226
Example 7.37	Schumann, *Papillons*, no. 1, mm. 1–8	227
Example 7.38	Beethoven, Sonata in F Major, op. 10 no. 2, i, mm. 19–22	227
	a. Printed score	227
	b. View of arm gesture from the frontal plane	227
Example 7.39	Beethoven, Sonata in C Minor, op. 13 ("Pathetique"), ii, mm. 9–12	228
	a. Printed score	228
	b. View of arm gesture from the frontal plane	228
Example 7.40	Chopin, Scherzo no. 4 in E Major, op. 54, mm. 961–967	228
Example 7.41	Debussy, "Reflets dans l'eau," from *Images*, Book 1, m. 23	229
	a. Printed score	229
	b. View of arm gesture from the frontal plane	229
Example 7.42	Grieg, "Wedding Day at Troldhaugen," from *Lyric Pieces*, op. 65 no. 6, mm. 49–50	229
Example 8.1	Intermediate: Schumann, "Hunting Song," from *Album for the Young*, op. 68 no. 7, mm. 1–12, 17–18, and 21–28	238
Example 8.2	Late Intermediate: Schumann, "Träumerei," from *Kinderszenen*, op. 15 no. 7, mm. 1–4, 6–7, 10–12, and 21–22	239
Example 8.3	Late Intermediate: Manuel de Falla, *Ritual Fire Dance* (for solo piano), mm. 40–43, 75–80, 115–118, 256–260, and 267–273	240
Example 8.4	Early Advanced: Grieg, "Wedding Day at Troldhaugen," from *Lyric Pieces*, op. 65 no. 6, mm. 49-56	242
Example 8.5	Advanced: Chopin, Ballade no. 1 in G Minor, Coda, mm. 208–212, 217, 240–242, 255–256, and 258–264	243
Example 8.6	Advanced: Rachmaninoff, Concerto no. 2 in C Minor, i, mm. 63–70	245

Figures

Figure 1.1	Impact of palm size on angles of finger abduction	3
Figure 1.2	Deahl's comparison of changes in horizontal dimensions	11
Figure 1.3	Deahl's comparison of changes in vertical dimensions	11
Figure 1.4	Stretching device recommended by Sigismond Thalberg	16
Figure 2.1	Relationship between load magnitude, repetition, and likelihood of injury	26
Figure 2.2	Surface EMG tracing of muscle activity	33
	a. Efficient dynamic muscle contraction	33
	b. Inefficient dynamic muscle contraction	33
Figure 2.3	The cardinal planes of motion	36
Figure 2.4	Neutral curves of the spine	38
Figure 2.5	Neutral seated position	39
Figure 2.6	Movements at the shoulder	40
Figure 2.7	Forearm rotation	43
Figure 2.8	Movements at the wrist	44
Figure 2.9	Neutral wrist and hand position while standing	46
Figure 2.10	Neutral wrist and hand position at the piano	46
	a. Viewed from above (transverse plane)	46
	b. Viewed from the side (sagittal plane)	46
Figure 2.11	Finger movements	47
Figure 2.12	Thumb movements	48
Figure 2.13	Bones and joints of the fingers	49
Figure 2.14	Finger extensors	50
Figure 2.15	The hand with the lumbricals activated	51
Figure 3.1	Basic oval gestures viewed from behind (frontal plane)	62
Figure 3.2	Forearm fan gestures viewed from above (transverse plane)	72
Figure 3.3	Passive spread	74
Figure 3.4	Horizontal versus vertical spanning between the second and third fingers	75
Figure 3.5	Selected vertical spans	76
	a. Vertical span between thumb and second finger	76
	b. Vertical span between second and fifth fingers	76
Figure 5.1	Ortmann's photographs illustrating the effects of fingering upon hand movements	109
Figure 7.1	Incremental crescendo	210

Tables

Table 1.1	Comparison of pianist's thumb to fifth finger span data in inches (centimeters)	6
Table 1.2	Proportions of adult pianists with "small" or "large" hands reported by Boyle, Boyle, and Booker	7
Table 1.3	Summary of Sakai's measurements of octave spans on historical instruments	9
Table 1.4	A comparison of nineteenth-century grand pianos with modern Steinway grand pianos	9
Table 1.5	Historical/reconstructed instruments measured by Deahl	10
Table 2.1	General work factors that impact piano playing	25
Table 3.1	Maladaptive tendencies for small-handed pianists to avoid	81
Table 3.2	Facilitative strategies for small-handed pianists to cultivate	81
Table 5.1	Traditional E-major scale fingering	136
Table 5.2	Alternate E-major scale fingering	136
Table 5.3	E-major scale fingering with simultaneous hand shifts	137
Table 8.1	Chords	232
Table 8.2	Arpeggiated passages or rolled chords	233
Table 8.3	Large leaps, hand shifts, and changes of direction	234
Table 8.4	Octaves and other harmonic double notes	235
Table 8.5	Broken intervals	236
Table 8.6	Legato	236
Table 8.7	Voicing	236

Index

abduction. *See also* finger abduction; stretches
defined, 247
of hand, 47
of thumb, 48*fig.*
acceleration, 31, 32, 247
accuracy
adaptive strategies for, 235*t*
increasing, 98–99
adduction, 47, 247
agonists, 29, 247
Albeniz, Isaac
"Prelude," from *Songs of Spain*, op. 232 no. 1, mm. 25–28, 183*ex.*
amplitude, 35, 49, 65–67, 247
anatomic neutral, 37, 38*fig.*, 39*fig.*, 41, 46*fig.*, 54, 101–2, 247
anatomy
as factor impacting piano playing, 25*t*
of the playing apparatus 37–52
antagonist(s), 29, 247
anthropometry, 24
arc-shaped movements, 62, 144, 227–28. *See also* curved movements; oval gestures; overarcs; underarcs
arm. *See also* forearm; upper arm; whole arm
in navigating distances, 71–72
in playing repeated chords and octaves, 182
arm impulses, 69–71, 154, 173, 182–85. *See also* impulse
arpeggiated accompaniments, legato in, 203–6
arpeggios
adaptive strategies for, 233–34*t*
eliminating stretches in, 86–94
encouraging compact hand positions in, 112–16
positional fingerings in, 137
releasing notes in, 174–77
Arrau, Claudio, 83
ascending pitches, vertical spanning in, 76, 77*ex.*

Bach, C. P. E.
fingering possibilities offered by, 107
Sonata in C Minor, Wq. 48 no. 4, iii, mm. 29–30, 134*ex.*
Bach, J. S.
Fugue in A Minor, BWV 865 from *Das Wohltemperierte Klavier*, Book 1, mm. 80–87, 215–16
Fugue in C Major from *Das Wohltemperierte Klavier*, Book I, BWV 846, mm. 24–27, 155*ex.*
Minuet in D Minor, BWV Anh. 132, mm. 9–12, 201
Prelude in C Minor, from *The Well-Tempered Clavier I*, BWV 847, mm. 1–2, 134*ex.*
Sinfonia no. 4 in D Minor, mm. 1–2, 102*ex.*
Banowetz, Joseph, 211, 212, 215
Bartók, Béla
"Absent Is My Sweetheart," from *For Children*, Sz. 42 no. 36, mm. 1–4, 117*ex.*
Bagatelle, op. 5 no. 5, mm. 1–4, 184*ex.*
"Buciumeana," from *Six Romanian Folk Dances*, mm. 17–18, 104*ex.*
"Free Variations," from *Mikrokosmos*, Book 6 no. 140, mm. 24–34, 217
"Six Dances in Bulgarian Rhythm," from *Mikrokosmos*, Book 6, 218

baseline, 33–35, 59, 71, 182, 247
Beethoven, Ludwig van
  "Appassionata" sonata, 108–9
  Arrangement of a Country Dance, WoO 114 no. 10, mm. 102, 201
  Concerto no. 5 ("Emperor"), op. 73, i, mm. 438–439, 96*ex.*
  Piano Trio in D Major, op. 70 no. 1 ("Ghost"), iii, piano part, mm. 35–49, 225*ex.*
  Rondo in C Major, op. 51 no. 1, mm. 17–20, 134*ex.*
  Rondo in C Major, op. 51 no. 2, mm. 1–3, 216*ex.*
  Sonata in A♭ Major, op. 110, ii, mm. 1–4, 85*ex.*
  Sonata in A♭ Major, op. 110, i, mm. 14–17, 117*ex.*
  Sonata in B♭ Major, op. 106, i, mm. 1–4, 96*ex.*
  Sonata in B♭ Major, op. 22, i, mm. 147–148, 89
  Sonata in C♯ Minor, op. 27, no. 2 ("Moonlight"), i, 86*ex.*, 147*ex.*
  Sonata in C♯ Minor, op. 27 no. 2 ("Moonlight"), iii, 157*ex.*, 187–88, 221
  Sonata in C Major, op. 2 no. 3, i, 178*ex.*, 223*ex.*
  Sonata in C Major, op. 2 no. 3, iv, mm. 1–4, 185*ex.*, 187
  Sonata in C Major, op. 53 ("Waldstein"), 96*ex.*, 142–43, 212
  Sonata in C Minor, op. 10 no. 1, i, mm. 21–22, 224*ex.*
  Sonata in C Minor, op. 10 no. 2, 184
  Sonata in C Minor, op. 111, 96*ex.*, 142–43
  Sonata in C Minor, op. 13 ("Pathetique"), i, 140, 151, 169*ex.*, 179*ex.*
  Sonata in C Minor, op. 13 ("Pathetique"), ii, 129*fig.*, 200, 207*ex.*, 228, 228*ex.*
  Sonata in D Minor, op. 31 no. 2 ("Tempest"), i, 95*ex.*, 212, 225*ex.*
  Sonata in E♭ Major, op. 14 no. 1, iv, mm. 68–69, 157*ex.*
  Sonata in E♭ Major, op. 31 no. 3, ii, 95*ex.*, 105
  Sonata in E♭ Major, op. 31 no. 3, i, mm. 1–4, 214*ex.*
  Sonata in E♭ Major, op. 37 no. 1, iii, mm. 21–24, 102*ex.*
  Sonata in E Major, op. 109, 142–43, 171*ex.*
  Sonata in E Major, op. 14 no. 1, i, mm. 5–7, 141
  Sonata in F Major, op. 10 no. 2, 185–86, 190, 227–28
  Sonatina in F Major, Anh. 5, i, mm. 19–26, 122*ex.*
Benda, Georg Anton
  Sonatina no. 3 in A Minor, i, mm. 1–4, 168*ex.*
biofeedback, 32, 37n22, 247
biomechanics, 24, 25*t*, 31, 35, 247
Bolcom, William
  Raggin' Rudi, mm. 5–8, 152*ex.*
Booker, Erica, 6–7
boundaries, 77–80
Boyle, Rhonda, 5–7, 13
Boyle, Robin, 5–7
Brahms, Johannes
  Capriccio in G Minor, op. 116 no. 3, m. 9, 132*ex.*
  Intermezzo in A Major, op. 118 no. 2, mm. 25–30, 90*ex.*
  Intermezzo in A Minor, op. 76 no. 7, mm. 9–12, 175*ex.*
  Intermezzo in E♭ Major, op. 117 no. 1, mm. 7–10, 165*ex.*
  Rhapsody in B Minor, op. 79 no. 1, 92*ex.*, 154*ex.*, 171*ex.*, 206*ex.*
  Rhapsody in G Minor, op. 79 no. 2, 129*fig.*, 152*ex.*, 205
  Waltz in A♭ Major, op. 39 no. 15, mm. 1–4, 119*ex.*
  Waltz in E Minor, op. 39 no. 4 (for solo piano), mm. 23–26, 162*ex.*
Brandfonbrener, Alice, 18
broken chords, 112–16, 143, 167–68, 174–77, 223
broken intervals, 149, 236*t*
broken octaves, 96, 177–78, 188
Burger, Ernst, 16–17
Burgmüller, Friedrich
  Etude in E Major, op. 105 no. 9, mm. 1–3, 184*ex.*
Busoni, Ferrucio
  Klavierübung, 135

cantabile melodies, legato in, 203–6
carpal tunnel syndrome, 45
changes of direction, adaptive strategies for, 234–35*t*
children
  adapting fingerings and means of tonal production for, 2
  and pedagogical misconceptions about small hands, 15
  small-handedness in, 8
Chopin, Frédéric
  Ballade no. 1 in G Minor, Coda, 243–44*ex.*
  Ballade no. 1 in G Minor, op. 23, 173*ex.*, 180*ex.*, 223–24
  Ballade no. 3, op. 47, mm. 213–216, 212*ex.*
  Ballade no. 3 in A♭ Major, mm. 165–66, 149*ex.*
  Concerto no. 2 in F Minor, op. 21, i, 149*ex.*, 167*ex.*
  Etude, op. 25 no. 11, 136
  Etude, op. 25 no. 1 ("Harp"), 175–76

Etude in A♭ Major ("Aeolian Harp"), op. 25 no. 1, 128*ex.*, 131*ex.*, 222*ex.*
Etude in A Minor, op. 10 no. 2, 188–89
Etude in C♯ Minor, op. 10 no. 4, 177*ex.*, 225–26
Etude in C Major, op. 10 no. 1, 115–16, 142, 159*ex.*
Etude in E♭ Major, op. 10 no. 11, mm. 1–3, 167*ex.*
Etude in F Major, op. 10 no. 8, 101*ex.*, 131*ex.*, 137*ex.*, 176*ex.*, 222
Etude in G♭ Major ("Black Key"), op. 10 no. 5, 131*ex.*, 149*ex.*, 197
on fingering, 106, 107–8
and finger repetition, 126
Nocturne in B♭ Minor, op. 9 no. 1, 127*ex.*, 207*ex.*
Nocturne in B Major, op. 9 no. 3, mm. 107–108, 126*ex.*
Nocturne in C♯ Minor, op. 27 no. 1, 94*ex.*, 99*ex.*, 127*ex.*, 140, 146*ex.*, 159*ex.*
Nocturne in C Minor, op. 48 no. 1, mm. 49–50, 199
Nocturne in D♭ Major, op. 27 no. 2, 146*ex.*, 208*ex.*
Nocturne in E♭ Major, op. 9 no. 2, 126*ex.*, 150*ex.*, 204, 220
Nocturne in F Major, op. 15 no. 1, 172*ex.*, 196, 198
Nocturne in G Major, op. 37 no. 2, mm. 1–2, 127*ex.*
Polonaise in C♯ Minor, op. 26 no. 1, 86–88, 120*ex.*, 195*ex.*
Polonaise in G Minor, B. 1, op. posth., mm. 23–26, 140
Prelude in A Major, op. 28 no. 7, 164*ex.*, 212
Prelude in C Minor, op. 28 no. 20, mm. 1–4, 154*ex.*
Prelude in D♭ Minor, op. 28 no. 15, mm. 40–43, 158*ex.*
Prelude in E Minor, op. 28 no. 4, mm. 16–19, 191–92
Scherzo no. 1 in B Minor, op. 20, 93*ex.*, 129*fig.*
Scherzo no. 2 in B♭ Minor, op. 31, mm. 118–132, 94*ex.*
Scherzo no. 4 in E major, op. 54, 136–37, 228–29
Sonata in B♭ Minor, op. 35, i, 119*ex.*, 139*ex.*, 163*ex.*, 165*ex.*
Sonata no. 2 in B♭ Minor, op. 35, i, mm. 9–13, 158*ex.*
Waltz in A♭ Major, op. 34 no. 1, mm. 17–40, 140
Waltz in A Minor, op. Posth., mm. 1–4, 112*ex.*
chord inversions, 61, 109, 111
chords. *See also* broken chords
adaptive strategies for, 232–33*t*
dividing or rolling, 160–67
eliminating stretches in, 86–94
encouraging compact hand positions in, 111–12
omitting and revising notes in, 190
playing staccato, 181–87
reducing size of shifts between, 98
releasing notes in, 168–73
vertical spanning to play large, 77
Chung, In-Seol, 45
Clementi, Muzio
Sonatina in C Major, op. 36 no. 3, i, mm. 10–12, 112*ex.*
cognitive factors and problems, 24, 25*t*
combining movements, 61–67
concentric contraction, 29n10, 247
consecutive divisions or rolls, 160, 161–66
consecutive double notes and octaves, 117–19
control, increasing, 98–99
Cortot, Alfred, 115–16, 136, 142n2
Crane, Frederick, 16
crescendos, 209–10
curved movements, 31, 32, 35, 60, 61. *See also* arc-shaped movements; oval gestures; overarcs; underarcs
Czerny, Carl, 107

damper pedal, 65, 87, 92, 97, 104, 108, 113, 119, 120, 127, 130, 132, 139, 146, 155, 156, 161–62, 164–66, 172–75, 180, 185–87, 194, 204–7, 209, 211–15, 220
Daquin, Louis-Claude
Le Coucou, mm. 1–12, 141
Deahl, Lora, 10–12, 11*fig.*
Debussy, Claude
"Danseuses des Delphes," from *Préludes*, Book 1, 103*ex.*, 153*ex.*, 155*ex.*
"Hommage à Rameau," from *Images*, Book I,98*ex.*, 139*ex.*
"La soirée dans Grenade," from *Estampes*, mm. 17–21, 218*ex.*
L'isle joyeuse,102*ex.*, 124*ex.*, 173*ex.*, 200
"Pagodes," from *Estampes*, 121*ex.*, 201
"Poissons d'or," from *Images*, Book II, mm. 14–15, 97*ex.*
"Prélude," from *Pour le Piano*, 85*ex.*, 99*ex.*, 102*ex.*, 214–15
"Reflets dans l'eau," from *Images*, Book I, 98*ex.*, 115*ex.*, 137–38, 229*ex.*
de Falla, Manuel
Ritual Fire Dance,240–41*ex.*
de Larrocha, Alicia, 213
Desai, Hina, 7

descending pitches, vertical spanning in, 77
De Smet, L., 18–19
diagnosing issues related to small-handedness, 232–37
direction, changes of, 234–35*t*
distal interphalangeal (DIP) joints, 48, 49*fig.*, 51
dividing chords, 160–67
double notes
- adaptive strategies for harmonic, 235*t*
- consecutive, 117–19
- legato in double-note passages, 206–8
- organizing series of, 64

double octaves, consecutive, 117–19
down-up / drop-lift motions, 62, 63–64, 67, 229, 247
dynamic loading, 34, 247
dynamic markings, 209–10

eccentric contraction, 29n10, 247
Eger, Patricia, 96
Ekier, Jan, 161n10, 165n11, 198n20
Elder, Dean, 212n12
electromyography (EMG), 32–33, 247
E-major scale fingering, traditional and alternate, 136–37*t*
environmental factors, impacting piano playing, 25*t*
Ergonomically Scaled Piano Keyboards, 12–14, 80
ergonomics, 24, 27, 247
ethnic groups, differing hand sizes among, 7
extension
- defined, 247
- of hand, 47
- of thumb, 48*fig.*
- of wrist, 44

extensors/extensor tendons, 50–51, 248

facilitative techniques
- addressing large reaches and distances, 71–77
- avoiding fatigue, 59–67
- to cultivate, 81–82*t*
- producing power, 67–71

factors impacting piano playing, 24–26
fan gestures, 144–46, 147–54, 248
Farias, J., 4, 19
fatigue
- acknowledging, 232
- adaptive strategies for, 232*t*, 233*t*, 234*t*, 235*t*, 236*t*
- avoiding, 59–67, 81*t*
- and injury, 27–28
- and maladaptive tendencies, 55–56, 81*t*
- and muscular contraction and release, 34
- and releasing notes, 167

feet, in maintaining balanced seating position, 39
females
- average hand span of adult, 5–7
- injury and hand size in, 18–19
- and pedagogical misconceptions about small hands, 14–15
- and 7/8-size keyboard, 13

fifth-finger melodies, 128–30, 179
finger abduction. *See also* stretches
- avoiding, 71
- impact of palm size on angles of, 3*fig.*
- and small-handed pianists, 108

fingering(s). *See also* refingering
- alternative approaches to, 110
- good, 106
- and hand size, 108–10
- and legato octaves and double-notes, 206–7
- non-consecutive, 121–23
- positional, 134–38, 249
- and rolling chords, 160
- standard, 107

finger repetitions, 124–27
fingers
- alignment of, 50
- leverage and segments of, 30
- moving longer over shorter, 131–33
- as part of playing apparatus, 47–52
- substituting, for another on same key, 138
- substituting thumb for, 133–34

finger-to-elbow alignment, 60–61
first-inversion chords, 86, 190, 199
fixation, 28, 52, 194, 248
flexion
- defined, 248
- of hand, 47
- of thumb, 48*fig.*
- of wrist, 44, 45

flexors, 50, 56, 248
force. *See also* power
- defined, 248
- and fortissimo playing, 208–11
- in Newtonian mechanics, 32
- transmitted through skeletal structures, 69

force load, and injury, 26, 27–28
forearm
- aligning, in skeletal position, 130–31
- and avoiding fatigue, 62
- fan gestures, 71–72, 73–74
- and grouping notes into larger gestures, 187
- maladaptive twisting of, 58
- as part of playing apparatus, 42–43

and traveling staccato chords and octaves, 184
forearm rotation, 42–43, 68–69, 72, 178–80
forearm staccato, 182, 183, 185, 248
fortissimo playing, 208–11
frontal plane, 35, 36*fig.*, 41, 62, 248
fulcrum, 28, 30, 130, 149, 182, 210, 248. *See also* leverage
fundamental principles. *See also* technique(s)
of movement, 28–37
parts of playing apparatus, 37–52
of work efficiency, 23–28

gestural shaping, 226–29
gestures, larger, 61–67, 187–88
Ghyselen, H., 18–19
Gieseking, Walter, 212n12, 214–15
Granados, Enrique
"May Song," from *Cuentos de la juventud*, op. 1 no. 3, mm. 15–17, 186*ex.*
Grieg, Edvard
"Gade," from *Lyric Pieces*, op. 57 no. 2, mm. 1–8, 105
Norwegian Dance, op. 35 no. 2 (solo piano), mm. 5–6, 125*ex.*
"Wedding Day at Troldhaugen," from *Lyric Pieces*, op. 65 no. 6, 85*ex.*, 100*ex.*, 111*ex.*, 127*ex.*, 128*ex.*, 194, 210–11, 229*ex.*, 242*ex.*
Griffes, Charles Tomlinson
Scherzo, op. 6 no. 3, mm. 29–33, 213*ex.*

Hallbeck, M. S., 4. 13
hand
aligning, in skeletal position, 130–31
encouraging compact, positions, 110–24
maintaining neutral position of, 101–2
as part of playing apparatus, 47–52
Händel, George Frideric
Suite in F Major, HWV 427, iv, m. 23, m. 26, and m. 42, 103*ex.*
"Hand Extender," 16
hand shifts
adaptive strategies for, 234–35*t*
facilitating, 95–98
hand size, fingering and, 108–10. *See also* small hands and small-handedness
hand span
anatomic dimensions affecting, 57
and passive spread, 72–75
and uncrossing parts, 84
and vertical spanning, 75–77
hand staccato, 69, 70–71, 182, 183, 185, 248
harmonic double notes, 235*t*
harmony, projecting, 104–5
Haydn, Joseph
Sonata in C Major, H. 50/L. 60, 211–12
Sonata in D Major, Hob. XVII: D1, iii, mm. 17–18, 123*ex.*
Sonata in D Major, L. 50/Hob. XVI: 37, i, mm. 7–8, 103*ex.*
Sonata in E Minor, L. 53/H. XVI: 34, i, mm. 30–32, 156*ex.*
Sonata in F Major, L. 2/H. XVI: 9, Scherzo, mm. 1–4, 174*ex.*
healthy practice behaviors, 20–21
Heller, Stephen
25 Etudes for Piano, op. 45 no. 2 ("Avalanche"), mm. 61–68, 190–91
25 Etudes for Piano, op. 45 no. 9 ("Celestial Voices"), 175*ex.*, 180*ex.*
25 Etudes for Piano, op. 45 no. 10, mm. 1–3, 171*ex.*
25 Etudes for Piano, op. 45 no. 25 ("Epilogue"), mm. 9–12, 185*ex.*
op. 45 no. 15 ("Warrior's Song"), 86*ex.*, 153*ex.*, 158*ex.*
Hillberry, Benjamin, 19
Hofmann, Josef, 14, 79
humerus, 41, 71n18, 248
hyperadduction, 48*fig.*
hyperextension, 44, 45, 62, 248
hyperflexion, 48*fig.*

impulse, 248. *See also* arm impulses
inertia, 31–32, 60, 248
injury. *See also* pain
factors impacting, 26
occupational, 44–45
relationship between load magnitude, repetition, and likelihood of, 26*fig.*
repetitive-strain, 18, 26–28
small-handedness and, 18–20, 54
and teaching healthy practice behaviors, 20–21
interossei muscles, 51
interphalangeal joints, 48, 49*fig.*, 51
ischial tuberosities, 37–38
isometric contraction, 29, 248

joint fixation, 28, 52, 194, 248
Joplin, Scott
The Entertainer, mm. 5–8, 191*ex.*

Kabalevsky, Dmitry
A Slow Waltz, op. 39 no. 23, mm. 1–8, 145*ex.*
Kalmosiri, Tretip, 4

Kaminsky, Pawel, 161n10, 165n11, 198n20
Kautsky, Catherine, 126n9, 199n21, 221n23, 237n1
keybed / keybedding, 57, 59, 69, 208, 248
key speed, 68–69, 208–9, 219
key surface, 248
kinematics, 24, 248
kinesiology, 24, 248
kinesthetic cues, 232
Kirnberger, J. Philipp
  Les Carillons, mm. 13–16, 146*ex.*
Kuhlau, Friedrich
  Prelude in G Major, mm. 1–3, 174*ex.*
  Sonatina in A Minor, op. 88 no. 3, iii, mm. 53–63, 105
  Sonatina in G Major, op. 55 no. 2, ii, mm. 18–19, 184*ex.*, 187
Kullak, Theodore, 173n13
Kwak, Jun Young,137

large reaches and distances
  adaptive strategies for, 234–35*t*
  addressing, 71–77, 82*t*
  covering, 143–54
  facilitating, 154–60
larger gestures, 61–67, 187–88
leaps, 95–98
legato
  adaptive strategies for, 236*t*
  and encouraging compact hand positions, 113
  maintaining, 102–3
  in octave and double-note passages, 206–8
  and passive spread, 155and vertical spanning, 76
  in widespread cantabile melodies and arpeggiated accompaniments, 203–6
legato octaves, 185–87
Leone, Carol, 13
leverage. *See also* fulcrum
  appropriate use of, 59–60
  and avoiding fatigue, 55
  common misperceptions regarding, 203
  and inertia, 31
  negotiating balance in, 52
  and playing repeated chords and octaves, 181–83
  as principle of movement, 28–30
  torso as source of, 39–40
Libin, Lawrence, 1
Liebermann, Lowell
  Gargoyles, 90–91
Ligeti, György
  Musica ricercata, vii, 159–60
limitations, physical and musical, 77–80
line, maintaining, 102–3
Liszt, Franz
  on crescendos, 209–10
  Étude en douze exercices, S. 136 no. 4, mm. 25–27, 88
  "Funérailles," from *Harmonies poétiques et religieuses*, mm. 1–6, 212*ex.*
  "Harmonies du soir," from the *Transcendental Etudes*, mm. 38–41, 133*ex.*
  La Campanella, 148
  Paganini Etude no. 6, Variation 3, mm. 68–71, 194*ex.*
  Piano Transcription of *Die Loreley*, S. 532, mm. 1–4, 204*ex.*
  positional fingerings proposed by, 135–36
  Sonnetto 104 del Petrarca, mm. 1–4, 198
  Transcendental Etude in F Minor, 184, 193
  Un Sospiro, 91*ex.*, 99*ex.*
load magnitude, 26*fig.*
lumbricals, 51
Lysens, R.,18–19

MacDowell, Edward
  "Sung Outside the Prince's Door," from *Forgotten Fairytales*, op. 4 no. 1, 164*ex.*, 214*ex.*
  "To a Wild Rose," from *Woodland Sketches*, op. 51, mm. 17–25, 90*ex.*
maladaptive tendencies
  to avoid, 81*t*
  fatigue, 55–56
  power, 56–57
  reach, 57–58
males
  average hand span of adult, 5–7
  and 7/8-size keyboard, 13
Martinu, Bohuslav
  "Columbine Remembers," from *Puppets*, Book II no. 3, H. 116, mm. 1–8, 138
  "Columbine Sings," from *Puppets*, Book II no. 5, H. 116, mm. 13–16, 163*ex.*
mass, 31, 32, 40, 41, 42, 68, 248
maximizing reach and power, 142, 201
  covering large distances, 143–54
  dividing or rolling chords, 160–67
  facilitating large reaches and distances, 154–60
  grouping notes into larger gestures, 187–88
  omitting and revising notes, 189–200
  playing staccato chords and octaves, 181–87
  releasing notes, 167–78
  taking advantage of skeletal support, 180–81
  using rotation for power, 178–80
melodic intervals
  encouraging compact hand positions in, 112–16
  releasing notes in, 174–77

melodic projection, 128–30
men. *See* males
Mendelssohn, Felix
  Praeludium in E Minor, mm. 9–10, 176*ex.*
  Rondo Capriccioso, op. 14, 114*ex.*, 118*ex.*, 128*fig.*, 129*fig.*
  Songs without Words, op. 19 no. 1, mm. 16–19, 89
  Songs without Words, op. 19 no. 2, mm. 1–21, 105
  Variations sérieuses, op. 54, Variation 3, 168*ex.*, 199
  "Venetian Boat Song," from *Songs without Words*, op. 30 no. 6, mm. 1–5, 159*ex.*
metacarpal bones, 48, 49*fig.*
metacarpophalangeal (MCP) joints, 48, 49*fig.*, 51, 56
momentum, 31, 32, 72, 179, 185, 225, 226, 228, 248
Mompou, Federico
  Fountain and Bell, mm. 22–24, 166*ex.*
motion, planes of, 35–36
movement, principles of. *See also* amplitude; combining movements
  additional characteristics of movement economy, 35
  leverage, 28–30
  muscular contraction and release, 32–35
  neutral position at piano, 37
  Newton's laws and, 31–32
  planes of motion, 35–36
Mozart, Wolfgang Amadeus
  Concerto in C Major, K. 467, i, mm. 238–240, 177*ex.*, 188
  Rondo in D Major, K. 485, 136
  Sonata in C Major, K. 279, i, mm. 22–24, 95*ex.*
  Viennese Sonatina no. 6 in C, iii, mm. 1–4, 169*ex.*
Muczynski, Robert
  Prelude op. 6 no. 1, mm. 61–65, 101*ex.*
multi-layered textures, releasing notes in, 168–73
muscle tension, 28–30, 33, 34, 37n22, 41, 54, 56
muscular bracing, 54, 248
muscular contraction and release, 32–35
muscular release, 34, 59, 168–69, 182, 249
musculoskeletal problems, 24

Nag, Anjali, 7
Nag, P. K., 7
neutralizers, 29, 248
neutral position, 37, 38*fig.*, 39*fig.*, 41, 46*fig.*, 54, 101–2, 247
Newton's laws, 31–32
non-consecutive divisions or rolls, 160, 161–66
non-consecutive fingerings, 121–23
notes
  grouping, into larger gestures, 187–88
  omitting and revising, 189–200
  voicing, 219–20

occupational injury, 44–45
octave passages, 118, 148, 206–8
octave(s). *See also* broken octaves
  adaptive strategies for, 235*t*
  as basic fingering referent, 112–13
  consecutive, 117–19
  playing staccato, 181–87
octave spans, measurement and analysis of, 8–12
older adults, small-handedness in, 8
Ortmann, Otto
  on appropriate use of levers, 60
  and defining small hand, 2–4
  on fingering, 106, 108–9
  on legato, 203, 205
  on muscular contraction and release, 34
  on playing staccato chords and octaves, 181
  on producing sufficient power, 56
  recommendations for small-handed pianists, 15
  on stretching hand, 142
  study of piano technique, 23–24
oval gestures, 62*fig.*, 65, 66, 144–47, 248. *See also* arc-shaped movements; curved movements; overarcs; underarcs
overarcs, 62, 65, 66, 71, 227–28, 248. *See also* arc-shaped movements; curved movements; oval gestures; underarcs
over-rotation, 248

pain. *See also* injury
  hand size as factor in development of, 18–19
  identifying and acknowledging, 21, 232
  ignoring, 20, 232
  reduced-size keyboards and, 13
palm size, impact on angles of finger abduction, 3*fig.*
Parakalis, James, 15
passive spread, 72–75, 154–55, 158–60, 248
pedaling. *See* damper pedal; sostenuto pedal
phalanges, of fingers, 48, 49*fig.*
Philipp, Isidore
  School of Technic for the Piano, 17fg.
physics, as factor impacting piano playing, 25*t*
piano(s)
  customized, 12–14
  evolution of modern, 8–14
  seating placement at, 38–39
  smaller size of earlier, 1
planes of motion, 35–36
playing apparatus, parts of
  forearm, 42–43
  hand and fingers, 47–52
  shoulder and upper arm, 40–41
  torso, 37–40
  wrist, 43–46

positional fingering, 134–38, 249. *See also* fingering(s)
power. *See also* force; maximizing reach and power
adaptive strategies for, 232*t*, 233*t*, 234*t*, 235*t*, 236*t*
and fortissimo playing, 208–11
increasing, 98–99
and maladaptive tendencies, 56–57, 81*t*
producing, 67–71, 82*t*
through arm impulses, 71
using forearm rotation for, 178–80
problem-solving skills, developing, 20–21
Prokofiev, Sergei
Sonata no. 2 in D Minor, i, mm. 85–90, 97*ex.*
Visions fugitives no. 9, mm. 1–2, 198
pronation, 42–43, 68, 73, 178, 249
proximal interphalangeal (PIP) joint, 48, 49*fig.*, 51
psychomotor problems, 24
psychosocial factors, impacting piano playing,25*t*

Rachmaninoff, Sergei
Concerto no. 2 in C Minor, i, mm. 63–70, 245*ex.*
Prelude in B♭ Major, op 23 no. 2, mm. 60–61, 196
"Prelude in C♯ Minor" from *Morceaux de fantasie*, op. 3 no. 2, 84*ex.*, 88–89, 179*ex.*, 192–93, 197
Prelude in G Minor, op. 23 no. 5, mm. 70–73, 201
radial deviation, 44, 45–46, 100, 174, 249
rags, 152
Ravel, Maurice
"Alborada del gracioso," from *Miroirs*, mm. 194–195, 163*ex.*
reach. *See also* maximizing reach and power
adaptive strategies for, 233*t*, 234*t*, 236*t*
and maladaptive tendencies, 57–58, 81*t*
reaction, 31, 32
Rebikov, Vladimir
"Une histoire triste qui finit bien," from *Tableaux pour enfants*, op. 37, mm. 21–24, 147*ex.*, 166*ex.*
redistribution, 83–84, 104–5
defined, 249
eliminating stretches in chords and arpeggios, 86–94
facilitating leaps or hand shifts, 95–98
facilitating trills and tremolos, 101–2
increasing accuracy, power, and control, 98–99
maintaining legato and line, 102–3
maintaining more neutral hand and wrist positions, 100–101
projecting harmony, 104–5
uncrossing parts, 84–85
reduced-size piano keyboards, 12–14
refingering, 106–8, 139–41. *See also* fingering
encouraging compact hand positions, 110–24
fingering and hand size, 108–10
moving longer fingers over shorter fingers, 131–33
optimizing sound production and voicing, 128–31
repeating same finger to recover space in hand, 124–28
retaking released notes silently, 138–39
substituting one finger for another on the same key, 138
substituting thumb for inner fingers, 133–34
using positional fingerings to achieve maximum speed, 134–38
release, 34, 59, 168–69, 182, 249
releasing / released notes, 138–39, 167–78, 207–8
Rempel, David, 45
repeated chords and octaves, 181–84
repertoire, 21, 80
repetitive-strain injuries, 18, 26–28
rests, 209
rhythmic inflection, 220–26
rolling / rolled chords, 86, 92, 147, 160–67, 233–34*t*
Rosen, Charles, 142
rotation
defined, 249
forearm, 42–43, 68–69, 72, 178–80
shoulder, 40*fig.*
rubato, 220–26

sagittal plane, 35, 36*fig.*, 44, 47, 62, 249
Sakai, Naotaka, 7, 8–9, 19
Scarlatti, Domenico
Sonata in C Minor, K. 116, mm. 108–115, 223*ex.*, 226*ex.*
Sonata in D Minor, K. 1, mm. 23–26, 148*ex.*
Schnabel, Artur, 142–43, 171–72
Schubert, Franz
Impromptu in G♭ Major, op. 90 no. 3, 132*ex.*, 170*ex.*, 205, 220
Valse Noble in A Minor, D. 969, op. 77 no. 9, mm. 1–4, 191*ex.*
Waltz in B Minor, op. 18 no. 6, mm. 31–32, 118*ex.*
Schultz, Arnold, 4

Schumann, Clara, 237n1
  "Notturno," from *Soireés musicales*, op. 6, mm. 1–6, 90
Schumann, Robert
  "Am Kamin," from *Kinderszenen*, op. 15 no. 8, mm. 1–16, 105
  "Aufschwung," from *Fantasiestücke*, op. 12 no. 2, mm. 1–3, 151*ex.*
  Davidsbundlertänze, mve. 3, mm. 24–32, 197
  "Der Dichter Spricht," from *Kinderszenen*, mm. 24–25, 161*ex.*
  "Glückes genug," from *Kinderszenen*, 121*ex.*, 205–6
  "Happy Farmer," from *Album for the Young*, mm. 1–3, 156*ex.*
  "Hunting Song," from *Album for the Young*, op. 68 no. 7, 98*ex.*, 133*ex.*, 153*ex.*, 238*ex.*
  injury of, 18
  "Kleine Romanze," from *Album for the Young*, op. 68, mm. 1–8, 140
  Kreisleriana, ii, mm. 1–2, 103*ex.*
  Kreisleriana, vi, 104*ex.*, 217*ex.*
  Kreisleriana, v, mm. 52–55, 92
  "Mignon," from *Album for the Young*, mm. 19–22, 125*ex.*
  "Northern Song," from *Album for the Young*, op. 68 no. 41, mm. 104, 181*ex.*
  Papillons no. 1, 186–87, 227
  "Träumerei," from *Kinderszenen*, op. 15, 162*ex.*, 172*ex.*, 239–40*ex.*
  Waltz, op. 124 no. 4, mm.1–8, 150*ex.*, 180*ex.*
  "Wild Rider," from *Album for the Young*, mm. 1–4, 114*ex.*
score, omitting and revising notes in, 189–200
seconds, playing, with thumb alone, 120–21
Serkin, Rudolf, 171–72
7/8-size keyboard, 12–13
shoulder, as part of playing apparatus, 40–41
Sibelius, Jean
  Romance, op. 24 no. 9, mm. 56–59, 192
simultaneous divisions or rolls, 160, 166–67
skeletal positionings, 130–31, 249
skeletal structures, transmitting force through, 69, 180–81, 209
slides, 124, 125–26
slurs, 62, 63–4
small hands and small-handedness
  accommodating, 20–21
  adversity due to, 245–46
  attempts to stretch, 15–18
  defining, 2–5
  demographics of, 5–8
  diagnosing issues related to, 232–37
  and injury, 18–20
  modern piano's relationship to, 8–14
  pedagogical misconceptions about, 14–18, 231
  as risk factor for injury, 54
  situations ideally suited for, 5
  understanding, 1–2, 21–22
social bias, 14–15
somatic training, somatics, 37n22, 249
sostenuto pedal, 215–18
sound production, optimizing, 128–31
speed
  adaptive strategies for, 233*t*, 234*t*, 235*t*, 236*t*
  positional fingerings for, 134–38
spinal alignment, 37–39
stabilizers, 29, 249
staccato, 49, 69, 70–71, 181–87
static loading, 34, 38, 41, 249
Steinbuhler, David, 4, 5, 6, 12
stretches. *See also* abduction; finger abduction; large reaches and distances
  damper pedal and alleviating problems of, 214
  eliminating, in chords and arpeggios, 86–94
stretching small hands, 15–18
stride-bass textures, 112
supination, 42–43, 249
swiveling, 58, 67, 159, 168, 249

tactile feedback, 232
task-specific factors, impacting piano playing, 25*t*
Tchaikovsky, Pyotr Ilyich
  "February," from *The Seasons*, op. 37a, 101*ex.*, 123*ex.*, 126*ex.*, 195*ex.*, 201
Tcherepnin, Alexander
  Bagatelle, op. 5 no. 1, m. 6, 195–96
  Bagatelle, op. 5 no. 9, mm. 20–23, 125*ex.*
  Bagatelle, op. 5 no. 10, mm. 94–102, 170*ex.*
technical problems, musical solutions for, 230
  fortissimo playing, 208–11
  gestural shaping, 226–29
  legato, 203–8
  pedaling, 211–18
  rhythmic inflection, tempo modification, and rubato, 220–26
  voicing, 219–20
technique(s), 54, 80–82. *See also* fundamental principles
  boundaries, 77–78
  facilitative, 58–77
  maladaptive tendencies, 55–58
tempo modification, 220–26
tension, 9, 28–30, 33, 34, 37n22, 41, 54, 56, 249
tenths, in consecutive and non-consecutive divisions and rolls, 165

Thalberg, Sigismond, stretching device recommended by, 16*fig.*
thumb
as part of playing apparatus, 47, 48*fig.*
playing seconds with, 120–21
repeated use of, to project melodies, 128–30
substituting, for inner fingers, 133–34
thumb melodies, 128–30, 179–80
thumb-to-fifth-finger spans, measurement and analysis of, 5–6
tonal intensity, 68–69
torso, as part of playing apparatus, 37–40
transverse plane, 35, 36*fig.*, 249
trapezius, 38, 41, 42, 56
traveling staccato chords and octaves, 184–85
tremolos, 101–2
trills, 101–2, 123–24
twisting, 58, 79, 133, 205, 249
two-note slur, 63

ulnar deviation, 44, 45–46, 174n14, 249
ultra-compact fingering, 129, 249
uncrossing parts, 84–85
underarcs, 62, 65, 66, 71, 227–28, 249. *See also* arc-shaped movements; curved movements; oval gestures; overarcs
upper arm
and grouping notes into larger gestures, 187
as part of playing apparatus, 40–41
and traveling staccato chords and octaves, 184

vector, 31, 249
vertical spanning, 75–77, 155–60, 176, 249
Villa-Lobos, Heitor
"O Policinelo," from *Prole de Bebe,* Bk. 1, mm. 1–2, 183*ex.*
voicing, 128–31, 219–20, 236*t*

Wagner, Christopher, 2, 5, 6
Walker, Alan, 209–10
waltz-bass textures, 112
whole arm, 41–2, 52, 130, 181–2, 249
Wieck, Alwin, 17
Wieck, Friedrich, 16–17
women. *See* females
work efficiency, principles of, 23–28. *See also* movement, principles of
work factors impacting piano playing, 24–26
wrist
alignment of, 108
and avoiding fatigue, 62, 65–7
maintaining neutral position of, 101–2
maladaptive swiveling of, 58, 67
as part of playing apparatus, 42, 43–46
and releasing notes, 174
Wristen, Brenda, 4, 12–13
wrist staccato. *See* hand staccato

Yoshimura, Eri,13, 19

The manufacturer's authorised representative in the EU for product
safety is Oxford University Press España S.A. of El Parque Empresarial
San Fernando de Henares, Avenida de Castilla, 2 - 28830 Madrid
(www.oup.es/en or product.safety@oup.com). OUP España S.A. also acts
as importer into Spain of products made by the manufacturer.
Printed and bound by CPI Group (UK) Ltd, Croydon, CR0 4YY
18/09/2026
02218912-0003